Maintenance 4.0 Technologies for Sustainable Manufacturing

Maintenance 4.0 Technologies for Sustainable Manufacturing

Guest Editor

Jasiulewicz-Kaczmarek Małgorzata

MDPI

Basel • Beijing • Wuhan • Barcelona • Belgrade • Novi Sad • Cluj • Manchester

Guest Editor
Jasiulewicz-Kaczmarek Małgorzata
Poznan University of Technology
Poznan
Poland

Editorial Office
MDPI AG
Grosspeteranlage 5
4052 Basel, Switzerland

This is a reprint of the Special Issue, published open access by the journal *Applied Sciences* (ISSN 2076-3417), freely accessible at: https://www.mdpi.com/journal/applsci/special_issues/Sus_Technologies.

For citation purposes, cite each article independently as indicated on the article page online and as indicated below:

Lastname, A.A.; Lastname, B.B. Article Title. *Journal Name* **Year**, *Volume Number*, Page Range.

ISBN 978-3-7258-2623-0 (Hbk)
ISBN 978-3-7258-2624-7 (PDF)
https://doi.org/10.3390/books978-3-7258-2624-7

Contents

applied sciences

MDPI

Editorial

Maintenance 4.0 Technologies for Sustainable Manufacturing

Małgorzata Jasiulewicz-Kaczmarek

Faculty of Management Engineering, Poznan University of Technology, 2, Prof. Rychlewskiego St., 60-965 Poznan, Poland; malgorzata.jasiulewicz-kaczmarek@put.poznan.pl

Citation: Jasiulewicz-Kaczmarek, M. Maintenance 4.0 Technologies for Sustainable Manufacturing. *Appl. Sci.* **2024**, *14*, 7360. https://doi.org/10.3390/app14167360

Received: 19 July 2024
Accepted: 18 August 2024
Published: 21 August 2024

1. Introduction

Manufacturing companies are navigating two pivotal trends that significantly impact their operations: sustainability and digitalization [1]. These trends present significant challenges, with sustainability becoming a global concern due to climate change and resource depletion [2]. In the sustainable development environment, manufacturing companies have been pressured to think beyond traditional economic measures and evaluate their business's environmental and social effects. They need not only to offer a return on investment but also to reduce the impact on the environment [3]. They must also constitute an attractive workplace for people and meet the requirements of stakeholders who can affect or be affected by the company. The second major trend, digitalization, is a cornerstone of the Industry 4.0 era, as described in the production literature [4]. In this era, manufacturing systems are empowered to monitor physical processes and make intelligent decisions through real-time communication and collaboration with humans, machines, sensors, and other elements. This evolution toward Industry 4.0 is pervasive, impacting all enterprise levels, including maintenance [5,6].

According to the authors of [7], maintenance can be defined as "the systematic execution of monitoring, repair, and replacement tasks designed to preserve or reinstate the desired functionality of a machine". Today, maintenance management is a very complex function involving technical and managerial skills and the flexibility to cope with the enormous dynamics of the business environment [8]. Maintenance management is key in modern production systems and requires proper attention [9]. Maintenance management should be treated as long-term strategic planning, integrating all stages of the product life cycle. This strategic planning must also anticipate changes in future social, economic, and environmental trends and incorporate innovative technologies into operations [10]. Moreover, according to [11–13], maintenance management is commonly considered the first step in an Industry 4.0 environment to have technical and economic advantages. Industry 4.0 technologies offer new possibilities for maintenance managers and support to improve maintenance strategies [14]. Moreover, according to the authors of [15], integrating I4.0 technology with maintenance can support the company in meeting the economic, environmental, and social challenges of sustainable manufacturing.

If technologies such as the Industrial Internet of Things (IIoT), Augmented Reality (AR), Virtual Reality (VR), Big Data Analytics (BDA), and AI are the main drivers of Industry 4.0, Maintenance 4.0 is the realization of these technologies. The ultimate objective of Maintenance 4.0 is to enhance and advance maintenance processes.

Over the last decade, several initiatives and approaches have been set up to support maintenance processes in adopting the principles and technologies of Industry 4.0 [16–18]. Modern technologies enabling the acquisition, integration, and analysis of numerous industrial data provide new possibilities for supporting maintenance processes. Predictive maintenance is a particularly interesting and wide-ranging field of application [19,20]. According to [21], "Predictive maintenance is an industrial science that utilizes condition-based monitoring technology to observe the health of machines, thereby enabling early detection of its deterioration via anomalies or faults; and provides a mechanism to plan and schedule maintenance actions to maximize its remaining useful-life". Many predictive maintenance methods have been developed with the development of big data methods and IoT technology [22,23]. In data-driven predictive maintenance strategies, machine learning (ML) can predict the system's operational status and remaining useful life (RUL) [24–26] Additionally, ML methods support the planning of maintenance activities via connected IoT technology to reduce downtime and maintenance costs and increase machine availability [27]. PdM can also be implemented using digital twins [28–30]. A "digital twin" is a dynamic, digital replica of a technical object, such as a physical system, device, machine, or production process. This solution is an integral part of Industry 4.0 and is particularly effective in sustainable production and maintenance [31,32]. Research on the application of DT in maintenance has been conducted by, among others, [33,34].

As predictive maintenance advances and is more widely adopted by companies, augmented reality can more effectively support machine inspection [35,36]. AR represents effective maintenance support, guiding operators through diagnostics, inspection, and training [37–41].

A literature analysis shows that implementing I4.0 technologies has many advantages [42–44]. However, despite the potential benefits, the implementation of I4.0 technologies in maintenance requires not only overcoming the technological challenges related to the integration with Industry 4.0. Other challenges, such as sustainability, must also be considered, including social, economic, and environmental challenges, to ensure sustainable manufacturing, which is greatly influenced by maintenance [45,46]. According to the authors of [47,48], the ability to anticipate, avoid, and quickly solve problems by applying Industry 4.0 technologies to maintenance practices is an invaluable tool for a company in achieving its sustainability goals

The objective of the following Special Issue (SI) is to present the latest advances and developments in new methods, techniques, systems, and tools dedicated to applying Maintenance 4.0 technologies to the economic, environmental, and social challenges of sustainable manufacturing. The present Special Issue captures the diversity of research focusing on the issues of Maintenance 4.0 and sustainability. The SI contains 13 articles, which are briefly described in the following chapter. The following Editorial encourages the reader to familiarize themselves with the articles and further develop the still topical issues of maintenance, Industry 4.0 technologies, and sustainable manufacturing.

2. Overview of Published Articles

Industry 4.0 is expected to revolutionize maintenance practices by reaching new predictive and prescriptive maintenance analytics levels. However, according to Nordal and El-Thalji (Contribution 1), justifying these new maintenance paradigms (predictive and prescriptive) is often difficult due to their multiple inherent trade-offs and hidden systems causalities. The prediction models in the literature can be considered as a "black box" that is missing the links between input data, analysis, and final predictions, which makes the industrial adaptability to such models almost impossible. The literature also omits modeling deterioration based on loading or considering technical specifications related to detection, diagnosis, and prognosis, which are all decisive for intelligent maintenance purposes. The authors propose a novel simulation model that enables estimation of the lifetime benefits of an industrial asset when an intelligent maintenance management system is utilized as a mixed maintenance strategy and predictive maintenance (PdM) is leveraged into opportunistic intervals

In the paper by Giacotto et al. (Contribution 2), the authors discuss how the maintenance technologies applicable to various machines need to be appropriately supported by a production environment, called an "ecosystem", that facilitates their integration within the process and their synergy with the operators. The authors tested the existing concepts of the Smart Prescriptive Maintenance Framework (SPMF) for introducing a prescriptive maintenance policy in an aviation assembly line.

Scheffer et al. (Contribution 3) propose an adaptive architectural framework aimed at shaping and structuring the process that provides operators with tailored support when using an augmented reality (AR) tool. It was found that the framework ensures that self-explanatory AR systems can capture the operator's knowledge, support the operator during maintenance activities, conduct failure analysis, provide problem-solving strategies, and improve learning capabilities. In the fourth article included in this Special Issue, Borro et al. (Contribution 4) present an example of the application of AR technology and wearable devices for the maintenance of bus fleets. The solution aims to improve the maintenance process by verifying the task checklist. The main contribution of the paper focuses on implementing prototypes at company facilities in an operational environment with real users and addresses the difficulties inherent in transferring the technology to a real work environment, such as a mechanical workshop.

The research from Nentwich and Reinhart (Contribution 5) and Xie et al. (Contribution 6) concerns an industrial robot (IR) monitoring system. Industrial robots are used in almost every industry, and their reliability is crucial to minimizing downtime and maximizing production. An IR includes various components (e.g., robot arms, body, arm, actuators, sensors, end effectors, switches, gears, and connections). Due to its complex nature, many faults can occur in a robotic system. Therefore, monitoring their condition is very important. This applies to both individual components, e.g., gears (Contribution 5) and the architecture of the monitoring system (Contribution 6).

Yin et al. (Contribution 7) propose a novel prediction scheme for the life prediction of equipment under multiple operating conditions based on morphological patterns and the symbolic aggregate approximation-based similarity measurement method (MP-SAX) and STM. The analysis and verification of public datasets of the turbofan engine from the NASA Ames Prognostics Data Repository proves that the proposed method can achieve life prediction only using original monitoring data without extracting the degradation trend of said data. In addition, the prediction result of the STM can be effectively improved by improving the STM's similarity measurement accuracy.

Kowalski and Waszkowski (Contribution 8) propose an innovative idea of taking environmental aspects into account when selecting loaders and haul trucks for excavated material transport tasks so that the amount of pollutants emitted by them in exhaust gases, e.g., the sum of hydrocarbons and nitrogen oxides (HC+NOx), is also taken into consideration when assigning the means of transport for particular tasks. Environmental aspects were also the subject of the paper by Cárcel-Carrasco et al. (Contribution 9). In this paper, the authors indicate that refrigeration production accounts for a significant proportion of electricity consumption in the main branches of the food industry. The authors state that regulating the power compressors' efficiency is a suitable way to save energy.

Research by Cárcel-Carrasco and Cárcel-Carrasco (Contribution 10) highlights the importance of knowledge in maintenance activities. The main objective of their study was to define the relationship of knowledge management within maintenance activities from the perspective of the technicians who work in these departments and extract the fundamental barriers and facilitators that these technicians consider for the creation, transmission, and use of this strategic knowledge.

The topic of the next two papers (Contribution 11 and Contribution 12) is the digital twin. The purpose of the paper by Rojek et al. (Contribution 11) was to present the results of research on the development of digital twins of technical objects, which involved data acquisition and their conversion into knowledge, the use of physical models to simulate tasks and processes, and the use of simulation models to improve the physical tasks and processes. The main goal of the paper by Pawlewski et al. (Contribution 12) was to demonstrate the research implications of a new trend in computer simulations using digital twin technologies to optimize intralogistics processes.

In the final paper included in this Special Issue, Bocewicz et al. (Contribution 13) consider the dynamic vehicle routing problem where a fleet of vehicles handles periodic deliveries of goods or services to spatially dispersed customers over a given time horizon.

The considered problem arises, for example, in systems in which garbage collection or DHL parcel deliveries, as well as preventive maintenance requests, are scheduled and implemented according to a cyclically repeating sequence. This is formulated as a constraint satisfaction problem implementing the ordered fuzzy number formalism, enabling the handling of the fuzzy nature of variables through an algebraic approach. The authors' computational results show that the proposed solution outperforms commonly used computer simulation methods.

3. Conclusions

The collection of articles on Maintenance 4.0 for sustainable production presented above covers numerous issues. It also points to the challenges companies face, regardless of the industry (e.g., food, rail, and aviation) and type of activity (manufacturing and services). Every company has assets (machines, devices) that require maintenance. Maintenance processes are fundamental to achieving company goals, including those directly related to sustainable development. Industry 4.0 technologies help transform maintenance processes into an intelligent and resilient system, supporting managers in achieving these goals.

Although submissions for the present Special Issue have already been closed, more detailed research is needed on Maintenance 4.0 technologies regarding the challenges of sustainable production. Analyzing the literature and observing business practices, it can be predicted that there will soon be great demand from different companies for new tools and methods offered by Maintenance 4.0 to predict, prevent, and reduce the impact of failure on all aspects of sustainable development.

Acknowledgments: This Special Issue would not have been possible without the contributions of the various talented authors, the hardworking and professional reviewers, and the dedicated editorial team of *Applied Sciences*. Congratulations to all authors—regardless of the final decisions on the submitted manuscripts, the feedback, comments, and suggestions from the reviewers and editors helped the authors to improve their papers. I would like to take this opportunity to express my sincere gratitude to all reviewers.

Conflicts of Interest: The author declares no conflicts of interest.

List of Contributions

1. Nordal, H.; El-Thalji, I. Lifetime Benefit Analysis of Intelligent Maintenance: Simulation Modeling Approach and Industrial Case Study. *Appl. Sci.* **2021**, *11*, 3487.
2. Giacotto, A.; Costa Marques, H.; Pereira Barreto, E.A.; Martinetti, A. The Need for Ecosystem 4.0 to Support Maintenance 4.0: An Aviation Assembly Line Case. *Appl. Sci.* **2021**, *11*, 3333.
3. Scheffer, S.; Martinetti, A.; Damgrave, R.; Thiede, S.; van Dongen, L. How to Make Augmented Reality a Tool for Railway Maintenance Operations: Operator 4.0 Perspective. *Appl. Sci.* **2021**, *11*, 2656
4. Borro, D.; Suescun, Á.; Brazález, A.; González, J.M.; Ortega, E.; González, E. WARM: Wearable AR and Tablet-Based Assistant Systems for Bus Maintenance. *Appl. Sci.* **2021**, *11*, 1443
5. Nentwich, C.; Reinhart, G. A Combined Anomaly and Trend Detection System for Industrial Robot Gear Condition Monitoring. *Appl. Sci.* **2021**, *11*, 10403
6. Xie, X.;Wu, X.; Hu, Q. Bi-Objective Optimization for Industrial Robotics Workflow Resource Allocation in an Edge–Cloud Environment. *Appl. Sci.* **2021**, *11*, 10066.
7. Yin, J.; Li, Y.; Wang, R.; Xu, M. An Improved Similarity Trajectory Method Based on Monitoring Data under Multiple Operating Conditions. *Appl. Sci.* **2021**, *11*, 10968.
8. Kowalski, A.; Waszkowski, R. Method of Selecting the Means of Transport of the Winning, Taking into Account Environmental Aspects. *Appl. Sci.* **2021**, *11*, 5512. 7.
9. Cárcel-Carrasco, J.; Pascual-Guillamón, M.; Salas-Vicente, F. Improve the Energy Efficiency of the Cooling System by Slide Regulating the Capacity of Refrigerator Compressors. *Appl. Sci.* **2021**, *11*, 2019.
10. Cárcel-Carrasco, J.; Cárcel-Carrasco, J.-A. Analysis for the Knowledge Management Application in Maintenance Engineering: Perception from Maintenance Technicians. *Appl. Sci.* **2021**, *11*, 703.
11. Rojek, I.; Mikołajewski, D.; Dostatni, E. Digital Twins in Product Lifecycle for Sustainability in Manufacturing and Maintenance. *Appl. Sci.* **2021**, *11*, 31.

12. Pawlewski, P.; Kosacka-Olejnik, M.; Werner-Lewandowska, K. Digital Twin Lean Intralogistics: Research Implications. *Appl. Sci.* **2021**, *11*, 1495.13.
13. Bocewicz, G.; Nielsen, P.; Jasiulewicz-Kaczmarek, M.; Banaszak, Z. Dynamic planning of mobile service teams' mission subject to order uncertainty constraints. *Appl. Sci.* **2020**, *10*, 8872.

References

1. Jasiulewicz-Kaczmarek, M.; Legutko, S.; Kluk, P. Maintenance 4.0 technologies–new opportunities for sustainability driven maintenance. *Manag. Prod. Eng. Rev.* **2020**, *11*, 74–87.
2. Malek, J.; Desai, T.N. A systematic literature review to map literature focus of sustainable manufacturing. *J. Clean. Prod.* **2020**, *256*, 120345.
3. Bastas, A. Sustainable manufacturing technologies: A systematic review of latest trends and themes. *Sustainability* **2021**, *13*, 4271. [CrossRef]
4. Felsberger, A.; Reiner, G. Sustainable industry 4.0 in production and operations management: A systematic literature review. *Sustainability* **2020**, *12*, 7982. [CrossRef]
5. Ahmed, U.; Carpitella, S.; Certa, A.; Izquierdo, J. A feasible framework for maintenance digitalization. *Processes* **2023**, *11*, 558. [CrossRef]
6. Werbińska-Wojciechowska, S.; Winiarska, K. Maintenance performance in the age of Industry 4.0: A bibliometric performance analysis and a systematic literature review. *Sensors* **2023**, *23*, 1409. [CrossRef]
7. Shao, Z.; Kumral, M. Logical analysis of data in predictive failure detection and diagnosis. *Int. J. Qual. Reliab. Manag*, 2024, *ahead of printing*.
8. Antosz, K.; Jasiulewicz-Kaczmarek, M.; Machado, J.; Relich, M. Application of Principle Component Analysis and logistic regression to support Six Sigma implementation in maintenance. *Maint. Reliab. Eksploat. I Niezawodn.* **2023**, *25*, 174603.
9. Shaheen, B.W.; Németh, I. Integration of Maintenance Management System Functions with Industry 4.0 Technologies and Features—A Review. *Processes* **2022**, *10*, 2173. [CrossRef]
10. Jasiulewicz-Kaczmarek, M.; Antosz, K.; Waszkowski, R.; Nielsen, I.; Čep, R. "Technology" as the fourth dimension of sustainable maintenance management. *IFAC-PapersOnLine* **2023**, *56*, 162–167.
11. Mosyurchak, A.; Veselkov, V.; Turygin, A.; Hammer, M. Prognosis of behaviour of machine tool spindles their diagnostics and maintenance. *MM Sci. J.* **2017**, *2017*, 2100–2104.
12. Forcina, A.; Introna, V.; Silvestri, A. Enabling technology for maintenance in a smart factory: A literature review. *Procedia Comput. Sci.* **2021**, *180*, 430–435.
13. Fasuludeen Kunju, F.K.; Naveed, N.; Anwar, M.N.; Ul Haq, M.I. Production and maintenance in industries: Impact of industry 4.0. *Ind.Robot. Int. J. Robot. Res. Appl.* **2022**, *49*, 461–475.
14. Mendes, D.; Gaspar, P.D.; Charrua-Santos, F.; Navas, H. Integrating TPM and Industry 4.0 to Increase the Availability of Industrial Assets: A Case Study on a Conveyor Belt. *Processes* **2023**, *11*, 1956.
15. Franciosi, C.; Voisin, A.; Miranda, S.; Iung, B. Integration of I4. 0 technologies with maintenance processes: What are the effects on sustainable manufacturing? *IFAC-PapersOnLine* **2020**, *53*, 1–6.
16. Dui, H.; Wu, X.; Wu, S.; Xie, M. Importance measure-based maintenance strategy optimization: Fundamentals, applications and future directions in AI and IoT. *Front. Eng. Manag.* **2024**. [CrossRef]
17. Keleko, A.T.; Kamsu-Foguem, B.; Ngouna, R.H.; Tongne, A. Artificial intelligence and real-time predictive maintenance in industry 4.0: A bibliometric analysis. *AI Ethics* **2022**, *2*, 553–577.
18. Patange, A.; Soman, R.; Pardeshi, S.; Kuntoglu, M.; Ostachowicz, W. Milling cutter fault diagnosis using unsupervised learning on small data: A robust and autonomous framework. *Maint. Reliab. Eksploat. I Niezawodn.* **2024**, *26*, 178274. [CrossRef]
19. Molęda, M.; Małysiak-Mrozek, B.; Ding, W.; Sunderam, V.; Mrozek, D. From corrective to predictive maintenance—A review of maintenance approaches for the power industry. *Sensors* **2023**, *23*, 5970. [CrossRef]
20. Mallioris, P.; Aivazidou, E.; Bechtsis, D. Predictive maintenance in Industry 4.0: A systematic multi-sector mapping. *CIRP J. Manuf. Sci. Technol.* **2024**, *50*, 80–103.
21. Siraskar, R.; Kumar, S.; Patil, S.; Bongale, A.; Kotecha, K. Reinforcement learning for predictive maintenance: A systematic technical review. *Artif. Intell. Rev.* **2023**, *56*, 12885–12947.
22. Gbadamosi, A.Q.; Oyedele, L.O.; Delgado, J.M.D.; Kusimo, H.; Akanbi, L.; Olawale, O.; Muhammedyakubu, N. IoT for predictive assets monitoring and maintenance: An implementation strategy for the UK rail industry. *Autom. Constr.* **2021**, *122*, 103486.
23. Zonta, T.; Da Costa, C.A.; da Rosa Righi, R.; de Lima, M.J.; da Trindade, E.S.; Li, G.P. Predictive maintenance in the Industry 4.0: A systematic literature review. *Comput. Ind. Eng.* **2020**, *150*, 106889.
24. Carvalho, T.P.; Soares, F.A.A.M.N.; Vita, R.; Francisco, R.D.P.; Basto, J.P.; Alcalá, S.G.S. A systematic literature review of machine learning methods applied to predictive maintenance. *Comput. Ind. Eng.* **2019**, *137*, 106024.
25. Welte, R.; Estler, M.; Lucke, D. A Method for Implementation of Machine Learning Solutions for Predictive Maintenance in Small and Medium Sized Enterprises. *Procedia CIRP* **2020**, *93*, 909–914.
26. Nunes, P.; Santos, J.; Rocha, E. Challenges in predictive maintenance–A review. *CIRP J. Manuf. Sci. Technol.* **2023**, *40*, 53–67.
27. Çınar, Z.M.; Abdussalam Nuhu, A.; Zeeshan, Q.; Korhan, O.; Asmael, M.; Safaei, B. Machine learning in predictive maintenance towards sustainable smart manufacturing in industry 4.0. *Sustainability* **2020**, *12*, 8211. [CrossRef]

28. Liu, Z.; Meyendorf, N.; Mrad, N. The role of data fusion in predictive maintenance using digital twin. *AIP Conf. Proc.* **2018**, *1949*, 020023.
29. Aivaliotis, P.; Georgoulias, K.; Chryssolouris, G. The use of Digital Twin for predictive maintenance in manufacturing. *Int. J. Comput. Integr. Manuf.* **2019**, *32*, 1067–1080.
30. Chen, C.; Fu, H.; Zheng, Y.; Tao, F.; Liu, Y. The advance of digital twin for predictive maintenance: The role and function of machine learning. *J. Manuf. Syst.* **2023**, *71*, 581–594.
31. Jiao, Z.; Du, X.; Liu, Z.; Liu, L.; Sun, Z.; Shi, G. Sustainable Operation and Maintenance Modeling and Application of Building Infrastructures Combined with Digital Twin Framework. *Sensors* **2023**, *23*, 4182. [CrossRef]
32. Singh, R.R.; Bhatti, G.; Kalel, D.; Vairavasundaram, I.; Alsaif, F. Building a digital twin powered intelligent predictive maintenance system for industrial AC machines. *Machines* **2023**, *11*, 796. [CrossRef]
33. Errandonea, I.; Beltrán, S.; Arrizabalaga, S. Digital Twin for maintenance: A literature review. *Comput. Ind.* **2020**, *123*, 103316.
34. Feng, Q.; Zhang, Y.; Sun, B.; Guo, X.; Fan, D.; Ren, Y.; Wang, Z. Multi-level predictive maintenance of smart manufacturing systems driven by digital twin: A matheuristics approach. *J. Manuf. Syst.* **2023**, *68*, 443–454.
35. Malta, A.; Farinha, T.; Mendes, M. Augmented reality in maintenance—History and perspectives. *J. Imaging* **2023**, *9*, 142 [CrossRef]
36. Haleem, A.; Javaid, M.; Singh, R.P.; Suman, R.; Khan, S. Management 4.0: Concept, applications and advancements. *Sustain. Oper. Comput.* **2023**, *4*, 10–21.
37. Scurati, G.W.; Gattullo, M.; Fiorentino, M.; Ferrise, F.; Bordegoni, M.; Uva, A.E. Converting maintenance actions into standard symbols for Augmented Reality applications in Industry 4.0. *Comput. Ind.* **2018**, *98*, 68–79.
38. Simon, J.; Gogolák, L.; Sárosi, J.; Fürstner, I. Augmented Reality Based Distant Maintenance Approach. *Actuators* **2023**, *12*, 302 [CrossRef]
39. del Amo, I.F.; Erkoyuncu, J.A.; Roy, R.; Wilding, S. Augmented Reality in Maintenance: An information-centred design framework. *Procedia Manuf.* **2018**, *19*, 148–155.
40. Ceruti, A.; Marzocca, P.; Liverani, A.; Bil, C. Maintenance in aeronautics in an Industry 4.0 context: The role of Augmented Reality and Additive Manufacturing. *J. Comput. Des. Eng.* **2019**, *6*, 516–526.
41. Runji, J.M.; Lee, Y.J.; Chu, C.H. Systematic literature review on augmented reality-based maintenance applications in manufacturing centered on operator needs. *Int. J. Precis. Eng. Manuf. -Green Technol.* **2023**, *10*, 567–585.
42. Bousdekis, A.; Lepenioti, K.; Apostolou, D.; Mentzas, G. A review of data-driven decision-making methods for industry 4.0 maintenance applications. *Electronics* **2021**, *10*, 828.
43. Tortorella, G.L.; Saurin, T.A.; Fogliatto, F.S.; Tlapa Mendoza, D.; Moyano-Fuentes, J.; Gaiardelli, P.; Seyedghorban, Z.; Vassolo, R.; Cawley Vergara, A.F.M.; Sunder, M.V.; et al. Digitalization of maintenance: Exploratory study on the adoption of Industry 4.0 technologies and total productive maintenance practices. *Prod. Plan. Control* **2024**, *35*, 352–372.
44. Kulshrestha, N.; Agrawal, S.; Shree, D. Spare parts management in industry 4.0 era: A literature review. *J. Qual. Maint. Eng.* **2024**, *30*, 248–283.
45. Samadhiya, A.; Agrawal, R.; Luthra, S.; Kumar, A.; Garza-Reyes, J.A.; Srivastava, D.K. Total productive maintenance and Industry 4.0 in a sustainability context: Exploring the mediating effect of circular economy. *Int. J. Logist. Manag.* **2023**, *34*, 818–846.
46. Suresh, M.; Dharunanand, R. Factors influencing sustainable maintenance in manufacturing industries. *J. Qual. Maint. Eng.* **2023**, *29*, 94–113.
47. El Kihel, Y.; El Kihel, A.; Bouyahrouzi, E.M. Contribution of Maintenance 4.0 in Sustainable Development with an Industrial Case Study. *Sustainability* **2022**, *14*, 11090. [CrossRef]
48. Mendes, D.; Gaspar, P.D.; Charrua-Santos, F.; Navas, H. Synergies between lean and Industry 4.0 for enhanced maintenance management in sustainable operations: A model proposal. *Processes* **2023**, *11*, 2691. [CrossRef]

applied sciences

MDPI

Article

Lifetime Benefit Analysis of Intelligent Maintenance: Simulation Modeling Approach and Industrial Case Study

Helge Nordal * and Idriss El-Thalji

Department of Mechanical and Structural Engineering and Materials Science, University of Stavanger, 4036 Stavanger, Norway; idriss.el-thalji@uis.no
* Correspondence: helge.nordal@uis.no

Abstract: The introduction of Industry 4.0 is expected to revolutionize current maintenance practices by reaching new levels of predictive (detection, diagnosis, and prognosis processes) and prescriptive maintenance analytics. In general, the new maintenance paradigms (predictive and prescriptive) are often difficult to justify because of their multiple inherent trade-offs and hidden systems causalities. The prediction models, in the literature, can be considered as a "black box" that is missing the links between input data, analysis, and final predictions, which makes the industrial adaptability to such models almost impossible. It is also missing enable modeling deterioration based on loading, or considering technical specifications related to detection, diagnosis, and prognosis, which are all decisive for intelligent maintenance purposes. The purpose and scientific contribution of this paper is to present a novel simulation model that enables estimating the lifetime benefits of an industrial asset when an intelligent maintenance management system is utilized as mixed maintenance strategies and the predictive maintenance (PdM) is leveraged into opportunistic intervals. The multi-method simulation modeling approach combining agent-based modeling with system dynamics is applied with a purposefully selected case study to conceptualize and validate the simulation model. Three maintenance strategies (preventive, corrective, and intelligent) and five different scenarios (case study data, manipulated case study data, offshore and onshore reliability data handbook (OREDA) database, physics-based data, and hybrid) are modeled and simulated for a time period of 20 years (175,200 h). Intelligent maintenance is defined as PdM leveraged in opportunistic maintenance intervals. The results clearly demonstrate the possible lifetime benefits of implementing an intelligent maintenance system into the case study as it enhanced the operational availability by 0.268% and reduced corrective maintenance workload by 459 h or 11%. The multi-method simulation model leverages and shows the effect of the physics-based data (deterioration curves), loading profiles, and detection and prediction levels. It is concluded that implementing intelligent maintenance without an effective predictive horizon of the associated PdM and effective frequency of opportunistic maintenance intervals, does not guarantee the gain of its lifetime benefits. Moreover, the case study maintenance data shall be collected in a complete (no missing data) and more accurate manner (use hours instead of date only) and used to continuously upgrade the failure rates and maintenance times.

Keywords: maintenance management; lifetime benefit; simulation modeling; multi-method simulation; industry 4.0; oil and gas; centrifugal compressor

Citation: Nordal, H.; El-Thalji, I. Lifetime Benefit Analysis of Intelligent Maintenance: Simulation Modeling Approach and Industrial Case Study. *Appl. Sci.* **2021**, *11*, 3487. https://doi.org/10.3390/app11083487

Academic Editor: Małgorzata Jasiulewicz-Kaczmarek

Received: 26 March 2021
Accepted: 10 April 2021
Published: 13 April 2021

1. Introduction

The opportunities within information and communication technology have revolutionized the industry by bringing the fourth industrial revolution, Industry 4.0, into reality. The main enablers of this new era are associated with the opportunities within emerging technologies such as the internet of things, big data, and cloud computing (including detection, diagnosis, and prognosis). These technologies are the fundamentals of Industry 4.0′s core concept, namely the cyber-physical-system that enables converging the physical space of equipment with cyberspace. Therefore, Industry 4.0 is considered as the future scenario of industrial production since it enables a new level of organizing and controlling the entire value chain within the product lifecycle, by creating a dynamic and real-time understanding of cross-company behaviors.

Several case studies [1] highlight the benefits of digital transformation in the oil and gas (O&G) sector. For example, reducing the upstream operations' finding and development costs by 5 percent; maintenance costs by 20 percent; overtime cost by 20 percent; downtime by 5 percent (mainly due to predictive maintenance (PdM)); inventory levels for spare parts by 20 percent; while boosting production by a conservative 3 percent in conventional land operations [1]. However, maintenance management and performance are complex aspects of an asset's operation that are difficult to justify because of their multiple inherent trade-offs and hidden systems causalities. Nevertheless, companies want to be capable of estimating the lifetime benefits in terms of improving availability and reducing the maintenance management workload, etc., by incorporating intelligent maintenance into the operation and maintenance of their engineering assets to demonstrate (1) how much to invest, (2) when to invest, and (3) the resulting expected lifetime benefits.

Therefore, the industry has really begun to appreciate the benefits of applying modeling and simulation methodologies as a supportive function to enable assessing the behavior and predicting the future outcome of, e.g., maintenance management. For example, Shoreline AS provides a simulation model that helps to simulate possible maintenance alternatives for offshore wind farms and select the most cost-effective by considering operational aspects such as weather forecast, accessibility, and resources i.e., technicians and type of vessel. Moreover, Miriam RAM studio simulates the availability and productivity of O&G installations based on reliability analysis. This helps designers to redesign or design out items to enhance availability and overall equipment effectiveness. These industrial simulation tools shall be enhanced until they capture and are able to estimate all the lifetime benefits of an intelligent maintenance management system. For example, industrial managers are looking forward to estimating the lifetime benefits of PdM and the potential opportunistic maintenance intervals in (1) reducing the corrective maintenance and unintended maintenance events, (2) reducing the preventive maintenance workload and minimizing the planned maintenance campaigns, (3) reducing the level of damage and repair, and (4) extending the lifetime of industrial assets. The desired simulation tool shall support estimating the scheduled maintenance workload (maintenance campaigns), potential corrective maintenance workload, PdM capabilities (effectiveness and earliness), and the planned and potential opportunistic maintenance intervals to perform intelligent maintenance. These basic functions shall enable industrial managers to (1) redesign their maintenance campaigns and potential corrective maintenance (with the help of intelligent maintenance) to fit opportunistic maintenance intervals, (2) reschedule maintenance campaigns at the utilization phase, and (3) redefine their loading and operating profile (optimal profile to produce as high as possible at a deterioration rate as low as possible) either for short-term tactical decisions as toleration to utilize the next potential opportunistic maintenance (avoid unintended maintenance visit) or for long-term strategic decisions to extend their assets' lifetime. For instance, Arun [2] illustrates how the change in loading profile (from stand-by redundancy to preschedule redundancy) extends the asset lifetime. In this case study, two out of three crude oil pumps were operating continuously, while one pump was in stand-by mode functioning as redundancy (triggered once one of the other two pumps fail). Following this, the company decided to change the operating policy and run each pump based on time, whereas each pump was operating for two months followed by one month in redundancy (monthly shift between the pumps to ensure that two pumps were running continuously).

The state-of-the-art of simulation models for maintenance practices shows three schools of thinking: discrete event, system dynamics, and agent-based modeling. The discrete-event simulation models for maintenance services and failure events are nicely summarized by Alabdulkarim, Ball et al. [3]. These models have the objective of independently simulating preventive maintenance and corrective maintenance events (due to probabilistic failures) and consider them as discrete events that return the asset to a state of "as good as new". First, maintenance practitioners and researchers [4,5] have noticed that preventive maintenance has a long-term effect on corrective maintenance and maintenance resources, and they defined the "shift the burden" phenomenon. Second, they noticed that preventive maintenance activities fix the symptoms of failures but might not fix the fundamental problem or cause of the propagating deterioration. Therefore, the system dynamics approach came as the second wave with its ability to model interactions (causalities) between maintenance policies (corrective and preventive) to enable the field of maintenance simulation to study the effect of several and mixed maintenance policies e.g., total productive maintenance [5], reliability centered maintenance [6], overall equipment effectiveness [7], and condition-based maintenance (CBM) [8–11]. In this context, simulation models involving maintenance policies using systems dynamics are nicely summarized by Linnéusson, Ng et al. [12]. In fact, system dynamics models are well known for their high level and abstractive representations (they consider the entire industrial system as one single system), which made maintenance practitioners and researchers search for another approach that models the individual behaviors (where they can decompose the system, but with traceable connections). Therefore, the third wave of maintenance simulation started with agent-based modeling where multi-agent models, multi-simulation models, and individual state-transition (statechart) were enabled. The agent-based models for maintenance simulation are still few, but rapidly growing [2,13–15].

The literature clearly introduces two research gaps. First, none of the existing simulation models have modeled the deterioration based on loading. Second, a model that includes the PdM capabilities of detection, diagnosis, and prediction processes is missing. In summary, to get the lifetime benefits of the referred simulation models and make them fit with the required above-mentioned functionalities (opportunistic maintenance intervals, PdM, and load-based deterioration), further contributions are required. In fact, the future simulation model required shall be able to consider: (1) the individual agent (physical component and failure modes), as Endrerud, Liyanage et al. [14] have done, besides, (2) modeling the PdM module, as Adegboye, Fung et al. [15] have done, (3) modeling the asset determination based on loading function as Arun [2] has done, and (4) modeling opportunistic maintenance intervals and leveraging PdM into these intervals in terms of intelligent maintenance. Table 1 highlights what is covered by the three latter studies and the missing scientific contribution (research gap) required to enable simulating intelligent maintenance operations.

Table 1. Existing research and required scientific contributions to satisfy intelligent maintenance operations.

Reference	Preventive and Corrective Maintenance Considering Agents	Modeling CBM and PdM	Load-Based Deterioration	Intelligent Maintenance (Leveraging PdM into Opportunistic Intervals)
[14]	X	-	-	-
[15]	X	X	-	-
[2]	X	X	X	-
This study	**X**	**X**	**X**	**X**

Therefore, the purpose and scientific contribution of this work is to develop a novel multi-method simulation model that enables estimating the lifetime benefits of an industrial asset, whereas intelligent maintenance is utilized as mixed maintenance strategies and the PdM is leveraged into opportunistic intervals.

Leveraging PdM requires an enhanced level of detection, diagnosis, and prognosis [16] with an integrated load-based deterioration model. To be more specific, the desired simulation model shall enable simulating the behavior of several maintenance strategies and fulfill specific industrial requirements to ensure its effectiveness, fitness to purpose, and adaptability. The desired model shall enable maintenance engineers to (1) allocate the scheduled maintenance campaigns for each component and differentiate between campaigns that lead to operational unavailability and not, (2) simulate the potential failure events and associated corrective maintenance events, and utilize their real historical failure and maintenance data or data extracted from the well-known failure database, i.e., the offshore and onshore reliability data handbook (OREDA) [17], or both, (3) assign "failure rate" and "mean time to repair" (MTTR) values for each maintainable item (component level) and associated failure modes, (4) simulate maintenance events that are triggered by CBM or PdM algorithms, (5) assign the capability level of condition monitoring techniques and prediction algorithms [16], and leverage the predicted failure events into opportunistic maintenance intervals in terms of intelligent maintenance, (6) simulate deterioration process and predict failure events based on realistic (fluctuating, seasonal patterns, stand-by operations, extreme loading intervals) loading and operating profiles. Thus, to build such a model and validate its structure and behavior, a case study of a centrifugal compressor used for natural gas transportation is purposefully selected.

The novel multi-method computational simulation model in this paper is decomposed into four sub-models (1) working state for operational availability and intelligent maintenance, (2) scheduled maintenance states (component level and equipment level) which also presents the opportunistic maintenance intervals, (3) failure states which represent failure modes and triggers for failure events, and (4) corrective maintenance states. Furthermore, to highlight the expected lifetime benefits of intelligent maintenance during 20 years of operation, two main use case scenarios shall be modeled: with and without intelligent maintenance. The latter use case scenario (without intelligent maintenance) has several sub-scenarios that also study the effectiveness of several possible data sources: (1) empiric case study data (experience), (2) manipulated empiric case study data, (3) the OREDA database [17], and (4) mixed data-input from both the empiric case study and the OREDA database. These four sub-scenarios along with the intelligent maintenance scenario result in a total of five simulated use case scenarios.

The six-step modeling and simulation methodology, presented in the following section, is applied to build the desired novel multi-method computational simulation model that combines agent-based modeling with system dynamics to simulate the five use case scenarios. In this case, the multi-method modeling software Anylogic 8 is utilized.

The rest of this section is organized as follows. First, Section 2 explains the materials and methodology of this study, which includes the entire six-step simulation modeling methodology adopted. Section 3 presents the simulated results obtained from the computational model. Section 4 discusses and validates the findings of this study. Finally, Section 5 offers some conclusions and makes recommendations for future work.

2. Materials and Methods

In this section, the adopted six-step simulation modeling methodology is presented. Thus, detailed descriptions of how the real-world case study was analyzed, conceptualized, and computerized into a simulation model are presented.

In fact, model-based representations in terms of process modeling and industrial simulation approaches have become a highly embraced tool with their growing complexity and capabilities [18]. Current literature presents several different methodologies that facilitate the successful development of a simulation model, with the most trusted modeling methodologies being that of [19–21], whereas the majority of literature relies on the methodology proposed by Sterman [21]. Nevertheless, the essence of the different methodologies is quite similar. This research adopts a six-step simulation modeling methodology that extends the essence of Sterman [21] by allocating additional emphasis on systems analysis and scenario modeling. The adopted six-step modeling and simulation methodology is as follows: (1) System analysis and project planning, (2) Conceptual modeling, (3) Computational modeling, (4) Scenario modeling, (5) Verification and validation, and (6) Visualization. In the following subsections, each step will be described in detail.

2.1. Step 1: System Analysis and Project Planning

The first step in the six-step modeling and simulation process starts with a system analysis addressing the needed fundamentals to attain an understanding of the system's behavior, i.e., structure, interfaces, processes, interactions, etc. To do so, relevant stakeholders must be addressed, including their needs and requirements to the system under study. Then, the model constraints must be defined by studying, e.g., system context, hierarchy, interface architecture, and functional and physical architecture in greater detail. In addition, other features posing a significance to the purpose must be identified, e.g., politics, market, technology.

The purpose of the system analysis step is to (1) identify the purpose and objective of the simulation, (2) analyze the case study data that is required to conceptualize the maintenance management practices (scheduled, corrective, condition monitoring, opportunistic intervals), especially, workflow, rules, conditions, and actions, (3) analyze the case study data that is required as inputs for the simulation model e.g., failure rates, maintenance service times, and (4) analyze the case study data that is required to validate the simulated behavior e.g., real availability and real corrective maintenance workload.

The purpose of the proposed multi-method simulation model is to simulate and estimate the potential lifetime benefits of implementing an intelligent maintenance management system in terms of availability and corrective maintenance workload during a time period of 20 years. Thus, the simulated outputs of the computational model address: (1) operational behavior, (2) maintenance event: timeline and workload, and (3) the occurrence of failures allocated at the component level. It is evident that operational availability is essential for the case company, as the end-user consumption is traceable to industrial operation and human welfare in Europe. Therefore, the operational behaviors including availability and unavailability caused by failures and the need for corrective maintenance are analyzed. This is easiest illustrated through a time-plot diagram showing continuous availability and unavailability as a function of time during operation. The maintenance event timeline of both scheduled maintenance and corrective maintenance is analyzed. First, a scheduled maintenance event timeline is analyzed as it introduces opportunistic maintenance intervals whereas future predicted failures can be allocated. Second, a corrective maintenance event timeline that demonstrates the corrective maintenance events required by the different use case scenarios is analyzed. This is especially interesting when it comes to comparing the corrective use case scenarios with the intelligent maintenance scenario. The maintenance workload is analyzed to demonstrate the allocation of maintenance management. The number of component failures occurring during operation is analyzed to compare different use case scenarios, which supports highlighting the number of corrective maintenance events that can potentially be replaced with intelligent maintenance. In addition, it addresses possible differences in input data originating from the empiric case study and the OREDA database.

To simulate the possible lifetime benefits of incorporating an intelligent maintenance system into the specific case study, data concerning failure rates and MTTR values of the specific case of interest is needed. In addition, data that enables determining the capabilities of detection, diagnosis, and prognosis of such a system is also needed. To do so, an analysis tool that has been developed by the authors on a previous occasion can be adapted [16]. A more detailed system analysis has already been performed and presented by the authors in [22].

2.2. Steps 2 and 3: Conceptualization and Computational Modeling

The conceptual modeling is all about synthesizing the developer's understanding of the real situation analyzed in the "system analysis and project planning" into a conceptual model. This is known as a time-consuming task in comparison to the other steps in the simulation modeling process [23]. In this context, the authors have already published a paper [24] where the conceptual model is described using a system dynamic approach. However, the authors later recognized that a multi-method modeling approach combining system dynamics with an agent-based modeling approach, whereas statecharts either triggered by rates, parameters, or conditions connected to system dynamics approaches are used, can enable better modeling of the maintenance policies. The statecharts in Figure 1 represent the maintenance management process in the case company, specifically for compressor equipment. The statechart is modeled using Anylogic Simulation package (8.5.1) and decomposed into the following four sub-models (1) Working state for operational availability and intelligent maintenance representing the daily operation and maintenance (including condition monitoring) activities that do not affect the operational behavior, (2) scheduled maintenance states (at component level and equipment level), requiring shutdown of the compressor equipment (presents the opportunistic maintenance intervals), (3) failure states, representing failure modes and triggers for failure events, and (4) corrective maintenance states, referring to the corrective maintenance needed to put the compressor equipment back in normal operation post-failure. These four sub-models are illustrated in Figures 2–4 and described in more detail in the following subsections, respectively.

2.2.1. Operational Availability and Intelligent Maintenance

The "working" state is considered as the mother state in Figure 2, which means that the equipment is available as long as the agent "Compressor" has not triggered a maintenance event requiring equipment stoppage. However, the equipment might be available and running normally in the "normal" state while the condition monitoring system is active, and the equipment health is being checked on a daily and monthly basis (rates). The daily and monthly checks have specific time amounts (i.e., timeout in Anylogic) and might trigger a maintenance event that results in equipment stoppage. The monthly monitoring checks are done by two stakeholders (1) condition monitoring providers and (2) the technical service provider. Moreover, there are minor and major scheduled maintenance work that is taking place while the equipment is running (which does not lead to production stoppage). Furthermore, the PdM might also trigger a maintenance event that can take place in the following opportunistic intervals. This state is named "Intelligent Maintenance" and does not lead to production stoppage as it utilizes potential opportunistic intervals. The time amount for these maintenance events specifically connected to the intelligent maintenance and is extracted based on OREDA data for MTTR.

Table 2 addresses the triggers of the transitions between the different states included in Figure 2. As seen, the transitions from the "normal" state to the states concerning (minor and major) "scheduled maintenance", and back again, are triggered by timeouts (specific time interval). The states of "condition monitoring" are triggered by rates. In this case, a conditional transition including a "randomTrue probability distribution" of the input failure data is used to demonstrate the probability of detection. This means that, if the condition is false, the condition monitoring system is not able to detect anything abnormal with the operation and enters the normal working state again. In contrast, if the condition is true, the condition monitoring system has detected abnormal behavior of the system and the presence of failure. The logic of these latter states concerning condition monitoring is not yet incorporated into the computational model, as the extracted failure rates used in this research only contain system failure, and therefore the impact of the monitoring system has already been taking into consideration, indirectly. However, the states are included in the computational model as they pose an impact on the model output in terms of maintenance workload. At last, the "IntelligentMaintenance" is triggered by the flow "OpportunisticMaintenance" from system dynamics and then back to the "normal" state again by a timeout function of 18 h that is extracted from the OREDA database [17] and traceable to the specific MTTR values of the failure modes monitored in this case study.

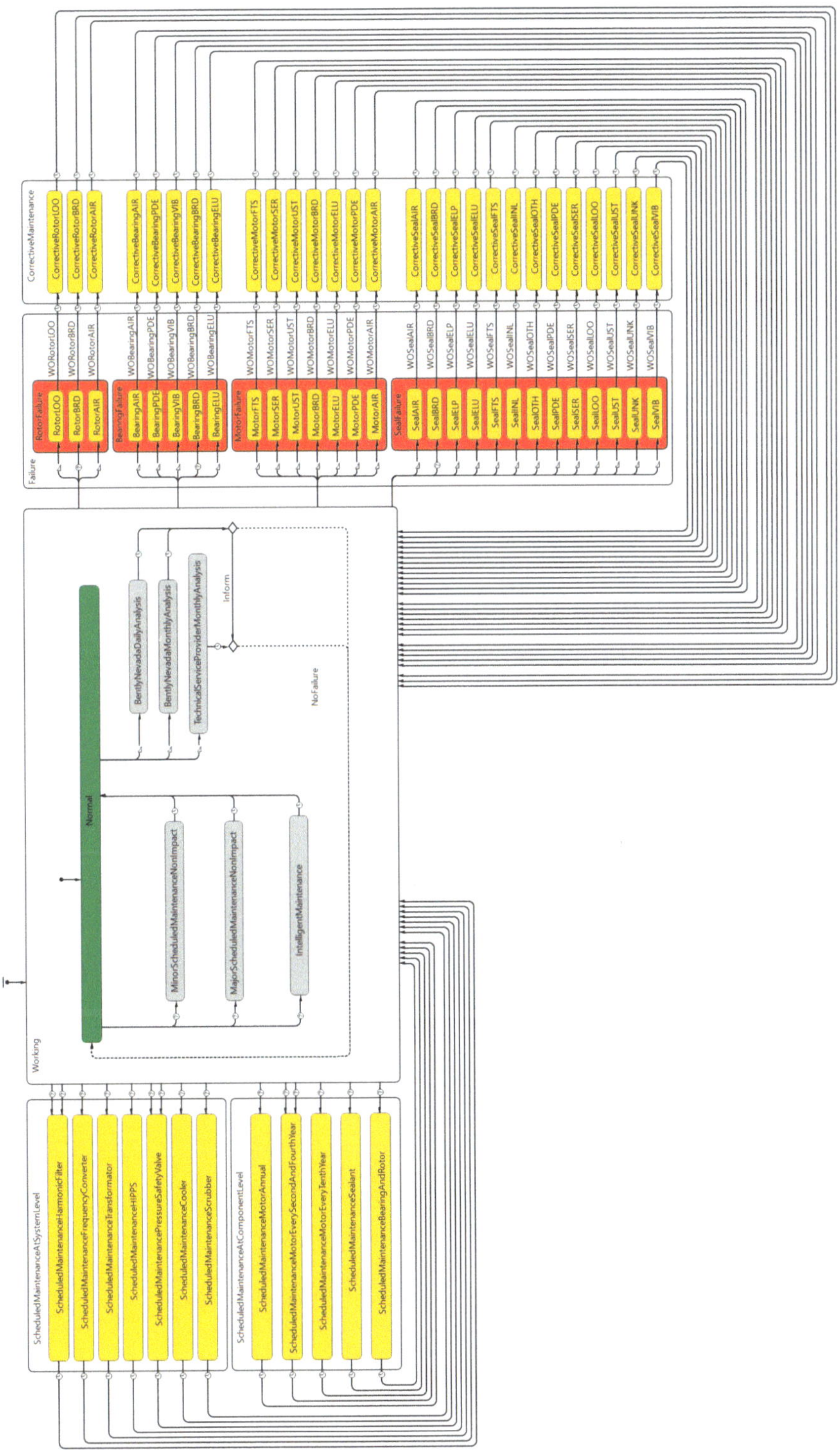

Figure 1. The novel computational model of intelligent maintenance.

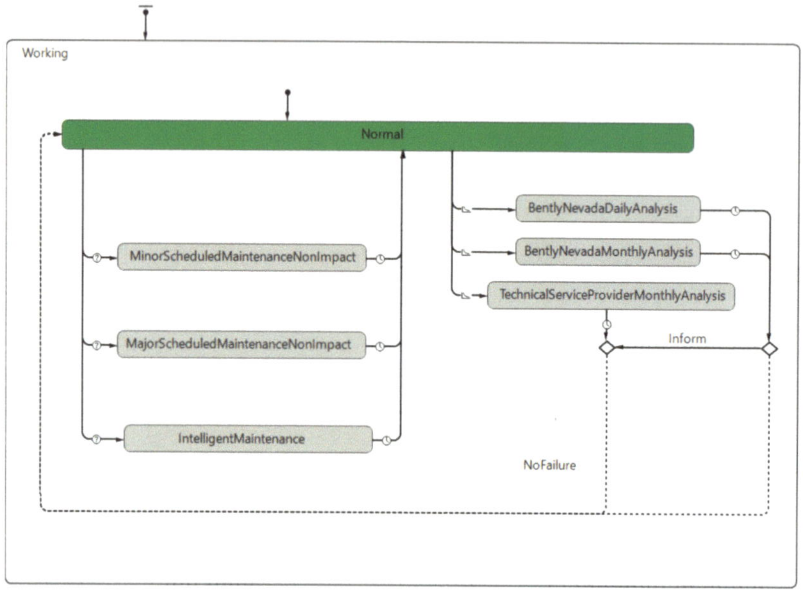

Figure 2. Operational availability and intelligent maintenance.

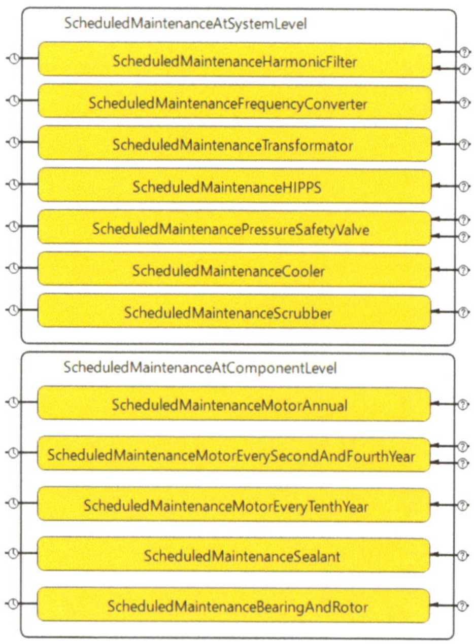

Figure 3. Scheduled maintenance state (at both system and component level) representing opportunistic maintenance intervals.

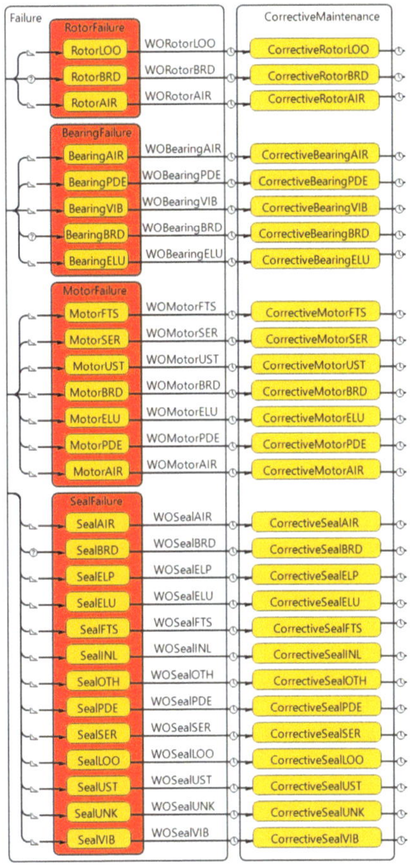

Figure 4. Failure events and corrective maintenance.

Table 2. The connection between states, triggers, logics, and data sources in Figure 2.

From State	To State	Trigger	Logic	Data Source
Normal	Scheduled maintenance non prod. Impact	Condition	Scheduled maintenance > 0	Case study
Scheduled maintenance non prod. Impact	Normal	Timeout	Triangular distr.	Case study
Normal	Condition Monitoring	Rates	Parameter	Case study
Condition Monitoring	Normal	Timeout & Condition	Parameter and randomTrue(failure rate) distr.	Case study
Normal	Intelligent Maintenance	Condition	Opportunistic Maintenance > 0	Calixto
Intelligent Maintenance	Normal	Timeout	18	OREDA

2.2.2. Scheduled Maintenance and Opportunistic Maintenance Intervals

The second sub-model highlights the scheduled maintenance at both system level and component level, as depicted in Figure 3. More specifically, the scheduled maintenance at both component and system level includes all the scheduled maintenance activities that require stoppage of the system under study. In this case, scheduled maintenance at the component level concerns maintenance activities directly connected to the components of the case study, while scheduled maintenance at the system level focuses on system-level (systems connected to the case study, e.g., scrubber, cooler)—hence, scheduled maintenance causing unavailability of any of these systems requires stoppage of the case study.

Table 3 highlights the triggers that are causing transitions between the different states present in the scheduled maintenance and opportunistic maintenance intervals. In this case, the scheduled maintenance states are triggered by rates that are extracted from the case company data. The durations of the states are, on the other hand, highlighted by triangular distributions either based on the case study data or the OREDA database [17] (dependent on use case scenario). Since the MTTR values of the scheduled maintenance activities are average values (assuming normal distribution), the triangular distribution is used to incorporate some variance in the data.

Table 3. The connection between states, triggers, logics, and data sources in Figure 3.

From State	To State	Trigger	Logic	Data Source
Normal	Scheduled maintenance at System Level	Condition	Scheduled maintenance > 0	Case study
Scheduled maintenance at System Level	Normal	Timeout	Triangular distr.	Case study
Normal	Scheduled maintenance at Component Level	Condition	Scheduled maintenance > 0	Case study
Scheduled maintenance at Component Level	Normal	Timeout	Triangular distr.	Case study

The main purpose of modeling the scheduled maintenance at both system and component levels is to address all planned maintenance activities that are causing production stops. These stops are decisive to address as they can be used as opportunistic maintenance intervals in which PdM can be leveraged in terms of intelligent maintenance. Hence, if the intelligent maintenance system enables detecting and predicting the future deterioration propagation of a failure, it can allocate the future required maintenance activity to a coming opportunistic maintenance interval, as long as this interval appears prior to the component fault.

2.2.3. Failure Events and Corrective Maintenance

The third sub-model concerns the occurrence of failure events and the associated corrective maintenance actions required to put the component back in operation. Figure 4 highlights all the failure modes that are associated with the case study based on both the empiric case study data and the OREDA database. The systems analysis step revealed some differences between the failure modes presented in the OREDA database and the ones presented in the case company notification system, as demonstrated in Table 4. Therefore, only the failure modes represented by the specific scenarios are assigned with values traceable to their specific data source, while the failure modes that do not appear in the specific use case scenario are assigned with a value of zero.

During simulation, the "failures" are triggered by either (1) failure rates that are either extracted from the empiric case study data or the OREDA database [17] (Scenarios 1–4) or (2) a condition based on deterioration rates supported by Calixto [25] (only valid for intelligent maintenance and thus Scenario 5). Then, the state of "corrective maintenance" is triggered by timeout functions including a triangular distribution of the MTTR values that are transparent with the specific use case scenario. Since the MTTR values of the corrective maintenance activities are average values (assuming normal distribution), the triangular distribution is used to incorporate some variance in the data. The connection between the states, triggers, values, and data source are summarized in Table 5.

Table 4. Differences in failure modes of centrifugal gas compressor presented by the empiric case study data and the offshore and onshore reliability data handbook (OREDA) database.

Component	Failure Mode	Failure Mode Abbreviation	Case Study	OREDA Database
Rotor	Abnormal instrument reading	AIR	X	-
	Breakdown	BRD	X	-
	Low output	LOO	-	X
Bearing	Abnormal instrument reading	AIR	X	X
	Breakdown	BRD	X	-
	Parameter deviation	PDE	-	X
	Vibration	VIB	-	X
	External leakage—Utility medium	ELU	X	-
Motor	Abnormal instrument reading	AIR	X	-
	Breakdown	BRD	X	-
	Parameter deviation	PDE	X	-
	External leakage—Utility medium	ELU	X	-
	Spurious stop	UST	X	X
	Fail to start on demand	FTS	-	X
	Minor in-service problems	SER	X	X
Seal	Abnormal instrument reading	AIR	X	X
	Breakdown	BRD	X	X
	Parameter deviation	PDE	X	X
	Vibration	VIB	-	X
	External leakage—Utility medium	ELU	X	X
	Spurious stop	UST	-	X
	Fail to start on demand	FTS	-	X
	Minor in-service problems	SER	-	X
	External leakage—Process medium	ELP	-	X
	Low output	LOO	-	X
	Internal leakage	INL	X	X
	Unknown	UNK	-	X
	Other	OTH	-	X

Table 5. The connection between states, triggers, logics, and data sources in Figure 4.

From State	To State	Trigger	Logic	Data Source
Working	Failure	Rates	Parameter	Case study/OREDA
Working [1]	Failure [1]	Condition [1]	Fault > 0 [1]	Calixto [1]
Failure	Corrective Maintenance	Timeout	Parameter	Fixed
Corrective Maintenance	Normal Working	Timeout	Triangular distr.	Case study/OREDA

[1] Only valid for the condition monitoring systems with detection and prediction capabilities. In this paper, this refers to intelligent maintenance.

2.3. Step 4: Scenario Modeling

This section is dedicated to scenario modeling, which facilitates simulating different use case scenarios, and furthermore attaining an understanding of sensitive data and influencing factors identified through the model outputs. To do so, four different use case scenarios are modeled with the purpose of highlighting the associated sensitiveness connected to the model input data i.e., failure rates and MTTR values extracted from either (1) the case study, (2) the well-known OREDA database, [17] which is highly applied in the O&G industry, or (3) both. In final, the last use case scenario (use case scenario 5) that concerns the loading and deterioration process of the case study is modeled. Its purpose is to highlight the connection between component deterioration, detection, diagnosis, and prognosis purposes in the context of implementing an intelligent maintenance management system into the case study. Therefore, this paper models in total five use case scenarios. The connection between the input data and use case scenarios is summarized in Table 6 and described in more detail in the following subsections.

Table 6. The five use case scenarios and associated input data.

Scenario	Input Data				
	Failure Rate		Deterioration Rate	MTTR	
	Case Study	OREDA	Calixto	Case Study	OREDA
1	X	-	-	X	-
2	X		-	X [1]	-
3	-	X	-	-	X
4	X	-	-	-	X
5	X	-	X	-	X

[1] Manipulated input data.

2.3.1. Scheduled and Corrective Maintenance Scenarios (1, 2, 3, and 4)

Scenario 1 includes failure rates and associated MTTR values that are extracted from the notification system of the case company. The data is extracted exactly how it is presented in the notification system.

Scenario 2 includes the same data as in the previous scenario. However, the difference in Scenario 2 is that all the values considered as "unreasonably extreme" are replaced with values the authors anticipate to be more reasonable when taking the connection between the specific failure and associated MTTR value into consideration.

Scenario 3 addresses input data involving both failure rates and MTTR values extracted from the OREDA database [17]. The OREDA database is in fact well-known and highly adopted by O&G companies in connection with analyses concerning risk and technical integrity. In practice, the OREDA database categorizes failure rates in terms of "lower", "mean", and "upper" failure rates, and MTTR values in terms of "mean" and "max". This research adopts the "upper failure rates" and the "max MTTR values", which experts claim to represent industrial experience the best.

The interesting context of this use case scenario is to highlight whether the case company experiences either higher or lower failure rates and MTTR values in comparison to the OREDA database. This will underpin whether the industry shall be recommended to support integrity assessments based upon their own empiric case study data or the OREDA database, dependent on the associated risk profile ("risk-averse", "risk-seeking", etc.).

The estimation of MTTR values originating from the empiric case study data is associated with the highest uncertainty as it depends on two different variables the maintenance personnel need to report (start of maintenance and end of maintenance). Therefore, Scenario 4 replaces the MTTR values from the empiric case study data with the ones presented in the OREDA database.

2.3.2. Intelligent Maintenance Based on Deterioration Modeling (Scenario 5)

One of the main issues of applying failure rates in connection with detection, diagnosis, and prognosis purposes is due to the straight lines in terms of pulses produced by the simulation. In more detail, such straight lines make it difficult, or even impossible, to justify the opportunity to detect, diagnose, and predict future deterioration evolution. The maintenance timeline concept based on failure events is not effective to enable CBM and PdM, as they require deterioration curves instead. Therefore, a deterioration model based on loading that addresses the deterioration curves for the individual component associated with the case study must be modeled.

The deterioration modeling process starts by first modeling the component deterioration using system dynamics, depicted in Figure 5. As seen from the loading model, it contains three different flows: (1) Loading, (2) Failure, and (3) Intelligent maintenance. Furthermore, one stock representing the "accumulated loading", and one parameter of "Opportunities" which represents the future opportunistic maintenance intervals defined by the scheduled maintenance requiring stops in operation.

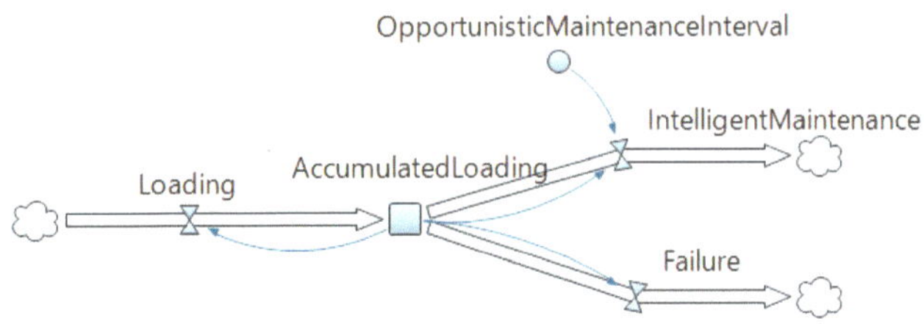

Figure 5. Loading model using systems dynamics.

The logic of each flow in the deterioration and intelligent maintenance module is described in more detail in Table 7.

Table 7. Description of the deterioration and intelligent maintenance module.

Element	Function	Logic
Loading	Estimates the loading rate per hour.	Described in Table 8.
Accumulated Loading	Accumulates the loading rate.	Integral value of loading.
Failure	Triggers failure and the need for corrective maintenance when the deterioration has reached a defined level.	rint(AccumulatedLoading) $\geq$ uniform_discr (100,100)? AccumulatedLoading = initial AccumulatedLoading value: 0
Opportunistic Maintenance Interval	Represents the opportunistic maintenance intervals defined by the scheduled maintenance.	Defines all scheduled maintenance events that are causing a stop in the operation in terms of pulseTrains.
Intelligent Maintenance	Triggers intelligent maintenance event when two conditions are satisfied: (1) opportunistic maintenance interval is available, and (2) the accumulated loading triggers detection or prediction alarm at the defined level.	OpportunisticMaintenanceInterval! = 0 && rint(AccumulatedLoading) $\geq$ uniform_discr (70,100)? AccumulatedLoading = initial AccumulatedLoading value: 0

Table 8. Connection between components, loading equations, and designed lifetimes.

Component	Loading Equation	Designed Lifetime [25]
Rotor	0.00013738 $\times$ AccumulatedLoading (initial AccumulatedLoading value = 0.5)	4.4 years (38,544 h)
Bearing	0.00010045 $\times$ AccumulatedLoading (initial AccumulatedLoading value = 0.5)	6.0 years (52,560 h)
Seal	0.00075232 $\times$ AccumulatedLoading (initial AccumulatedLoading value = 2×10^{-14})	4.7 years (41,172 h)

In more detail, the "Loading" flow expresses the entire deterioration process and includes the loading equation of the specific component under study. Such an equation can be established by first identifying a failure distribution that demonstrates the evolution of a specific failure through a deterioration curve. To do so, there exist several failure distributions applied to demonstrate the degradation evolution from a healthy component to a faulty one [26–28]. Some of the most applied failure distributions concerning aging equipment are, e.g., "traditional view", "bathtub curve", and "slow aging" (linear deterioration) [29]. However, concerning component deterioration, the distribution of either exponential distribution or power-law distribution is most frequently adopted.

Second, a designed load case that assumes constant loading from the beginning of the operation until it fails must be addressed. Such time to failure can for instance be based on recommendations from the component vendor, estimated through equations offered by the manufacturer (e.g., [30]) or other reliable data sources (e.g., Calixto [25] or OREDA [17]). In final, the suitable deterioration curve identified must be fitted with the time to failure through an iterative simulation process that highlights the entire deterioration process from a healthy component to a faulty one.

In practice, the condition monitoring system monitors the deterioration process of the component during normal operation. If the monitoring system is not capable of detecting and predicting the deterioration, the accumulated deterioration level reaches the designed lifetime (100% deterioration) that triggers the "Failure" flow in system dynamics (Figure 5), which furthermore triggers the associated "failure" state in the agent-based computational model (shown in Figure 1). However, if the condition monitoring system is able to detect and predict the level of deterioration propagation prior to component failure, it tries to leverage the PdM event into a coming opportunistic maintenance interval represented by the "OpportunisticMaintenanceInterval" parameter connected to the "IntelligentMaintenance" flow. In this case, the opportunity of leveraging a predicted failure event to a future opportunistic maintenance interval is based on two criteria: (1) when the future opportunistic maintenance intervals appear and (2) the capabilities offered by the specific monitoring system, i.e., levels of detection and prediction that are demonstrated in detail in [16]. Illustratively, if the condition monitoring system is able to detect component deterioration one week prior to failure and the next opportunistic maintenance interval appears first after four weeks, there exists no opportunity to leverage the PdM into an opportunistic maintenance interval, and corrective maintenance is thus required. In contrast, if the deterioration is detected five weeks prior to component failure and the next opportunistic maintenance interval appears after four weeks, the PdM can be leveraged into the future opportunistic maintenance interval in terms of "intelligent maintenance". Therefore, exploiting these opportunistic maintenance intervals to perform intelligent maintenance will thus reduce the unplanned operational unavailability and cost (since corrective maintenance is replaced by intelligent maintenance).

At last, this paper presents an illustrative example of how an intelligent maintenance system that enables detecting, diagnosing, and predicting the specific failure mode of breakdown (BRD) of the rotor, bearing, and seal. It is important to emphasize that the transparency between failure modes, and detection, diagnosis, and prognosis processes shall be analyzed individually for the specific condition monitoring system applied. In this context, the authors recommend the future readers perform the analysis presented in [16] to determine these specific capabilities of an associated condition monitoring system of interest.

The final use case scenario, Scenario 5, is dedicated to the deterioration modeling of the components associated with the case study, i.e., rotor, bearing, and seal. Modeling component deterioration is required to highlight the capabilities of implementing an intelligent maintenance management system, i.e., levels of detection, diagnosis, and prognosis [16]. This paper develops individual loading equations of the components of interest, based on the plot presented by Calixto [25]. Since Calixto only represents the deterioration curves and not the specific deterioration equations, the associated loading equations presented in Table 8 are replications.

The individual loading equations developed are then incorporated into the "Loading model" (shown in Figure 5) and are simulated and optimized to fit the designed lifetime presented by Calixto [25] using Anylogic, as depicted in Figure 6. As seen, the deterioration curves highlight the entire deterioration process from when the specific component starts operating until a fault is present at the designed lifetime. The deterioration curves also demonstrate that the component deterioration propagates differently. Clearly, this affects the opportunities of detecting, diagnosing, and predicting the future behavior of the associated component deterioration process. For example, the deterioration of the seal appears with a steeper slope in comparison to the two other components, which thus reduces the opportunities of performing intelligent maintenance as it is more difficult to detect and predict the occurrence of seal deterioration. In contrast, the deterioration curve of bearing introduces the gentlest slope, which increases the opportunities to perform intelligent maintenance as it is possible to detect and predict the occurrence of bearing deterioration at an early stage.

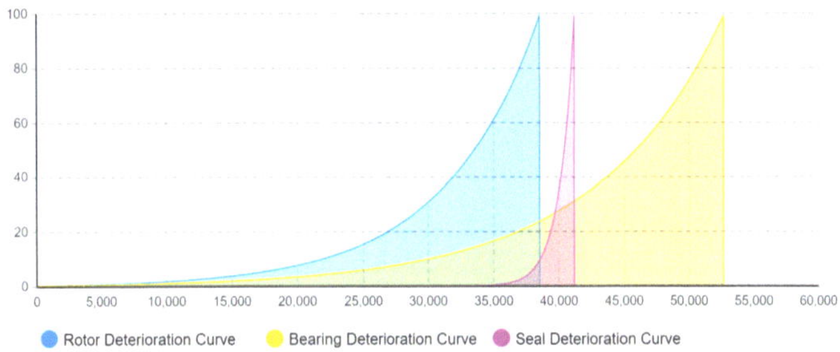

Figure 6. The deterioration curves modeled for each component, individually.

It is important to highlight that this research adopts a detection and prediction level of 70% for Scenario 5. The main justification of this selection is based on recommendations from the literature [31] and experts in the field.

2.4. Steps 5 and 6: Verification, Validation, and Visualization

The fifth step in the simulation modeling methodology concerns the verification and validation of the simulation. In this case, all the applied data, i.e., operation and maintenance including condition monitoring (Figure 2), scheduled maintenance plans (Figure 3), and experienced failures and following corrective maintenance (Figure 4) including failure modes, failure rates, and MTTR values are extracted from the notification system of the case company and incorporated into the computational model. The applied data is also validated through several discussions with engineers and experts in the field represented by the case company and stakeholders for verification and validation purposes to attain a correct description of the case study, to increase the reliability of the results obtained from the simulations. In final, to improve the reliability of the simulations even more, similar data, i.e., failure modes, failure rates, and MTTR values are extracted from the well-known OREDA database [17] and also compared to the real-time data extracted from the notification system of the case company.

The computational model is considered generic in that sense the future adopter can fit the model to their own purposes. In more detail, this means that the future adopters can replace the components with the ones of interest. Furthermore, the associated scheduled maintenance causing operational unavailability and thereby representing opportunistic maintenance intervals, and corrective maintenance data including failure modes, failure rates, and MTTR values can be replaced. This means that all data can be replaced by the ones of interest, however, the logic of the model must be kept, i.e., triggers and equations.

3. Results

The simulated results for the five scenarios are presented in accordance with the model outputs presented in Section 2.1. The following results are presented for each use case scenario: (1) operational behavior, (2) maintenance event: timeline and workload, and (3) component failure during a time period of 20 years (175,000 h). In addition, one subsection is dedicated to comparing the corrective maintenance (Scenario 4) with the intelligent maintenance scenario (Scenario 5). At last, one subsection demonstrates the effect of proliferating PdM capabilities in connection with intelligent maintenance.

3.1. Operational Behavior and Availability

The operational behavior highlights the operational availability and unavailability for each scenario, shown in Figures 7–11 and summarized in Table 9. In the figures, the y-axis shows the operational behavior in percentage as a function of time in hours represented by the x-axis.

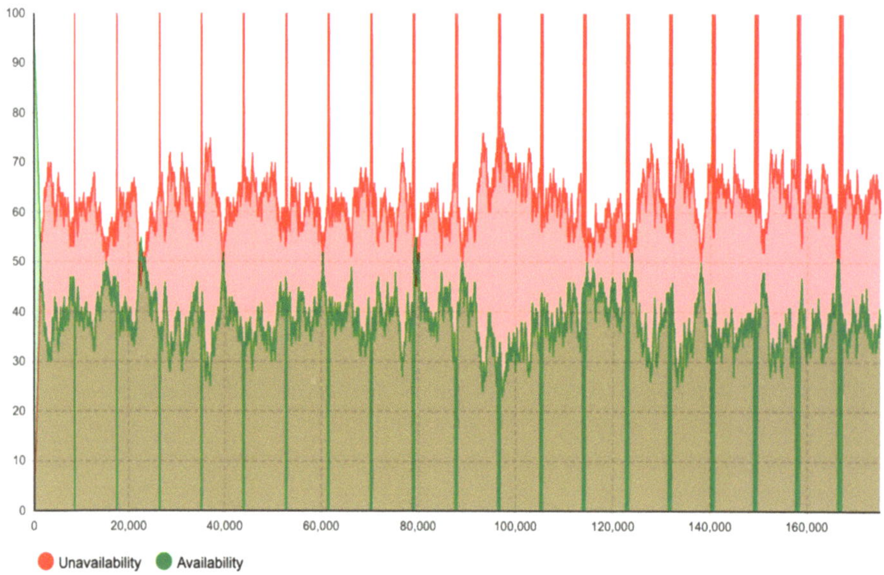

Figure 7. Operational behavior of Scenario 1 during 20 years of operation.

Figure 8. Operational behavior of Scenario 2 during 20 years of operation.

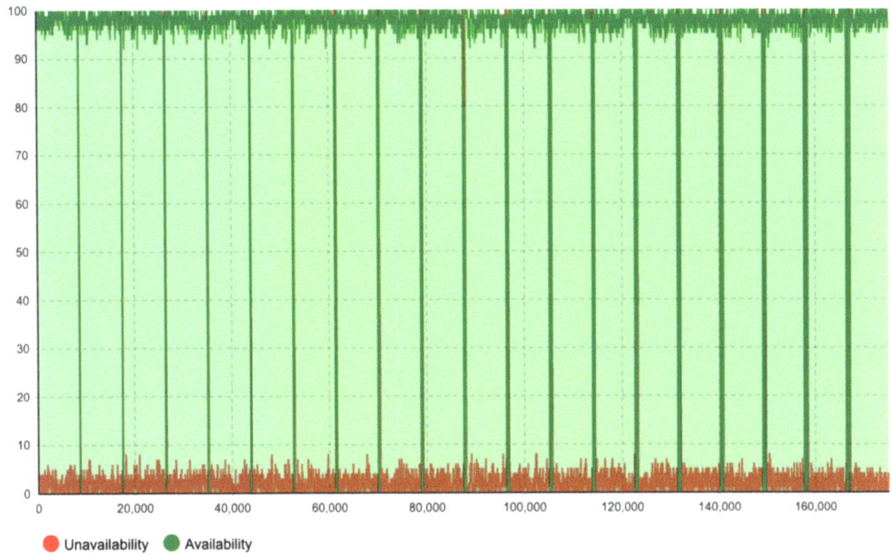

Figure 9. Operational behavior of Scenario 3 during 20 years of operation.

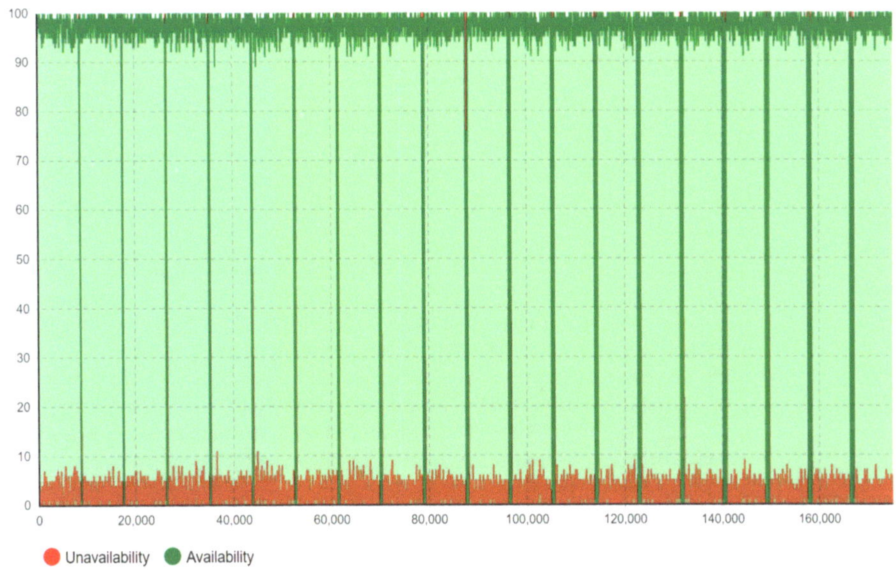

Figure 10. Operational behavior of Scenario 4 during 20 years of operation.

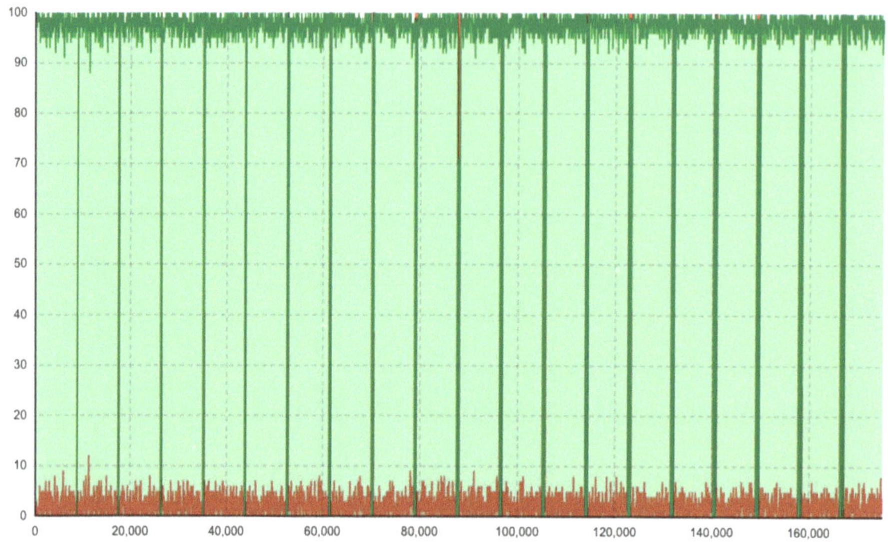

Figure 11. Operational behavior of Scenario 5 during 20 years of operation.

Table 9. Summary of operational behavior of the different simulated use case scenarios.

Scenario	Operational Behavior [%]		
	Availability	Unavailability	Total
1	37.756	62.244	100
2	73.384	26.616	100
3	96.220	3.780	100
4	95.750	4.250	100
5	96.018	3.982	100

The results highlight that Scenario 1, which is based on case study historical failure data, offers the lowest operational availability of approximately 37.756% availability during 20 years of operation, which is not valid. Scenario 2 provides more valid numbers of availability with 73.384%, as the unreasonably extreme data values (closed dates for maintenance events) caused by human factors in the notification process are manipulated and corrected. Scenario 3, which is based on OREDA failure rates and MTTR values, highlights the availability of 96.220%, which is a significant improvement compared to scenarios 1 and 2. Scenario 4, which is a mixed scenario based on case study failure rates and OREDA MTTR values, shows the availability of 95.750%. Scenarios 3 and 4 demonstrate the effect of failure rates on availability, where the availability decreased 0.470% (96.220–95.750%) when case study failure rates are used. Scenarios 2 and 4 highlight the effect of MTTR on availability, where the availability increased 22.366% (95.750–73.384%) when OREDA MTTR values are used. Scenarios 3 and 4 are valid scenarios when compared with real availability numbers. However, it shows that OREDA is more reliable and valid in terms of MTTR data, as the used unit is hours, not days like in case study data. In final, Scenario 5 addresses a total of operational availability of 96.018%, which corresponds to an improvement of 0.268% (96.018–95.750%) in comparison to Scenario 4, which includes the same input data but without an intelligent maintenance system.

3.2. Maintenance Event: Timelines and Workloads

This subsection presents the maintenance event timelines during 20 years of operation that include the associated workloads of (1) scheduled maintenance policy, (2) corrective maintenance policy with five different scenarios, and (3) intelligent maintenance policy.

3.2.1. Scheduled Maintenance Event Timeline

The scheduled maintenance plan provided by the case study company covers two categories: (1) downtimes that lead to production stoppage, and (2) downtimes that have no effect on production. In this context, the scheduled maintenance intervals that lead to production stoppages are especially interesting as they represent opportunistic intervals, as shown in Figure 12. In the figure, the x-axis represents the simulated time in hours and the y-axis defines the state of the operation that is either characterized by 0 or 1. Furthermore, 0 defines an operational state when the equipment operates as normal and 1 defines an operational state when the equipment is out of operation due to scheduled maintenance. As highlighted by the figure, there are in total 19 opportunistic maintenance intervals occurring during 20 years of operation, which PdM can be leveraged into in terms of intelligent maintenance.

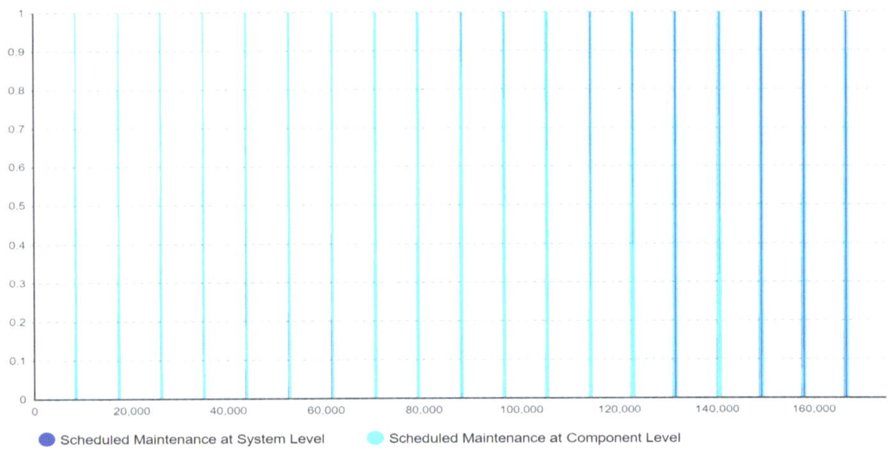

Figure 12. Scheduled maintenance representing opportunistic maintenance intervals during 20 years of operation.

3.2.2. Corrective Maintenance Event Timeline

The corrective maintenance event timelines for the five modeled use case scenarios are depicted in Figures 13–17. In the figures, the x-axis represents the simulated time in hours and the y-axis defines the state of the operation that is either characterized by 0 or 1. Furthermore, 0 defines an operational state when the equipment operates as normal and 1 defines an operational state when the equipment has failed and is out of operation due to corrective maintenance. The planned maintenance timeline, presented in Figure 12, is also included in all these corrective scenarios. The corrective maintenance event timeline shows the failure and corrective maintenance events for the most important components, i.e., bearing, rotor, seal, and motor. The corrective maintenance timeline of Scenario 1, Figure 13, clearly highlights the effect of incomplete maintenance data, where some maintenance events have quite a long maintenance time interval (due to the closing date either being missing or considered as an unreasonable extreme value). This issue was enhanced in Scenario 2, shown in Figure 14, where the closing dates were manipulated. Thus, the availability has changed from 37.756% in Scenario 1 to 73.384% in Scenario 2, which means that 35.629% of the availability in Scenario 1 is just related to incomplete maintenance dates. However, the availabilities obtained for Scenarios 1 and 2 do not match the real availability figures.

One issue that shall be highlighted is that the time unit used for maintenance time (MTTR) in the case company is days. It means the maintenance time for any maintenance event will have a minimum duration of 24 h, even though it might take 4 h in reality. This issue can clearly be illustrated when OREDA MTTR values are used, as in Scenario 4 (Figure 16), where the availability increases from 73.384% (days used in Scenario 2) to 95.750% (hours used in Scenario 4). Thus, the time unit for the MTTR values is of significant importance which may lead to an error of 22.366% (95.750–73.384%) in availability if days unit is used instead of hours.

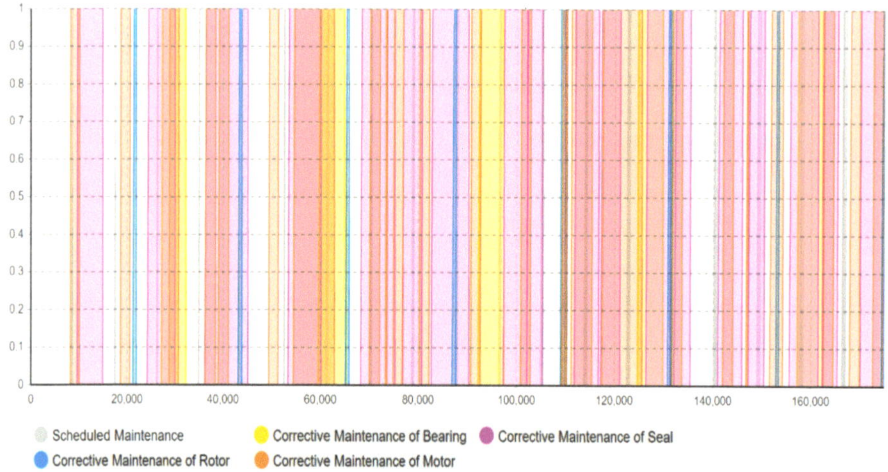

Figure 13. Corrective maintenance event timeline of Scenario 1 during 20 years of operation.

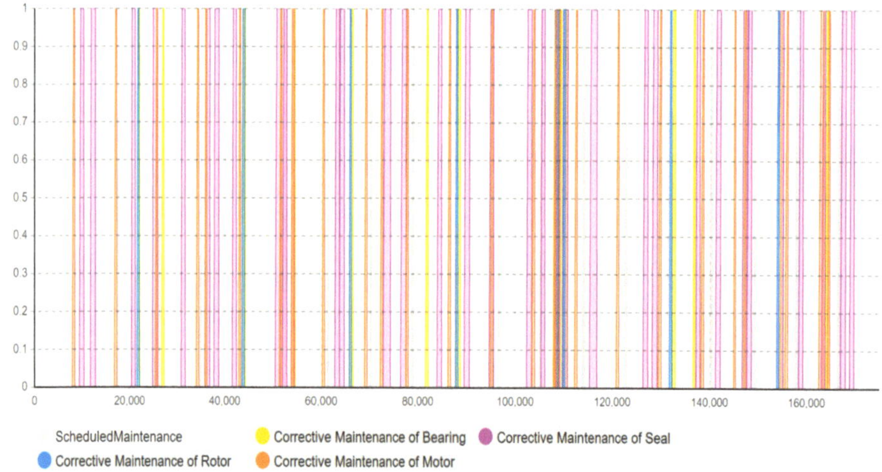

Figure 14. Corrective maintenance event timeline of Scenario 2 during 20 years of operation.

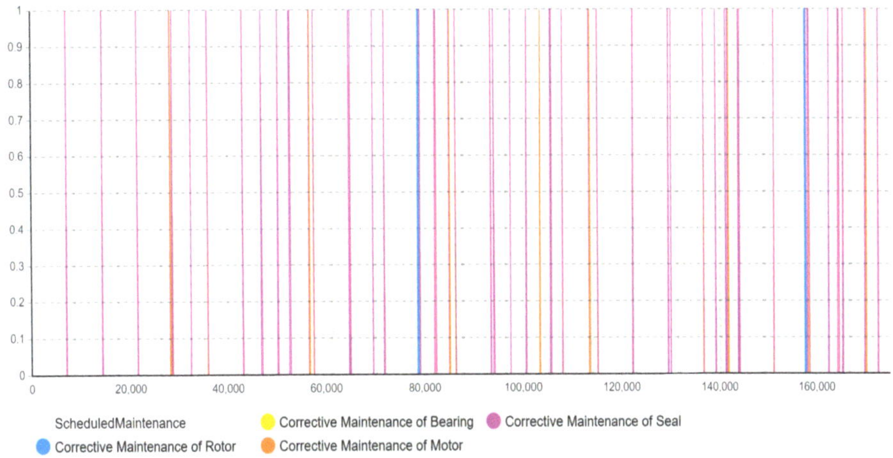

Figure 15. Corrective maintenance event timeline of Scenario 3 during 20 years of operation.

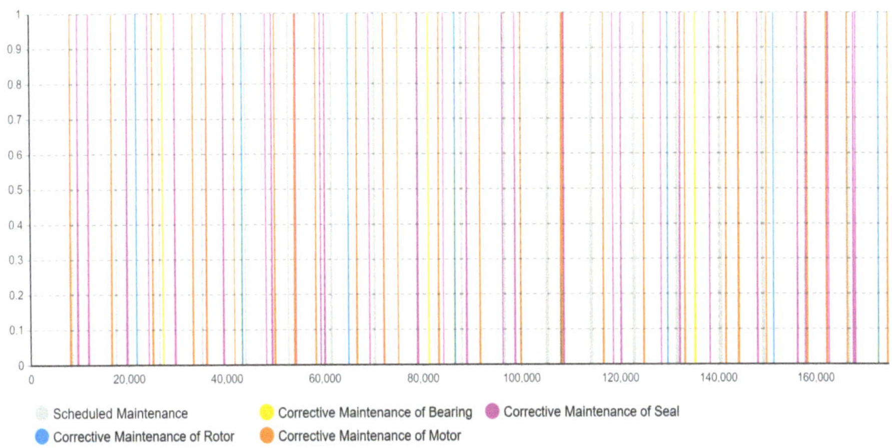

Figure 16. Corrective maintenance event timeline of Scenario 4 during 20 years of operation

The maintenance timeline for Scenario 3 (Figure 15), which is based on OREDA failure rates and MTTR values, shows different numbers (at the component level) and locations of corrective maintenance events compared to Scenario 4 (which is based on case study failure rates and OREDA MTTR values). This is caused by the fact that the scenarios include different failure rates at the component level. However, the number of corrective maintenance events in total (at equipment level) and availability are almost the same. The corrective maintenance timeline for Scenario 5 (Figure 17) provides a smaller number of corrective maintenance events compared to Scenario 4 (Figure 16), due to the lower failure rates presented by the deterioration curves. For example, bearing deterioration curves estimate the bearing to fail every six years (around three times during 20 years of operation), while the OREDA database anticipates that the bearing will fail 26 times during 20 years of operation.

In summary, the maintenance timelines (presented in Figures 13–17) visualize the number of corrective maintenance events with their associated maintenance time intervals. To get more insight into the maintenance workload for each scenario, Table 10 is provided. In addition, Table 11 presents the number of maintenance events for each scenario in more detail.

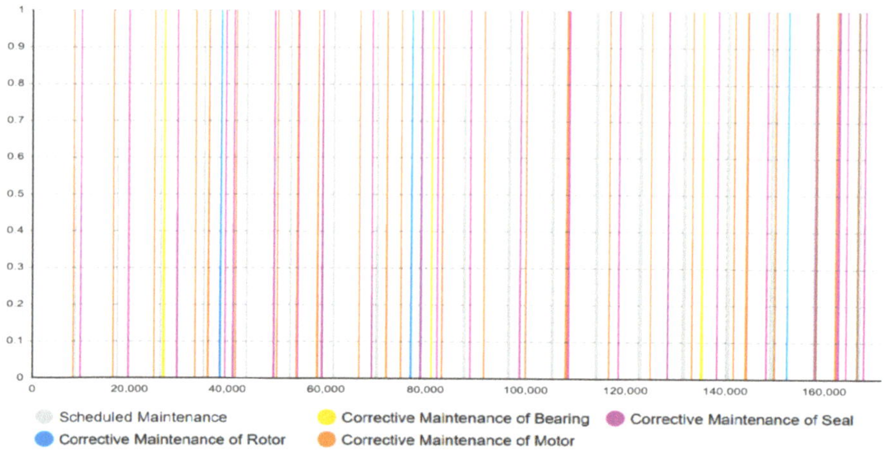

Figure 17. Corrective maintenance event timeline of Scenario 5 during 20 years of operation.

Table 10. Summary of the maintenance workload of the five simulated use case scenarios.

Scenario	Maintenance Workload [h]				
	Corrective Maintenance	Scheduled Maintenance		Intelligent Maintenance	Total
		with Impact	without Impact		
1	107,700	1311	3425	0	112,436
2	44,055	2578	6682	0	53,316
3	3385	3237	8763	0	15,385
4	4183	3262	8736	0	16,181
5	3724	3251	8765	360	16,101

Table 11. Failures occurring during 20 years of operation based on the simulated results.

Component	Number of Failures during 20 Years of Operation				
	Scenario				
	1	2	3	4	5
Rotor	8	8	2	8	3
Bearing	13	14	26	13	5
Seal	19	35	45	29	22
Motor	24	27	7	26	26
Total	64	84	80	76	56

By comparing the number of failures for Scenarios 1, 2, and 4 (which is based on case study historical failure data) and for Scenario 3 (based on OREDA data), it can be observed that the OREDA data is underestimating the number of failures for some components (e.g., rotor and motor), while it is overestimating the number of failures for some components (e.g., bearing and seal). That might be due to the mean issue (as the OREDA database presents the mean and upper failure rates of several failure events occurring based on several installations, where some failure events might even not be registered). By comparing the number of failures occurring in Scenarios 1, 2, and 4 with Scenario 5, it can be concluded that Scenario 5 (based on well-known deterioration curves) underestimates the number of failures. Let us take rotor as an example, the mean time between failures based on the deterioration curve is 4.4 years for rotor (see Table 8), while the mean time between failures based on case study historical failure data is 2.5 years (see Table 11). This can be justified as the available deterioration curves that authors could find only cover three out of twenty-eight failure modes considered in Scenario 1, 2, and 4.

3.3. Intelligent Maintenance and the Effect of Proliferating Detection and Prediction Levels

This subsection demonstrates the maintenance events based on prediction leveraged intelligent maintenance at several detection and prediction levels (50, 60, 70, 80, 90, and 100%), as depicted in Table 12. In contrast to the other use case scenarios, these results are only based on designed lifetimes presented by Calixto specifically (see Table 8). Therefore, the sole purpose of this subsection is to study if the assumption of increasing levels of detection and prediction offers increasing flexibilities of leveraging PdM into an opportunistic maintenance interval in terms of intelligent maintenance. Not surprisingly, the table underpins the connection between increasing detection and prediction levels, and the opportunity of performing intelligent maintenance. Furthermore, it highlights that the detection and prediction levels of 50% and 60% provide the same opportunities in this case (in total eight intelligent maintenance events), while the detection levels of 70, 80, and 90% include five, four, and three intelligent maintenance events, respectively.

Table 12. The effect of proliferating detection and prediction levels.

Detection and Prediction Level	Corrective Events	Opportunistic Events	Corrective Maintenance Reduction [%]
50%	3	8	72.727
60%	3	8	72.727
70%	6	5	45.454
80%	7	4	36.363
90%	8	3	27.272
100%	11	0	0

4. Discussion and Validation

4.1. Data Collection

Data in terms of scheduled maintenance plans and experienced corrective maintenance including failure modes, failure rates, and MTTR values were extracted from the notification system of the case company and incorporated into the computational model. In addition, several discussions with engineers have been conducted to attain a correct description and understanding of the case study and its data. In final, to improve the reliability of the simulations, data including failure modes, failure rates, and MTTR values were extracted from the OREDA database and compared with the experienced case study data.

4.2. Human Factors in Notification Processes

The data collection process of the empiric case study data became a lot more time-consuming than first anticipated. Its sole reason is traced back to human factors present in the notification processes that evidentially reduced the quality of the data significantly. This issue is clearly demonstrated by the simulated results. Use case Scenarios 1, 2, and 4 include the same failure rates extracted from the case study but with different MTTR values. In more detail, Scenario 1 includes MTTR values as presented in the notification process, Scenario 2 includes manipulated MTTR values considered as unreasonable extremes in the previous scenario, and Scenario 4 includes MTTR values extracted from the OREDA database [17]. Therefore, the volatile changes in operational behavior and significant differences in maintenance workload between these three scenarios are solely traced back to the MTTR values. Respectively, the maintenance workload (in hours) devoted to corrective maintenance for Scenarios 1, 2, and 4 are 107,700, 44,055, and 3262 with associated operational availabilities of 37.756, 73.384, and 95.750%.

From the authors' perspective, the main issue of incorporating human factors in the notification processes is traced back to the maintenance personnel's opportunity of developing a notification that is solely based on subjective perceptions, without any associated requirements concerning the level of details of the individual notifications. Following this, the simulated results also justify why the O&G companies keep using the OREDA database [17] and not their own empiric data in connection with analysis related to, e.g. technical integrity and risk. However, it is a paradox that case-specific data do not express the case of interest the best. Therefore, for the future, it shall be recommended that the notification processes avoid incorporating human factors, at least, reducing its impact by making the notification process (partial) automatic or based on a templated questionary with pre-defined alternatives the maintenance personnel is required to answer before the notification is considered as complete.

At last, it is also important to emphasize that the failure data originating from the case study includes several components of one component, i.e., rotor, bearing, and seal However, due to difficulties in differentiating between these specific components, the failure rates presented in this research do not take into consideration the number of each component. Illustratively, this means that this research estimates one failure rate composing all the failures associated with one type of component, without taking its population into consideration. Therefore, the failure rate assumes that failure of, for instance, one bearing results in failure of all the bearings present in the case study at the same time.

4.3. Intelligent Maintenance (Scenario 5) vs. Corrective Use Case (Scenario 4)

The final results of this research clearly demonstrate tempting lifetime benefits during 20 years of operation. In comparison, the intelligent maintenance system is expected to improve the operational availability by 0.268% by replacing 2.721% ((4183/16181) − (3724/16101) = 2.721%) of the corrective maintenance workload with intelligent maintenance. In workload, it equals replacing 459 h of corrective maintenance which corresponds to a reduction of 11% ((4183 − 3724)/4183 = 11%) of the total corrective maintenance workload. Specifically, the intelligent maintenance system reduced the unintended corrective maintenance visits by 20 (26.316%), whereas a reduction of 5 (62.500%), 8 (61.538%), and 7 (24.138%) corrective maintenance visits are traced back to the rotor, bearing, and seal, respectively. Following, these 20 corrective maintenance events were replaced by intelligent maintenance which leverages the PdM capabilities into opportunistic maintenance intervals and thereby does not affect the operational availability.

4.4. Additional Lifetime Benefits of Intelligent Maintenance in Industry 4.0

There exist some aspects that can improve the lifetime benefits even more, which are not presented in this paper. First, reducing component loading to extend the remaining useful life estimation and by this reach an opportunistic maintenance interval that was initially not reachable. Second, the expected improvements in terms of maintenance performance and in reducing the level or repair. In fact, enabling detecting, diagnosing, and predicting the future behavior of component deterioration is expected to support developing detailed work orders and ensure that the necessary spare parts and resources are available at the time of intelligent maintenance. However, since the proposed intelligent maintenance system remains to be implemented, it is difficult to justify the realistic values of these improvements. Nevertheless, this can be implemented in a future stage after obtaining operational experience post the implementation, which is traceable back to the MTTR values presented in, e.g., the notification system.

4.5. Intelligent Maintenance vs. Maintenance 4.0

There exists a large number of terminologies that are supposed to define maintenance management in Industry 4.0 such as, e-Maintenance [32], intelligent maintenance [33,34], smart maintenance [35], deep digital maintenance [36], and Maintenance 4.0 [37]. However, this paper adopts the terminology of intelligent maintenance, which intentionally differs from other terminologies e.g., smart maintenance [35], e-maintenance [32], as the focus is not primarily based on data analysis i.e., detection, diagnosis, and prognosis. However, this paper extends the scope to also consider enterprise-level data e.g., spare part management, seasonal loadings, available resources, in order to provide a solid foundation for the maintenance decision management that shall ensure that the right maintenance takes place at the right time. Furthermore, the term Maintenance 4.0 might bring the question about other technologies like robotics, augmented reality, additive manufacturing, i.e., 3D printed spare parts.

4.6. From a Case-Specific Computational Model into a Generic Computational Model

Although this paper develops a computational model based on a case study and presents simulated results associated with the case-specific data, it is important to emphasize that the computational model is easily converted to other cases of interest as the paper adopts a generic research methodology. To do so, the future adopter solely needs to incorporate general information from the specific case of interest including failure modes, scheduled maintenance plans representing opportunistic maintenance intervals, failure rates, and MTTR values. In this context, the authors recommend future adopters apply the PdM assessment matrix [16] to identify associated failure modes, failure mechanisms, and to determine the levels of detection, diagnosis, and prognosis associated with the specific condition monitoring system included in the case of interest. The only requirement is that the computational model presented in this research retains its model structure, triggers, and logic.

5. Conclusions

The simulated results obtained from the multi-method computational model developed in this paper clearly show the ability to estimate the lifetime benefits of applying several maintenance strategies (preventive, corrective, predictive, and opportunistic) on an industrial asset. Simulating preventive, corrective, and opportunistic maintenance is already done in literature (discussed in the introduction). The novelty and scientific contribution of this computational model is mainly traced back to its ability to (1) simulate and estimate CBM and PdM behaviors and their lifetime benefits, (2) leverage PdM into opportunistic maintenance in terms of intelligent maintenance, and (3) estimate and quantify the maintenance workload and determine the specific maintenance event timeline.

Simulating CBM and PdM behaviors was enabled by the deterioration timeline concept where a deterioration curve based on loading profile is simulated, and detection and prediction levels are incorporated. In fact, most of the existing simulation models utilize the failure timeline concept generating pulse train curve, which is useless in order to incorporate detection and prediction levels. It can be concluded that the load-based deterioration curve, shown in Figure 6, is an effective concept to enable the lifetime benefits estimation of CBM and PdM. Definitely, this is a challenging issue since there are some components that either have an unknown deterioration curve or random failure curve (undetectable or unpredictable). For example, only deterioration curves for the rotor, bearing, and seal were available for this case study.

The developed multi-method simulation model enables leveraging PdM capabilities into potential opportunistic intervals in terms of intelligent maintenance. It enables studying if the designed PdM specifications support gaining the lifetime benefits by utilizing potential opportunistic intervals or not. It is a core aspect to consider whether the maintenance system is intelligent or not. Intelligence in this context means that the maintenance management system is able to use detection, prediction, and scheduling analytics to optimize the maintenance events and utilize opportunistic intervals. It can be concluded based on Table 11 that the corrective maintenance events were reduced by earlier detection level or farther predictive horizon, e.g., detection and prediction at 60% of a component lifetime offers increased lifetime benefits (72.727% reduction in corrective maintenance events related to bearing, seal, and rotor) compared with the corrective maintenance reduction percent (27.272%) at 90% of asset lifetime. Please note that PdM at 90% of a component lifetime is capable of detecting sudden failures before their occurrence, however, the opportunistic intervals will not be utilized due to the short time notice. It is important to enable maintenance engineers to determine the optimal technical specifications, i.e., detection and predictive capabilities, and be able to revise and optimize such technical specifications at the design phase.

Moreover, this model has adopted the "timeline" concept to estimate and quantify the maintenance workload amount (how much) in the specific timeline (when), rather than just the accumulated workload amount for the entire lifetime. As shown in Figures 13–17, the corrective maintenance events are time-specific. The timeline concept is required and highly useful for maintenance scheduling purposes, especially, to utilize opportunistic maintenance (based on usage or season) in an intelligent manner. Regarding the quantification of lifetime benefits of intelligent maintenance, the developed simulation model mainly covers two aspects of lifetime benefits (1) operational behavior and (2) maintenance workload For example, the intelligent maintenance system for this case study at 70% detection and prediction level (able to detect failures after 70% of the asset lifetime), is estimated to improve the operational availability by 0.268% (shown in Table 9) and reduce the maintenance workload devoted to corrective maintenance by 459 h (based on Table 10) which equals 11% during 20 years of operation. Furthermore, intelligent maintenance management is also estimated to reduce the scheduled maintenance workload (that leads to downtime) by 0.339% ((3262–3251)/3262 = 0.339%), however, it will increase the scheduled maintenance workload (that does not lead to downtime) by 0.333% ((8765–8736)/8736 = 0.333%).

In summary, the developed simulation model has shown the ability to estimate the lifetime benefits in terms of operational availability and its reduction of corrective maintenance workload. The authors claim that the lifetime benefits of intelligent maintenance will become even greater than what is anticipated in this paper, once other lifetime benefit aspects, which are not covered by this research, are considered. This includes lifetime benefit aspects, i.e., increasing both number and levels of detection and prediction of failure modes, improving maintenance performance by reducing the level of repair, reducing scheduled maintenance workload, enhancing asset performance, lifetime extension measures for tactical and strategical decisions, and health, safety, and environmental issues, and capital allocations. Definitely, the simulation model shall be developed further to estimate all these lifetime benefits.

The structure of the developed simulation model is valid as it was extracted and validated based on experts from the case study. The structure illustrated in the statechart (Figure 1) represents (1) maintenance policy type (corrective and scheduled) and decision making (trigger and condition to get notifications), and (2) failure modes. The statechart represents how the system in this specific case company generates failure or maintenance notification and how it can trigger maintenance events. It is important to highlight that this state chart is valid for other O&G companies operating in the Norwegian Continental Shelf. Regarding the failure modes, the statechart considers all standardized failure modes (based on ISO14224) matching the well-known OREDA database. Thus, the authors claim that the presented state chart is generic for O&G compressors, while the methodology is generic for any equipment of interest.

The model inputs are also analyzed in a pragmatic manner, i.e., several data sources (historical data records from the case study, OREDA, and physics-based deterioration curves). The historical data related to failure and corrective maintenance events provide valid and reliable failure rates and MTTR values, as long as the incomplete data (e.g., maintenance ending date) are manipulated. Failure rates and MTTR values extracted from the OREDA data are well known and accepted in the Norwegian O&G industry as a valid and reliable source of information. The deterioration curves extracted from Calixto [25] are also valid and reliable curves.

The model outputs, i.e., simulated behaviors and estimated key performance indicators have been validated by comparing them to real-world data (case study historical data). The simulated availability and corrective maintenance timelines were validated with case company experts and numbers originating from case study literature [1]. It can be concluded that the computational model is quite effective in terms of computation time. This simulation model uses hours as time-unit, which means it simulates failure rate per hour and checks all conditions (triggers) every time unit. It takes on average around 48 h (where a "normal computer" is used) to provide results at equipment level, i.e., compressor. However, for future simulations, the authors recommend days as time-unit, especially once the model is scaled up to system-level, i.e., compression section and plant-level. In addition, it is recommended that the failure rates are simulated based on years instead of hours.

The computational model is easily generalized to fit any condition monitoring system of interest. In this context, future adopters solely need to incorporate general information from the specific case of interest, i.e., failure modes, scheduled maintenance representing opportunistic maintenance intervals, failure rates, and MTTR values. In fact, the authors recommend applying the PdM assessment matrix [16] to identify associated failure modes, failure mechanisms, and to determine the levels of detection, diagnosis, and prognosis associated with the specific condition monitoring included in the case of interest. The only requirement is that the computational model presented in this retains its model structure, triggers, and logic.

Regarding scenarios (Table 6), it is recommended to use Scenario 4 for further simulation as the failure rates are quite reliable in the case study historical data, while the MTTR values presented by the OREDA database are most reliable and accurate (presented in hours in comparison to the case study presenting the MTTR in days).

At last, besides the quantifiable results presented in this research, it also addresses the sensitiveness and challenges concerning incorporating human factors into the failure notification processes. From the authors' perspective, the main issue of incorporating the human factors in the notification processes is traced back to the maintenance personnel's opportunity of developing a notification that is solely based on subjective perceptions, without any associated requirements to the level of detail for the individual notification. Following this, the simulated results also justify why O&G companies keep using the OREDA database [17] and not the company's own empiric data in connection with analysis related to, e.g., technical integrity and risk. Therefore, for the future, it shall also be recommended that the notification processes avoid incorporating human factors, at least, reducing its impact by making the notification process (partially) automatic or based on a templated questionary with pre-defined alternatives that the maintenance personnel are required to answer before the notification is considered as complete.

Author Contributions: H.N.—Conceptualization, Data curation, Formal analysis, Investigation, Methodology, Project Administration, Resources, Software, Validation, Visualization, Writing—original draft, Writing—review and editing; I.E.-T.—Conceptualization, Funding acquisition, Investigation, Methodology, Resources, Software, Supervision, Validation, Visualization, Writing—original draft, Writing—review and editing. Both authors have read and agreed to the published version of the manuscript.

Funding: This research including APC was funded by the University of Stavanger, grant number IN-11626.

Institutional Review Board Statement: Not applicable.

Informed Consent Statement: Not applicable.

Acknowledgments: The authors acknowledge the University of Stavanger for funding this research (grant number: IN-11626). It would not have been possible without your financial support. In addition, the authors express their gratitude to the case study company for their continuous support in providing interesting industrial data.

Conflicts of Interest: The authors declare no conflict of interest.

References

1. World Economic Forum. *Digital Transformation Initiative Oil and Gas Industry*; World Economic Forum: Geneva, Switzerland, 2017
2. Joshi, A.S. Evaluating the Benefit of PdM on Critical Equipment by Use of Modelling and Simulation: A Case Study. Master's Thesis, University of Stavanger, Stavanger, Norway, 2020.
3. Alabdulkarim, A.A.; Ball, P.D.; Tiwari, A. Applications of simulation in maintenance research. *World J. Model. Simul.* **2013**, *9*, 14–37.
4. Crespo Marquez, A.; Usano, R. Learning Maintenance Management through System Dynamics: A Continuous Improvement Process. In Proceedings of the 12th International Conference of the System Dynamics Society, Stirling, Scotland, 1994; Available online: https://proceedings.systemdynamics.org/1994/proceed/papers_vol_2/cresp020.pdf (accessed on 10 January 2021).
5. Thun, J.-H. Maintaining preventive maintenance and maintenance prevention: Analysing the dynamic implications of Total Productive Maintenance. *Syst. Dyn. Rev.* **2006**, *22*, 163–179. [CrossRef]
6. Khorshidi, H.A.; Gunawan, I.; Ibrahim, M.Y. Reliability centered maintenance using system dynamics approach. In Proceedings of the IEEE International Conference on Industrial Technology (ICIT), Seville, Spain, 17–19 March 2015; pp. 1932–1936.
7. Zuashkiani, A.; Rahmandad, H.; Jardine, A.K.S. Mapping the dynamics of overall equipment effectiveness to enhance asset management practices. *J. Qual. Maint. Eng.* **2011**, *17*, 74–92. [CrossRef]
8. Honkanen, T. Modelling Industrial Maintenance Systems and the Effects of Automatic Condition Monitoring. Ph.D. Thesis, Helsinki University of Technology Information and Computer Systems in Automation, Espoo, Finland, 2004.
9. Jokinen, T.; Ylén, P.; Pyötsiä, J. *Dynamic Model for Estimating the Added Value of Maintenance Services*; Technical Research Centre of Finland, VTT: Espoo, Finland, 2011.
10. Ferri, G. *Simulation Tool Development to Support Customer-Supplier Relationship for CBM Services*; Politecnico di Milano: Milan, Italy, 2012
11. Linnéusson, G. Towards Strategic Development of Maintenance and Its Effects on Production Performance: A Hybrid Simulation-Based Optimization Framework. Ph.D. Thesis, University of Skövde, Skövde, Sweden, 2018.
12. Linnéusson, G.; Ng, H.C.A.; Aslam, T. Towards strategic development of maintenance and its effects on production performance by using system dynamics in the automotive industry. *Int. J. Prod. Econ.* **2018**, *200*, 151–169.
13. Borshchev, A. *The Big Book of Simulation Modelling: Multimethod Modelling with Anylogic 6*; AnyLogic North America: Chicago, IL, USA, 2013.
14. Endrerud, O.-E.V.; Liyanage, J.P.; Keseric, N. Marine logistics decision support for operation and maintenance of offshore wind parks with a multi method simulation model. In Proceedings of the 2014 Winter Simulation Conference, Savannah, GA, USA, 7–10 December 2014.
15. Adegboye, M.A.; Fung, W.K.; Karnik, A. Recent Advances in Pipeline Monitoring and Oil Leakage Detection Technologies: Principles and Approaches. *Sensors* **2019**, *19*, 2548. [CrossRef] [PubMed]
16. Nordal, H.; El-Thalji, I. Assessing the Technical Specifications of Predictive Maintenance: A Case Study of Centrifugal Compressor. *Appl. Sci.* **2021**, *11*, 1527. [CrossRef]
17. OREDA. Offshore and Onshore Reliability Data Handbook—Topside Equipment. In *Sintef NTNU*, 6th ed.; DNV GL: Trondheim, Norway, 2015; Volume 1.
18. McKee, D.W.; Clement, S.J.; Ouyang, X.; Xu, J.; Romanoy, R.; Davies, J. The Internet of Simulation, a Specialisation of the Internet of Things with Simulation and Workflow as a Service (SIM/WFaaS). In Proceedings of the 2017 IEEE Symposium on Service-Oriented System Engineering (SOSE), San Francisco, CA, USA, 6–9 April 2017; pp. 47–56.
19. Forrester, J.W. System Dynamics, Systems Thinking, and Soft OR. *Syst. Dyn. Rev.* **1994**, *10*, 245–256. [CrossRef]
20. Randers, J. *Elements of the System Dynamics Method*; Cambridge University Press: Cambridge, UK, 1980.
21. Sterman, J.P. *Business Dynamics: Systems Thinking and Modeling for a Complex World with CD-ROM*; McGraw-Hill Education: New York, NY, USA, 2000.
22. Nordal, H.; El-Thalji, I. Modeling a predictive maintenance management architecture to meet industry 4.0 requirements: A case study. *Syst. Eng.* **2020**, *24*, 34–50. [CrossRef]
23. Tako, A.A.; Robinson, S. Model building in System Dynamics and Discrete-event Simulation: A quantitative comparison. In Proceedings of the 2008 International Conference of the System Dynamics Society, Athens, Greece, 20–24 July 2008.
24. Nordal, H.; El-Thalji, I. Operations dynamics of gas centrifugal compressor: Process, health and performance indicators. *Linköping Electron. Conf. Proc.* **2019**, *170*, 229–235.
25. Calixto, E. Chapter 3-Reliability and Maintenance. In *Gas and Oil Reliability Engineering*, 2nd ed.; Calixto, E., Ed.; Gulf Professional Publishing: Boston, MA, USA, 2016; pp. 159–267.
26. *IEC 61649. Weibull Analysis*; International Electrotechnical Commission (IEC): Geneva, Switzerland, 2008.
27. *ISO 13381-1. Condition Monitoring and Diagnostics of Machines—Prognostics—Part 1: General Guidelines*; International Organization for Standardization: Geneva, Switzerland, 2015.
28. Eren, Ş.; Enaras, A. Structure and Behavior in System Dynamics: A Case Study in Logistic. *J. Bus. Res. Turk.* **2017**, *9*, 321–340. [CrossRef]
29. Moubray, J. *Reliability-Centered Maintenance*, 2nd ed.; Industrial Press Inc.: New York, NY, USA, 1997.

30. SKF Bearing Rating Life. Available online: https://www.skf.com/group/products/rolling-bearings/principles-of-rolling-bearing-selection/bearing-selection-process/bearing-size/size-selection-based-on-rating-life/bearing-rating-life (accessed on 12 January 2021).
31. Li, H.; Li, Y.; Yu, H. A Novel Health Indicator Based on Cointegration for Rolling Bearings' Run-To-Failure Process. *Sensors* **2019**, *19*, 2151. [CrossRef] [PubMed]
32. Candell, O.; Karim, R.; Söderholm, P. eMaintenance—Information logistics for maintenance support. *Robot. Comput. Integr. Manuf.* **2009**, *25*, 937–944. [CrossRef]
33. Liyanage, J.P.; Lee, J.; Emmanouilidis, C.; Ni, J. Integrated E-maintenance and Intelligent Maintenance Systems. *Handb. Maint. Manag. Eng.* **2009**, 499–544.
34. Lee, J.; Ni, J.; Singh, J.; Jiang, B.; Azamfar, M.; Feng, J. Intelligent Maintenance Systems and Predictive Manufacturing. *J. Manuf. Sci. Eng.* **2020**, *142*, 1–40. [CrossRef]
35. Marhaug, A.; Schjølberg, P. Industry 4.0 and Smart Maintenance: From Manufacturing to Subsea Production Systems. In *International Workshop of Advanced Manufacturing and Automation (IWAMA 2016)*; University of Manchester: Manchester, UK, 2016.
36. Rødseth, H.; Schjølberg, P.; Marhaug, A. Deep digital maintenance. *Adv. Manuf.* **2017**, *5*, 299–310. [CrossRef]
37. Martinetti, A.; Demichela, M.; Singh, S.; Soares, M.; Silva, J. Maintenance 4.0: Where Are We? A Systematic Literature Review. *Appl. Chall. Maint. Saf. Eng. Ind.* **2020**, 1–30. [CrossRef]

applied sciences

MDPI

Article

The Need for Ecosystem 4.0 to Support Maintenance 4.0: An Aviation Assembly Line Case

Alessandro Giacotto [1], Henrique Costa Marques [1,*], Eduardo Afonso Pereira Barreto [1] and Alberto Martinetti [2]

1 Logistics Engineering Laboratory, Aeronautics Institute of Technology, São José dos Campos, SP 12.228-900, Brazil; agiacott@ita.br (A.G.); barretoe@ita.br (E.A.P.B.)
2 Design, Production and Management Department, University of TWENTE, 7522 NN Enschede, The Netherlands; a.martinetti@utwente.nl
* Correspondence: hmarques@ita.br; Tel.: +55-12-3947-5763

Featured Application: The proposed prescriptive maintenance framework supports the maintenance teams in the Industry 4.0 ecosystem, enhancing the availability of diverse machinery and team capabilities.

Abstract: Manufacturing and assembling aircraft require hundreds of different machines for various process applications. The machines have different complexity and often different ages; however, they have to ensure a higher precision than other industrial fields. Recent technology advancement in maintenance approaches offers a wide range of opportunities to provide performance and availability. The paper discusses how the maintenance technologies applicable to the various machines need to be appropriately supported by a production environment, called "ecosystem", that allows their integration within the process and their synergy with the operators. (1) A background analysis of the aircraft production environment is offered. (2) A possible framework for designing a proper ecosystem 4.0 for integrating maintenance activities with design solutions and data gathering is provided. (3) A case study based on the assembly line of specific aircraft is adopted for testing the validity of the framework. (4) Finally, a discussion highlights the critical points of the research, underlying future work.

Keywords: ecosystem 4.0; maintenance 4.0; aviation; aircraft manufacturing; assembly line

Citation: Giacotto, A.; Costa Marques, H.; Pereira Barreto, E.A.; Martinetti, A. The Need for Ecosystem 4.0 to Support Maintenance 4.0: An Aviation Assembly Line Case. *Appl. Sci.* **2021**, *11*, 3333. https://doi.org/10.3390/app11083333

Received: 12 March 2021
Accepted: 5 April 2021
Published: 8 April 2021

1. Introduction

Large commercial, executive, or defence aircraft assemblers have manufacturing lines composed of many machines for various purposes. Nevertheless, human labour is still responsible for the vast majority of the assembly and finishing phases of different parts. This is partly due to the large size of aeronautical parts that require large and heavy machines, which has hampered the development of automation in the sector. More demanding than the automotive industry, the aeronautical sector also has higher precision and quality requirements, further increasing technological development for manufacturing through autonomous machines [1]. Bogue [2] mentioned that robots' use is far more limited in the aerospace industry than in the automotive sector. The most significant development of automation in the sector consists of drilling and riveting panels for wings and fuselage and transporting large parts for the assembly of aircraft wings and fuselage sections [1,3,4]. Robotic cells have been used in this sector, with high availability, precision, and quality, also becoming a point of attention for producing the input for several other workshops on the assembly line [3].

The challenge in an aircraft assembly company's production line maintenance is to have different types of machinery at different maintenance paradigms. There are machines built decades ago that only have a maintenance plan (scheduled maintenance interventions) based on obsolete information. In contrast, newer machines already have predictive maintenance or preventive maintenance approaches in a Total Productive Maintenance (TPM) policy. Adjusting the corrective, preventive, and predictive maintenance plans to be carried out by a small group of maintainers becomes a significant challenge as the machines degrade and the degradation model is not defined. This happens due to a lack of manufacturer's information, ageing of the machinery, the absence of sensors, or a combination of them [4]. However, due to the appearance of corrective maintenance or a change in the demand for the manufacture of new aircraft, there may be a change in parameters and a need to redesign several machines' maintenance plans in the line. Such a situation generates rework and, possibly, a not optimal result for the production line's production and maintenance. The loss of expected productivity and sub-optimal maintenance costs may reduce the company's revenue. A flexible plan is built each year of operation so that there is a statistical base that will support maintenance for decision-makers.

Big Data analytics and Digital Twins came to support this approach, but only monitored machines or processes may take advantage of the technique. The Digital Twins' concept is the most comprehensive in terms of using technologies aimed at Industry 4.0. The work of Qi [5] declares that the main difficulty is to establish the right technologies and tools to use the approach appropriately. The result of He [6] brings the evidence of acquiring situation awareness from digitized processes, captured data, and performed prognosis throughout the virtualized system, providing timely information to the various actors and decision-makers. Gao [7] pursues the challenge to provide an effective and efficient way to enhance productivity using Big Data Analytics, keeping the economy within the established boundaries and promoting the value-added product. In all these works, the maintenance costs are relevant and should be considered in the total cost.

This research paper seeks to identify the Industry 4.0 technologies' current opportunities to generate an "ecosystem" that facilitates information acquisition on the processes being executed and the existing assets to obtain the best possible results. It is intended to build a framework to support situational awareness to decision-makers and to suggest the optimal use of the current maintenance workforce being allocated, favouring a smooth and more extensive adoption of the prescriptive maintenance approach. Prescriptive maintenance uses the same knowledge in terms of data from predictive maintenance, integrating it with advice related to the maintenance window of opportunity and the tasks the workers need to perform. It allows generating better overall performance optimizing the allocations of resources, such as work floor surface and assets, for each task at the right moment. Therefore, the output of a prescriptive maintenance process must be a dynamic schedule of preventive maintenance according to the maintenance teams' adversities, spare parts, equipment availability, and tools in the identified maintenance window of opportunity, and the correct maintenance procedure manual. The primary purpose is to keep the machinery's high availability during the expected time of its utilization, given the production demand at the manufacture and assembly lines. The work in Choubey [8] brought the evidence that prescriptive maintenance is still in its infancy and lacks real-world implementations and lessons learned to empower its usage on a large scale. Marques [9] established the Smart Prescriptive Maintenance Framework (SPMF) utilized in the present work to provide a path to prescriptive maintenance implementation in any industry using the ecosystem 4.0 capabilities.

Considering that the window of opportunity's prediction is the foremost approach to optimise the entire maintenance process, the present study identified possible techniques to establish this purpose. Several methods are being utilised in literature to predict the window of opportunity based on production planning and product quality. The work in Shamsaei [10] exposed how limiting the problem size could be depending on the solver and algorithmic approach being used. The authors managed to increase the problems' size using a hybrid capacity planning approach based on non-cyclic maintenance (NCMP) and cyclic maintenance (CMP), showing that the solution could have exponential possibilities. The authors utilized a Mixed-Integer Programming (MIP) solver to such problems and could find the solution in a reasonable time, but they did not work with multiple machines. Matyas [11] proposed a procedural approach using multivariate data analysis and simulation tools to identify data correlations and real-time failures and implement it in a real-world scenario in the automotive manufacturing industry. The authors considered a scenario of production planning with multiple machines' maintenance planning but did not consider a dynamic production environment. Koops [12] presented an analytical process for prescriptive maintenance based on data analysis and Monte Carlo simulation; however, no discussion was mentioned about ecosystem 4.0 as an enabler. Kerin [13] discussed in depth the enablers and main paradigms of the Industry 4.0 remanufacturing and proposed framework but did not present simulations or applications in the prescriptive maintenance field. Similarly, Navas [14] presented a smart maintenance framework, discussed the technological enablers, and asserted that maintenance 4.0 is an uncontestable trend but did not present a simulation or case study.

To capture the dynamicity of the environment is necessary to apply the Industrial Internet of Things (IIoT) and understand the patterns in the captured data as already declared in Gao's work [7]. IIoT comprehends sensors that provide data to support Big Data Analytics and the pattern recognition effort that uses machine learning algorithms as presented in the work of Doce [15]. Diez's work [16] exposed the main contributions to state-of-the-art descriptive, predictive and prescriptive maintenance using optimization and machine learning algorithms. The authors proposed trends and perspectives about prescriptive maintenance, including metrics variability and conflicting objectives such as productivity and reliability that may affect the design of efficient solvers for problem resolution. Based on the considerations about the ecosystem 4.0 opportunity to contribute to the prescriptive maintenance approach and the trends exposed, the present work provides a novel framework that integrates different techniques, combining MIP and a real world dynamic production environment case study as the main contribution to the prescriptive maintenance implementation.

2. Materials and Methods

The Smart Prescriptive Maintenance Framework (SPMF), introduced in Marques [9], was developed and tested on a study case to support the introduction of prescriptive maintenance. The present work assembly line presents complex production systems (robots and other equipment), a specific operating environment that provides production requirements or production levels, and a well-defined maintenance capability constituted by maintenance labour, tooling, tribal knowledge and infrastructure suitable for the implementation of the SPMF.

The SPMF is built on three domains of interest captured through data fusion methodologies and integrated by artificial intelligence approaches. The SPMF's domains of interest are:

1. the system's Reliability, Availability, Maintainability and Safety (RAMS) factors;
2. the operating environment in which the system is being deployed;
3. the organization's maintenance capabilities [9].

As restrictions treated in de Mello [3], time is considered during the maintenance schedule that cannot be postponed or anticipated. Cost is also considered, although indirectly, in the restriction that imposes a limit on the available maintenance man-hours and by the objective function, which aims to minimize the maintenance man-hours spent, as presented in Section 3.2.

According to Marques [9], although the SPMF was conceived to implement the prescriptive maintenance on a commercial jet fleet, it can be generalized and used to support any complex system. Each domain was selected because it contains essential system information and performance requirements that feed the prescriptive maintenance framework algorithm, so the best group of maintenance actions is defined.

Figure 1 presents the framework which unifies ecosystem 4.0 and the SPMF applied to the aviation assembly line. The bottom right box shows the idealized assembly line constituted by robots, tools, equipment, workers, and logistics activities. The whole production process and resources are scheduled by the product demand [17].

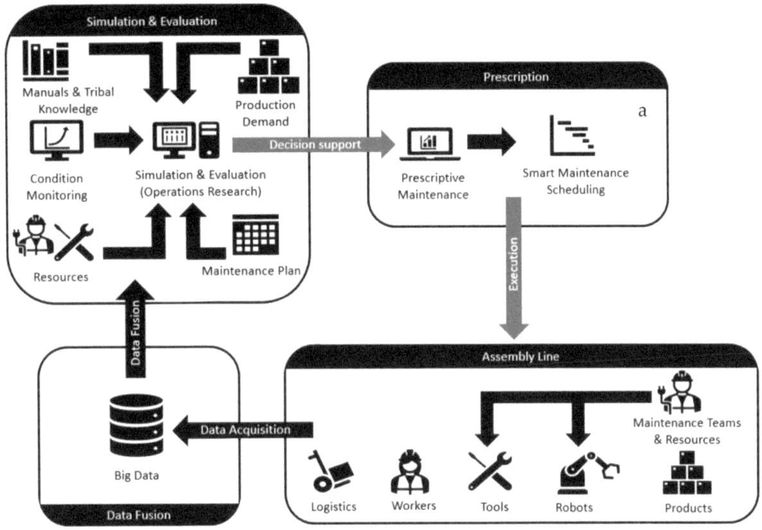

Figure 1. SPMF for integrating different Ecosystem 4.0 in the assembly line.

The framework was built considering the three SPMF's domains previously mentioned:

1. RAMS information is represented by the technical publications, manuals, specs, maintenance plan, and the data gathered from the condition monitoring;
2. The operating environment and requirements are essentially defined by the production demand, robots, tools, and workforce; and
3. The organization's maintenance capabilities are described by the maintenance resources such as tools, labour available, and maintenance tribal knowledge.

Wired or wireless networks and sensors continuously collect temperature, position, humidity, vibration sensors, Radio Frequency Identification (RFID) and internal processes such as built-in-test capabilities, performance, health status and equipment usage. The acquired data sets are successively stored in a centralized big data warehouse. The information is standardized and structured for the simulation and evaluation stage to prescribe and schedule maintenance tasks in real time and keep the assembly line functioning at the required performance through data fusion.

The simulation process aims to develop the best maintenance course of action to maximize the Overall Equipment Effectiveness (OEE) and minimize maintenance man-hours while ensuring production level.

In the simulation and evaluation stage, all the knowledge-base, parameters, system characteristics, and requirements are considered to generate prescriptive maintenance recommendations.

The information considered by the simulation algorithm within the SPMF's framework is listed below:

- Maintenance manual and maintenance team's "tribal" knowledge;
- Equipment condition monitoring to support Prognostics and Health Management, including Remaining Useful Life (RUL) evaluation;
- Available resources such as maintenance labour and tools;
- The maintenance plan, including tasks and their interval; and
- Production demand.

The methodology adopted for the simulation is the Mixed-Integer Linear Programming or Mixed-Integer Programming (MIP), well known in the Operations Research field The choice of MIP is related to its intrinsic characteristic to optimize a complex organization's operational efficiency while considering demands, capacity and other business rules as a constraint of the system. Shamsaei [10] mentioned that MIP is simpler, faster and more effective than other methodologies such as heuristics and meta-heuristics for industrial systems problems, like the proposed case study. This approach is also strengthened by Schrotenboer [18], who demonstrated that MIP could be successfully used to optimize maintenance equipment under operational uncertainties.

In the top-right box, identified with the letter "a", in Figure 1, a real-time updated prescriptive maintenance plan is presented to the maintenance engineer who takes the final decision about how to maintain the assembly line. This decision support system is often constituted by a "smart" maintenance scheduling that provides, also in real time and in Gantt form, a dynamic maintenance plan for each piece of equipment.

The plan is then executed on the assembly line by the maintenance team. In a feedback loop form, the OEE parameters and maintenance labour performance are measured. The system assesses its performance, continuously improves the algorithm, and adjusts parameters according to product demand and assembly resources in a non-stop, continuous improvement cycle.

3. Case Study

The case study's goal is to determine whether the adoption of SPMF improves OEE and decreases the maintenance man-hour of a commercial aircraft wing assembly line. The methodology to build the case study included semi-structured interviews with maintenance engineers who have been responsible for the maintenance of more than 90 machines of an aerospace manufacturer assembly line for 20 years. The field exploration helped to define parameters such as the monthly wing demand in a production rump up scenario, the number of production cells, maintenance team size, the meantime to repair (MTTR), and the maintenance strategy usually adopted. It is essential to mention that some parameters could not be collected or were not available from the field exploration, thus were identified through a literature review. The limitation arising from adopting this literature's parameters is that the data are different from the field's information. However, the magnitude of the possible incongruency does not invalidate the study since the product (aircraft wing) and the assembly line type (organized in levels) are the same observed in both scenarios: in the field and the literature. Table 1 presents the parameters collected through field exploration, the ones identified from the literature review, and both.

Table 1. Parameters' sources

Parameter or Assembly Line Characteristics	Field Exploration	Literature
Monthly production	x	
Assembly level		[19]
Number of robots	x	
Riveting per robot, per wing		[19]
Mean Time Between Failures	x	[20–23]
Production rate		[19]
Robot Maintenance Class	x	[24]
Robot Maintenance Strategy	x	
Maintenance Tasks Types and Intervals	x	[21]
Mean Time to Repair	x	
Team size and expertise	x	

3.1. Assembly Line Description

The assembly line mentioned in de Mello [3] is considered and shown in Figure 2, capable of assembling a three-piece wing box as an example. This assembly operation is divided into four steps, as demonstrated in Figure 2:

1. First assembly level: parts like skins, stringers, stiffeners, and doublers are joined to form upper and lower panels, as well as spar, ribs, and bulkhead subassemblies;
2. Second assembly level: these subassemblies are joined to form left, right and centre rib/spar grid structures;
3. Third assembly level: upper and lower panel subassemblies are joined with the rib/spar to form left, right and centre boxes;
4. Fourth assembly level: three wing boxes are joined to form the final wing.

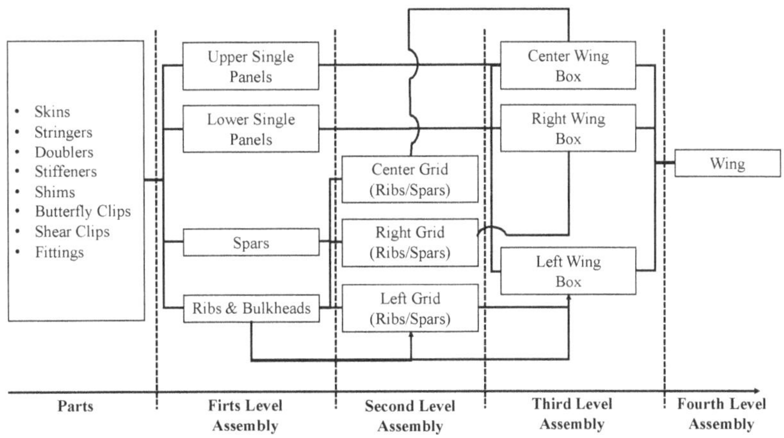

Figure 2. Wing flow assembly, adapted from Diez-Olivan [15].

As shown in Table A1 of Appendix A, 57% of all fasteners and rivets needed to assemble the wing are used at the first level, making it easier to automatize. Joining skin panels to the grid in the third level requires 34% of all rivets and fasteners, and even though automatization here is more challenging than the first level, it is still worth it. Although the final level only requires 3% of the fasteners, the time needed to install these larger fasteners without the aid of robots would be a magnitude higher than the time required to install fasteners in first level assembly; thus, automatization is highly recommended in this level, and it was adopted for the case study analysed in this paper. The second level only requires 6% of the total rivets/fasteners. Finally, manual installation is easy and acceptable in this step, thus turning the use of robots uneconomical. As a result, the second level's assembly tasks were assumed to be manually executed in the model proposed. The line was modelled according to the parameters listed in Table A2 of Appendix A. The parameters were acquired from the semi-structured interviews during the field exploration and confirmed through the literature review, as previously presented in Table 1.

3.2. Mathematical Model

Given that the preventive maintenance of an assembly line is an unavoidable part of the operation, robots must be stopped for regular maintenance activities. In this regard, time spent in maintenance is unproductive, reducing the availability of KPI's in the assembly process. The aim of the proposed model is therefore to reduce this unproductive time brought on by preventive maintenance. Khatab [25] presents a similar objective function, however, focusing on the cost incurred by the labour hours. Since this paper focuses on the overall efficiency of the assembly line, the repair teams are considered to have a fixed cost to be on standby for maintenance, thus this cost is not relevant at his point. The assembly line described in Section 3.1 was mathematically modelled according to the parameters variables and formulation presented below:

Parameters

I: set of maintenance activities;
J: set of maintenance teams;
R: set of robots;
K: set of robot types;
T: total number of maintenance activities required;
W: total working hours of maintenance teams in a period.
Robot_Vector: vector that lists the robot types and their position in the assembly line
Experience_Matrix: defines the *J* teams' experience and knowledge in maintaining each of the K robot types. In alignment with the observations in the assembly line during the field exploration, it was adopted as a system of five experience levels as suggested by the International Labour Organisation [26] as explained below:

- Level 1—shallow experience: performs 25% slower than the MTTR for a team to execute the maintenance tasks;
- Level 2—low experience: performs 10% slower than MTTR for a team to execute the maintenance tasks;
- Level 3—average experience: it takes a time equal to the MTTR to execute the maintenance task;
- Level 4—high experience: performs 10% faster than the MTTR to execute the maintenance task; and
- Level 5—very high experience: performs 25% faster than the MTTR to execute the maintenance task.

Experience_Factor: vector that lists the experience levels used to calculate how long it takes for the team to perform a maintenance task.
$Average_{ir}$: average time to perform maintenance task *i* for robot *r*.
t_{ijr}: the time required by team *j* to perform maintenance activity *i* on robot *r*.
Decision variables
z_{ijr}: decision variable that assumes the value one (1) if team *j* performs maintenance activity *i* on robot *r*.
C: number of cancelled maintenance activities
Objective Function

$$\min \sum_{i\in I}\sum_{j\in J}\sum_{r\in R} t_{ijr} z_{ijr} \tag{1}$$

Restrictions

$$\sum_{j\in J} z_{ijr} \leq 1 \;\forall\, i \in I,\ \forall\, r \in R \tag{2}$$

$$\sum_{r\in R} t_{ijr} z_{ijr} \leq W \;\forall\, i \in I,\ \forall\, j \in J \tag{3}$$

$$C + \sum_{i\in I}\sum_{j\in J}\sum_{r\in R} z_{ijr} = T \tag{4}$$

$$z_{ijr} \in \{0,1\} \;\forall\, i \in I,\ \forall\, j \in J,\ \forall\, r \in R \tag{5}$$

The objective function, Equation (1), aims to minimize the robot downtimes by reducing maintenance man-hours over the 15 years of operation considered. Here, t_{ijr} is defined in Equation (6):

$$t_{ijr} = Average_{ir} \times Experience\_Factor\left(ExperinceMatrix_{j,RobotVector_r}\right) \tag{6}$$

Equation (2) is a restriction that imposes that maintenance activity must only be done once by one team. Equation (3) determines that the maintenance man-hours cannot exceed the established maximum work hours of a team in a certain period, which, in this case, is

16 man-hours/day in seven days, meaning there are a total of 112 man-hours each seven days. Variable C in Equation (4) represents the number of tasks that could not be performed because of unavailable labour. Equation (5) defines variable z_{ijr} as binary, equal to 1 if maintenance i, for robot r, is performed in time t and equal to 0 otherwise.

A more global parameter is used to better compare the different cases presented in this work. This parameter is the OEE that takes into account availability, performance, and quality. For all effects in this work, the performance and quality are considered constant; thus, the availability can be more efficiently compared.

Equations (7) and (8) define the measured OEE and availability for the tested cases:

$$OEE = Availability \times Performance \times Quality \quad (7)$$

$$Availability = \frac{AvailableHours - Downtime}{AvailableHours} \quad (8)$$

Due to the scenario's structure tested in the following cases, the amount of maintenance activity hours in some periods is purposefully superior to the available labour hours. Thus, each case presents several cancelled maintenance activities in total planning. These cancelled activities are considered as downtime for the robots. Except for case 2, all cases were solved using the mixed-integer linear programming model presented previously. The system used to run the simulations was a laptop computer with 8 GB of RAM and operating system macOS Big Sur version 11.1 [27] running the open-source GUROBI solver [28] in MATLAB version R2020b [29].

Introduction to the different scenarios

Within the assembly line described in Section 3, four maintenance scenarios were modelled to evaluate how the SPMF can improve the OEE and the equipment availability over an assembly line life-cycle of 15 years. This time period was selected to include over-haul and heavy maintenance, that, according to field exploration, literature review, and historical data, happen every 12 to 15 years for the considered equipment. These four different scenarios are relevant because they model, under some limitations, real-life maintenance environments and how the gradual adoption of prescriptive maintenance can improve efficiency. Both topics will be discussed in Section 4.

The scenarios also model the actual industry assembly line environment and evaluate the potential of substituting an expert maintenance engineer with an algorithm supported by Ecosystem 4.0. For example, actual robots' capabilities of drilling, MTBF, quantity, the maintenance schedule intervals and type, teams' size and skills, these are all parameters considered in the scenarios and the engineer's ability to prescribe the best team for a particular maintenance task.

3.2.1. Scenario 1: Man-Hour Expertise Considered Constant

Case 1 uses the formulation presented previously to simulate an experienced maintenance planner with knowledge of available teams to reduce maintenance downtimes. By doing this, the best team is prescribed for each maintenance activity. However, this case does not consider any evolution of team skills in the lifetime of the robots.

The SPMF algorithm can read the scheduled maintenance that has to be performed in period i, on equipment r, and select the best maintenance team based on its expertise (constant for the life-cycle considered 15 years). Performance and quality are also regarded as constant, equal to 0.85 and 0.95, respectively. Table 2 shows the results:

Table 2. Scenario 1 results

Total Maintenance Man-Hour (h)	Cancelled Tasks	Total Downtime (h)	Availability	OEE
9676.5	19	11,804.50	86.49%	69.84%

3.2.2. Scenario 2: Man-Hour Expertise Considered Constant and Randomized Team Assignment

In this simulation, SPMF can read the scheduled maintenance that has to be performed in period i, on equipment r, but does not select the best maintenance team based on its expertise. It is the scenario in which team expertise is unknown. Thus, the teams are assigned randomly; therefore, the formulation presented previously is not used to solve this case differently from the other ones. Performance and quality are again considered constant, equal to 0.85 and 0.95, respectively. Results are presented in Table 3:

Table 3. Scenario 2 results.

Total Maintenance Man-Hour (h)	Cancelled Tasks	Total Downtime (h)	Availability	OEE
10,868	19	12,996	85.12%	68.74%

3.2.3. Scenario 3: Man-Hour Expertise Considered Variable

In this simulation, the SPMF algorithm plans the preventive maintenance, which has to be performed in period i, on equipment r, and selects the best maintenance team based on its expertise, which is now considered variable. In other words, the specialization of labour is simulated due to practice over the years. As proof of concept, it is assumed that expertise levels are updated yearly through machine learning. This assumption can be later altered by an ecosystem 4.0 that can monitor real-time expertise through sensors and algorithms that can track and evaluate individual performance directly or indirectly by monitoring equipment OEE. Table 4 shows the results:

Table 4. Scenario 3 results

Total Maintenance Man-Hour (h)	Cancelled Tasks	Total Downtime (h)	Availability	OEE
8153.8	19	10,281.8	88.23%	71.25%

3.2.4. Scenario 4: Man-Hour Expertise Considered Variable and Tasks Rescheduling

This case study adds to the SPMF algorithm the capability of maintenance task rescheduling on top of team assignment based on variable expertise. It is assumed that the information about the tasks, robots, and teams is provided in real-time by the IoT infrastructure. Here, the possibility of anticipating maintenance activities that were cancelled due to insufficient labour is presented in an attempt to reduce downtime further. Table 5 shows the results:

Table 5. Scenario 4 results

Total Maintenance Man-Hour (h)	Cancelled Tasks	Total Downtime (h)	Availability	OEE
8426.1	15	10,106.1	88.43%	71.41%

4. Discussion

Comparing the simulations and assuming a similar utilisation date for all the robots, a convergence was observed in the number of heavy maintenance tasks. It resulted in the optimisation of the maintenance process. This finding denoted that only an increase in working hours (or maintenance man-hours), achievable primarily by increasing the team size, could decrease the cancelled tasks. This hypothesis was tested in scenario 4: as shown in Figure 3, the team size was increased until the point that no tasks were cancelled, and, as a result, higher robots' availability and OEE were obtained.

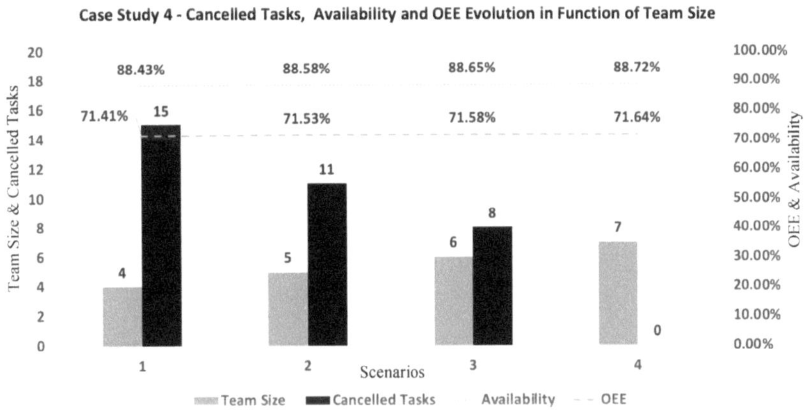

Figure 3. Case Study 4—Cancelled tasks, availability, and OEE evolution in team size function.

The study was limited by the following assumptions:

- No corrective maintenance was considered;
- Robots remaining useful life and degradation curve were not modelled; and
- Production demand, product quality, and robots' MTBF were considered constant.

It meant that the performance of the supporting capability of a complex system at its end of life-cycle, when more corrective maintenance takes place, or scheduled maintenance is anticipated, could not be assessed. As mentioned in the conclusion, future work will expand the simulation up to the inclusion of such situations.

Comparing the OEE and the availability obtained in each simulation, it was observed that the best results were obtained in the fourth scenario, as expected, since it featured not only assignments according to teams' expertise and increasing expertise over time, but also task rescheduling. The second scenario, characterized by randomized teams' assignment and constant expertise, presented the lowest OEE overall, confirming that not considering workforce skills in maintenance tasks' assignment is not a good strategy. Scenario 1, which considers workforce skills, presents improvements compared to scenario 1, while scenario 3, which in addition to teams 'skills, considers skill improvements over time, as considered in the fourth scenario, further improves the results obtained in scenario 1. Figure 4 summarizes the comparisons of the results.

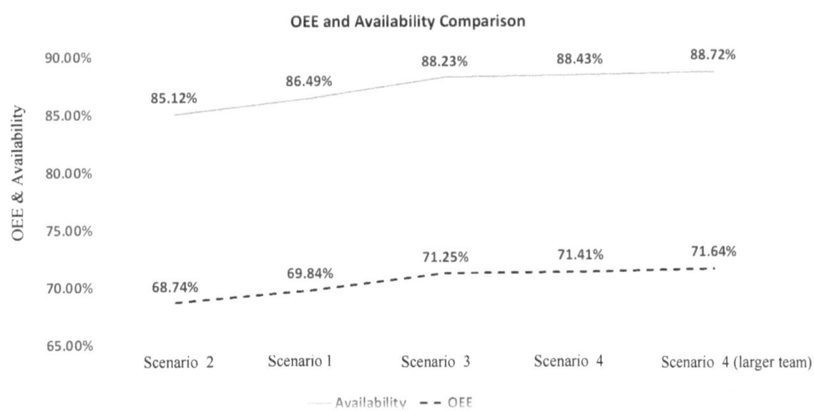

Figure 4. OEE and availability comparison among simulations.

5. Conclusions

The paper tests the existing concepts of the SPMF for introducing prescriptive maintenance policy in an aviation assembly line composed of operators and 55 robots, 36 of which presented 25-year preventive maintenance plans and smart monitoring possibilities. Four different scenarios, characterised by different combination of man-hour expertise, team assignment and task rescheduling, were considered. This approached allowed identifying the most efficient situation to adopt. In the first simulation, the case study focused on assigning the best team to complete the maintenance task relative to a specific robot type in the scheduled interval. In the second case study, teams were randomly assigned to perform the maintenance tasks. The third simulation expertise was assumed variable, and the SPMF algorithm capable again of assigning the best team. The fourth study case focused on assigning the best team in a scenario of variable expertise and smart rescheduling maintenance. It was recognized as the most efficient with the highest OEE and lowest cancelled maintenance tasks due to unavailable labour.

Analysing each simulation result, it was observed that the number of tasks that could not be performed (cancelled tasks) was converging, indicating that, assuming a similar usage starting date for all the robots, maintenance rescheduling was not accurate enough to avoid cancelled tasks without increasing the size of the maintenance team.

The results demonstrated that the SPMF concepts' effectiveness helped the maintenance specialist decision in an Ecosystem 4.0 supported assembly line, reducing the human effort for maintenance schedule significantly.

Future work will focus on expanding the simulation model to include equipment degradation and RUL as triggers for maintenance (condition-based maintenance), different equipment usage start date, tooling and materials availability, product demand, robots MTBF and quality variability, as well as "tribal" knowledge and historical maintenance data as sources of lessons learned to improve the algorithm.

Author Contributions: Conceptualization, H.C.M. and A.G.; formal analysis, E.A.P.B.; methodology, A.G. and E.A.P.B.; software, A.G. and E.A.P.B.; writing—original draft preparation, H.C.M.; writing—review and editing, A.M.; All authors have read and agreed to the published version of the manuscript.

Funding: This study was financed in part by the Coordenação de Aperfeiçoamento de Pessoal de Nıvel Superior—Brazil (CAPES)—Finance Code 001.

Institutional Review Board Statement: Not applicable.

Informed Consent Statement: Not applicable.

Data Availability Statement: Not applicable.

Conflicts of Interest: The authors declare no conflict of interest.

Abbreviations

CBM	Condition-Based Maintenance
CMP	Cyclic Maintenance
IIoT	Industrial Internet of Things
MIP	Mixed Integer Programming
MTBF	Meantime Between Failures
MTTR	Meantime to Repair
NCMP	Non-Cyclic Maintenance
OEE	Overall Equipment Effectiveness
RAMS	Reliability, Availability, Maintainability, and Safety
RUL	Remaining Useful Life
SPMF	Smart Prescriptive Maintenance Framework
TPM	Total Productive Maintenance

Appendix A

Appendix A presents the tables containing the number of rivets considered in the assembly line and the study case parameters for all 55 machines considered.

Table A2 shows the parameters of the assembly line described in Section 3.1. The columns description follows:

- 1st column: the equipment was modelled in 12 different robot types distributed in 42 production cells along with 13 cells equipped with human labour and tools;
- 2nd column: it lists the assembly level according to the description of Section 3.1;
- 3rd column: it presents the riveting capacity per wing. It is essential to add that the monthly wing demand was considered equal to 43;
- 4th column: the MTBF was identified for each robot type;
- 5th column: here the production rate, in riveting per minute, was defined for each robot according to the information collected during the interviews and the literature review, as described in Table 1;
- 6th column: for each piece of equipment, a maintenance class was selected. According to Gopalakrishnan [24], Class A represents the assembly line's most important robots, which, in turn, receive the most resources and more sophisticated maintenance strategies in an attempt to minimize downtime since an inactive Class A robot is highly uneconomic for management. Class B equipment is less critical than Class A but more fundamental than Class C. Usually, Class B equipment receives fewer resources than Class A but is still within the equipment set that presents a maintenance strategy that aims to minimize downtime preventive maintenance. Class C robots usually do not receive any resources to prevent unforeseen failures since downtime is economically acceptable for this equipment class. Being less costly than Class A and B, often these Class C equipment are substituted upon failure by the maintenance team;
- 7th column: as a function of the class, a maintenance strategy is assigned to each robot. Total Production Maintenance (TPM) combined with Condition Based Maintenance (CBM) are some examples. The strategy can also be constituted only by TPM or just letting the equipment fail, that is, on condition;
- 8th–15th columns: the last eight columns list the four considered maintenance tasks, their respective intervals, and Mean Times to Repair (MTTR).

Table A1. Number of rivets/fasteners for a three-piece wing [19].

Assembly Level	Assembly Strategy	Structural Component	Number of Rivets/Fasteners per Joining Operation					
			Stringer/Caps to Skin/Web	Stiffeners to Web	Shear & Butterflies/Clips to Skins and Stringers	Spar Caps to Panels	Box to Box	Total
1st Level	Robot	Panels Spars Ribs/Bulkhead	24.300 14.400	7.800 1.400				60.500
2nd Level	Human + Tool	Ribs/Bulkhead/Spars-Grid		6.000				6000
3rd Level	Robot	Left Box Right Box Center Box			9.000 9.000 2.300	7200 7200 1800		36,500
4th Level	Robot						2700	2700
		Total	38.700	27,800	20,300	16,200	2700	105,700

Table A2. Assembly Line Model Parameters.

Robot	Assembly Level	d_i (Riveting/Wing)	MTBF (hs)	Production Rate (Riveting/Minute)	Maint. Class	Maint. Strategy	Maintenance Task							
							Visual Inspection		Battery Servicing		Overhaul		Refurbishment	
							Interval (Week)	MTTR (hs)	Interval (Month)	MTTR (hs)	Interval (Month)	MTTR (hs)	Interval (Years)	MTTR (hs)
Robot Type 1	1st level	10,084	570	10	Class A	TPM + CBM	weekly	0.1833	12	2	36	12	120	96
Robot Type 2	1st level	10,084	370	10	Class B	TPM	weekly	0.1833	13	2	48	12	170	96
Robot Type 3	1st level	10,083	470	10	Class B	TPM	weekly	0.1833	12	2	50	12	180	96
Robot Type 1	1st level	10,083	570	10	Class B	TPM	weekly	0.1833	12	2	36	12	120	96
Robot Type 2	1st level	10,083	370	10	Class B	TPM	weekly	0.1833	13	2	48	12	170	96
Robot Type 3	1st level	10,083	470	10	Class B	TPM	weekly	0.1833	12	2	50	12	180	96
Human + Tool	2nd level	465	501	0.5	Class C	On Condition	N/A	N/A	N/A	N/A	N/A	N/A	N/A	N/A
Human + Tool	2nd level	465	501	0.5	Class C	On Condition	N/A	N/A	N/A	N/A	N/A	N/A	N/A	N/A
Human + Tool	2nd level	465	501	0.5	Class C	On Condition	N/A	N/A	N/A	N/A	N/A	N/A	N/A	N/A
Human + Tool	2nd level	465	501	0.5	Class C	On Condition	N/A	N/A	N/A	N/A	N/A	N/A	N/A	N/A
Human + Tool	2nd level	460	501	0.5	Class C	On Condition	N/A	N/A	N/A	N/A	N/A	N/A	N/A	N/A
Human + Tool	2nd level	460	501	0.5	Class C	On Condition	N/A	N/A	N/A	N/A	N/A	N/A	N/A	N/A
Human + Tool	2nd level	460	501	0.5	Class C	On Condition	N/A	N/A	N/A	N/A	N/A	N/A	N/A	N/A

Table A2. *Cont.*

Robot	Assembly Level	d_i (Riveting/Wing)	MTBF (hs)	Production Rate (Riveting/Minute)	Maint. Class	Maint. Strategy	Maintenance Task							
							Visual Inspection		Battery Servicing		Overhaul		Refurbishment	
							Interval (Week)	MTTR (hs)	Interval (Month)	MTTR (hs)	Interval (Month)	MTTR (hs)	Interval (Years)	MTTR (hs)
Human + Tool	2nd level	460	501	0.5	Class C	On Condition	N/A	N/A	N/A	N/A	N/A	N/A	N/A	N/A
Human + Tool	2nd level	460	501	0.5	Class C	On Condition	N/A	N/A	N/A	N/A	N/A	N/A	N/A	N/A
Human + Tool	2nd level	460	501	0.5	Class C	On Condition	N/A	N/A	N/A	N/A	N/A	N/A	N/A	N/A
Human + Tool	2nd level	460	501	0.5	Class C	On Condition	N/A	N/A	N/A	N/A	N/A	N/A	N/A	N/A
Human + Tool	2nd level	460	501	0.5	Class C	On Condition	N/A	N/A	N/A	N/A	N/A	N/A	N/A	N/A
Human + Tool	2nd level	460	501	0.5	Class C	On Condition	N/A	N/A	N/A	N/A	N/A	N/A	N/A	N/A
Robot Type 1	3rd level	1141	570	1.2	Class B	TPM	weekly	0.1833	12	2	36	12	120	96
Robot Type 2	3rd level	1141	370	1.2	Class B	TPM	weekly	0.1833	13	2	48	12	170	96
Robot Type 3	3rd level	1141	470	1.2	Class B	TPM	weekly	0.1833	12	2	50	12	180	96
Robot Type 4	3rd level	1141	430	1.2	Class B	TPM	weekly	0.1833	12	2	53	12	171	96
Robot Type 5	3rd level	1141	400	1.2	Class B	TPM	weekly	0.1833	14	2	37	12	180	96
Robot Type 6	3rd level	1141	440	1.2	Class B	TPM	weekly	0.1833	12	2	41	12	173	96
Robot Type 7	3rd level	1141	510	1.2	Class B	TPM	weekly	0.1833	12	2	39	12	138	96
Robot Type 8	3rd level	1141	517	1.2	Class B	TPM	weekly	0.1833	11	2	53	12	123	96
Robot Type 9	3rd level	1141	319	1.2	Class B	TPM	weekly	0.1833	12	2	51	12	180	96
Robot Type 10	3rd level	1141	289	1.2	Class B	TPM	weekly	0.1833	12	2	41	12	132	96
Robot Type 1	3rd level	1141	570	1.2	Class B	TPM	weekly	0.1833	12	2	36	12	120	96
Robot Type 2	3rd level	1141	370	1.2	Class B	TPM	weekly	0.1833	13	2	48	12	170	96
Robot Type 3	3rd level	1141	470	1.2	Class B	TPM	weekly	0.1833	12	2	50	12	180	96
Robot Type 4	3rd level	1141	430	1.2	Class B	TPM	weekly	0.1833	12	2	53	12	171	96
Robot Type 5	3rd level	1141	400	1.2	Class B	TPM	weekly	0.1833	14	2	37	12	180	96
Robot Type 6	3rd level	1141	440	1.2	Class B	TPM	weekly	0.1833	12	2	41	12	173	96
Robot Type 7	3rd level	1141	510	1.2	Class B	TPM	weekly	0.1833	12	2	39	12	138	96
Robot Type 8	3rd level	1141	517	1.2	Class B	TPM	weekly	0.1833	11	2	53	12	123	96
Robot Type 9	3rd level	1141	319	1.2	Class B	TPM	weekly	0.1833	12	2	51	12	180	96
Robot Type 10	3rd level	1141	289	1.2	Class B	TPM	weekly	0.1833	12	2	41	12	132	96
Robot Type 1	3rd level	1140	570	1.2	Class B	TPM	weekly	0.1833	12	2	36	12	120	96
Robot Type 2	3rd level	1140	370	1.2	Class B	TPM	weekly	0.1833	13	2	48	12	170	96

Table A2. *Cont.*

Robot	Assembly Level	d_i (Riveting/Wing)	MTBF (hs)	Production Rate (Riveting/Minute)	Maint. Class	Maint. Strategy	Maintenance Task							
							Visual Inspection		Battery Servicing		Overhaul		Refurbishment	
							Interval (Week)	MTTR (hs)	Interval (Month)	MTTR (hs)	Interval (Month)	MTTR (hs)	Interval (Years)	MTTR (hs)
Robot Type 3	3rd level	1140	470	1.2	Class B	TPM	weekly	0.1833	12	2	50	12	180	96
Robot Type 4	3rd level	1140	430	1.2	Class B	TPM	weekly	0.1833	12	2	53	12	171	96
Robot Type 5	3rd level	1140	400	1.2	Class B	TPM	weekly	0.1833	14	2	37	12	180	96
Robot Type 6	3rd level	1140	440	1.2	Class C	On Condition	N/A	N/A	N/A	N/A	N/A	N/A	N/A	N/A
Robot Type 7	3rd level	1140	510	1.2	Class C	On Condition	N/A	N/A	N/A	N/A	N/A	N/A	N/A	N/A
Robot Type 8	3rd level	1140	517	1.2	Class C	On Condition	N/A	N/A	N/A	N/A	N/A	N/A	N/A	N/A
Robot Type 9	3rd level	1140	319	1.2	Class C	On Condition	N/A	N/A	N/A	N/A	N/A	N/A	N/A	N/A
Robot Type 10	3rd level	1140	289	1.2	Class C	On Condition	N/A	N/A	N/A	N/A	N/A	N/A	N/A	N/A
Robot Type 3	3rd level	1140	470	1.2	Class C	On Condition	N/A	N/A	N/A	N/A	N/A	N/A	N/A	N/A
Robot Type 4	3rd level	1140	430	1.2	Class A	TPM + CBM	weekly	0.1833	12	2	53	12	171	96
Robot Type 11	4th level	675	489	1	Class A	TPM + CBM	weekly	0.1833	10	2	48	12	144	96
Robot Type 11	4th level	675	489	1	Class A	TPM + CBM	weekly	0.1833	10	2	48	12	144	96
Robot Type 12	4th level	675	511	1	Class A	TPM + CBM	weekly	0.1833	12	2	39	12	151	96
Robot Type 12	4th level	675	511	1	Class A	TPM + CBM	weekly	0.1833	12	2	39	12	151	96
	Total	105,700												

References

1. Schoenberger, R. From Cars to Planes—Aerospace Manufacturing and Design. Available online: https://www.aerospacemanufacturinganddesign.com/article/amd1114-aerospace-assembly-automation/ (accessed on 22 March 2021).
2. Bogue, R. The Growing Use of Robots by the Aerospace Industry. *Ind. Robot Int. J.* **2018**, *45*, 705–709. [CrossRef]
3. de Mello, J.M.G.; Trabasso, L.G.; Reckevcius, A.C.; Palmeira, A.L.O.A.; Reiss, P.; Caraca, W. A Novel Jigless Process Applied to a Robotic Cell for Aircraft Structural Assembly. *Int. J. Adv. Manuf. Technol.* **2020**, *109*, 1177–1187. [CrossRef]
4. Duffuaa, S.; Raouf, A. *Planning and Control of Maintenance Systems: Modelling and Analysis*, 2nd ed.; Springer International Publishing: Cham, Switzerland, 2015; ISBN 978-3-319-19802-6.
5. Qi, Q.; Tao, F.; Hu, T.; Anwer, N.; Liu, A.; Wei, Y.; Wang, L.; Nee, A.Y.C. Enabling Technologies and Tools for Digital Twin. *J. Manuf. Syst.* **2019**. [CrossRef]
6. He, B.; Bai, K.-J. Digital Twin-Based Sustainable Intelligent Manufacturing: A Review. *Adv. Manuf.* **2020**. [CrossRef]
7. Gao, R.X.; Wang, L.; Helu, M.; Teti, R. Big Data Analytics for Smart Factories of the Future. *CIRP Ann.* **2020**, *69*, 668–692. [CrossRef]
8. Choubey, S.; Benton, R.G.; Johnsten, T. A Holistic End-to-End Prescriptive Maintenance Framework. *Data-Enabled Discov. Appl.* **2021**, *4*, 11. [CrossRef]
9. Marques, H.; Giacotto, A. *Prescriptive Maintenance: Building Alternative Plans for Smart Operations*; Swedish Society of Aeronautics and Astronautics (FTF): Stockholm, Sweden, 2019; pp. 231–236.
10. Shamsaei, F.; Van Vyve, M. Solving Integrated Production and Condition-Based Maintenance Planning Problems by MIP Modeling. *Flex. Serv. Manuf. J.* **2017**, *29*, 184–202. [CrossRef]
11. Matyas, K.; Nemeth, T.; Kovacs, K.; Glawar, R. A Procedural Approach for Realizing Prescriptive Maintenance Planning in Manufacturing Industries. *CIRP Ann.* **2017**, *66*, 461–464. [CrossRef]
12. Koops, L.G. Optimized Maintenance Decision-Making—A Simulation-Supported Prescriptive Analytics Approach Based on Probabilistic Cost-Benefit Analysis. *PHM Soc. Eur. Conf.* **2020**, *5*, 14. [CrossRef]
13. Kerin, M.; Pham, D.T. Smart Remanufacturing: A Review and Research Framework. *J. Manuf. Technol. Manag.* **2020**, *31*, 1205–1235. [CrossRef]
14. Navas, M.A.; Sancho, C.; Carpio, J. Disruptive Maintenance Engineering 4.0. *Int. J. Qual. Reliab. Manag.* **2020**, *37*, 853–871. [CrossRef]
15. Munín-Doce, A.; Díaz-Casás, V.; Trueba, P.; Ferreno-González, S.; Vilar-Montesinos, M. Industrial Internet of Things in the Production Environment of a Shipyard 4.0. *Int. J. Adv. Manuf. Technol.* **2020**, *108*, 47–59. [CrossRef]
16. Diez-Olivan, A.; Del Ser, J.; Galar, D.; Sierra, B. Data Fusion and Machine Learning for Industrial Prognosis: Trends and Perspectives towards Industry 4.0. *Inf. Fusion* **2019**, *50*, 92–111. [CrossRef]
17. Chang, H.-M.; Huang, C.; Torng, C.-C. Lean Production Implement Model for Aerospace Manufacturing Suppliers. *Int. J. Innov. Manag. Technol.* **2013**. [CrossRef]
18. Schrotenboer, A.H.; Ursavas, E.; Vis, I.F.A. Mixed Integer Programming Models for Planning Maintenance at Offshore Wind Farms under Uncertainty. *Transp. Res. Part C Emerg. Technol.* **2020**, *112*, 180–202. [CrossRef]
19. Sarh, B. *Wing Structural Assembly Methodology*; SAE International: Long Beach, CA, USA, 1998; p. 982156.
20. Kampa, A.; Gołda, G.; Paprocka, I. Discrete Event Simulation Method as a Tool for Improvement of Manufacturing Systems. *Computers* **2017**, *6*, 10. [CrossRef]
21. Kampa, A. The Review of Reliability Factors Related to Industrial Robo. *Robot. Autom. Eng. J.* **2018**, 3. [CrossRef]
22. Dhillon, B.S.; Yang, N. Reliability Analysis of a Repairable Robot System. *J. Qual. Maint. Eng.* **1996**, *2*, 30–37. [CrossRef]
23. Dhillon, B.S.; Aleem, M.A. A Report on Robot Reliability and Safety in Canada: A Survey of Robot Users. *J. Qual. Maint. Eng.* **2000**, *6*, 61–74. [CrossRef]
24. Gopalakrishnan, M.; Skoogh, A. Machine Criticality Based Maintenance Prioritization: Identifying Productivity Improvement Potential. *Int. J. Product. Perform. Manag.* **2018**, *67*, 654–672. [CrossRef]
25. Khatab, A.; Diallo, C.; Venkatadri, U.; Liu, Z.; Aghezzaf, E.-H. Optimization of the Joint Selective Maintenance and Repairperson Assignment Problem under Imperfect Maintenance. *Comput. Ind. Eng.* **2018**, *125*, 413–422. [CrossRef]
26. ILO Benchmarking of Qualifications Frameworks: A Report on Potential Comparability between the Bangladesh Technical and Vocational Qualifications Framework and Qualifications Frameworks of Other Countries of Origin and Destination of Migrant Workers—Skills and Employability Network. Available online: http://apskills.ilo.org/resources/benchmarking-of-qualifications-frameworks-a-report-on-potential-comparability-between-the-bangladesh-technical-and-vocational-qualifications-framework-and-qualifications-frameworks-of-other-countries-of-origin-and-destination-of-migrant-workers (accessed on 11 March 2021).
27. Apple MacOS Big Sur. Available online: https://www.apple.com/macos/big-sur/ (accessed on 11 March 2021).
28. Gurobi Software Downloads and License Center. Available online: https://www.gurobi.com/downloads/ (accessed on 11 March 2021).
29. Mathworks MATLAB & Simulink. Available online: https://www.mathworks.com/products/matlab.html (accessed on 11 March 2021).

applied sciences

MDPI

Article

How to Make Augmented Reality a Tool for Railway Maintenance Operations: Operator 4.0 Perspective

Sara Scheffer *, Alberto Martinetti, Roy Damgrave, Sebastian Thiede and Leo van Dongen

Department of Design, Production and Management, Faculty of Engineering Technology, University of Twente, De Horst 2, 7522 LW Enschede, The Netherlands; a.martinetti@utwente.nl (A.M.); r.g.j.damgrave@utwente.nl (R.D.); s.thiede@utwente.nl (S.T.); l.a.m.vandongen@utwente.nl (L.v.D.)

* Correspondence: s.e.scheffer@utwente.nl; Tel.: +31-53-489-2875

Abstract: In the last few decades, several initiatives and approaches are set up to support maintenance procedures for the railway industry in adopting the principles of Industry 4.0. Contextualized maintenance technologies such as Augmented Reality (AR) overlay can integrate virtual information on physical objects to improve decision-making and action-taking processes. Operators work in a dynamic working environment requiring both high adaptive capabilities and expert knowledge. There is a need to support the operators with tailor-based information that is customized and contextualized to their expertise and experience. It calls for AR tools and approaches that combine complex methodologies with high usability requirements. The development of these AR tools could benefit from a structured approach. Therefore, the objective of this paper is to propose an adaptive architectural framework aimed at shaping and structuring the process that provides operators with tailored support when using an AR tool. Case study research is applied within a revelatory railway industry setting. It was found that the framework ensures that self-explanatory AR systems can capture the knowledge of the operator, support the operator during maintenance activities, conduct failure analysis, provide problem-solving strategies, and improve learning capabilities. This study contributes to the necessity of having a human-centered approach for the successful adaption of AR technology tools for the railway industry.

Keywords: augmented reality; railway maintenance; Operator 4.0

Citation: Scheffer, S.; Martinetti, A.; Damgrave, R.; Thiede, S.; van Dongen, L. How to Make Augmented Reality a Tool for Railway Maintenance Operations: Operator 4.0 Perspective. *Appl. Sci.* **2021**, *11*, 2656. https://doi.org/10.3390/app11062656

Academic Editor: Jasiulewicz-Kaczmarek Małgorzata

Received: 17 February 2021
Accepted: 12 March 2021
Published: 16 March 2021

1. Introduction

Maintenance, repair, and overhaul (MRO) are critical areas in railway asset management and are crucial for industry growth and seamless railway operations. Asset management, also referred to as engineering management, ensures proper management of assets throughout their entire lifecycle [1]. Having a holistic approach to asset management establishes an important link between operational performance and management practices [2]. This holistic approach is key when managing risks and opportunities to achieve the desired balance between cost, risk, and performance. Attention should be paid to align asset management strategies with the overall organizational strategy across different levels of decision-making to contribute to practice-oriented and technical aspects [3]. In light of the above, maintenance plays an important role in physical asset management and contributes to operational performance.

1.1. AR Technology

Globally, the railway industry is focusing on attaining efficiency in MRO by accelerating the maintenance and repair activities through industry specialized augmented reality (AR) technology solutions [4]. The global augmented reality industry for the MRO market was valued at $403.3 million in 2018 and is expected to reach $3319 million by 2024 [4]. AR-based innovations have received a lot of traction in the last few years.

In just over a decade, AR has matured and proven to be an innovative and effective solution to help solve a number of the critical problems to assist and improve maintenance processes before they are implemented [5]. This technology is based on human-computer interaction and overlays digital virtual information in the real-world environment. The information display and image overlay are context-sensitive and user-dependant, meaning that they depend on the perceived objects in combination with the user's expectations. This technology enables railway companies to examine, monitor, and analyze rolling stock components with great effectiveness and efficiency. The evolution of technologies such as cloud computing, cognitive computing, and machine learning is paving the way for the growth of AR in MRO [6].

1.2. AR Application Fields

Besides maintenance, other AR application fields can be specified like medical, military, robotics, education, and geospatial [7]. AR assists in standardizing and making workflows more user-friendly and efficient by contextualizing and personalizing information. Although AR has great potential, the hardware is not yet user-friendly [8]. Recent work introduced the use of AR as a tool for processing and visualization of imaging data which can be subtracted from medical devices such as MRI scanners, simulation of surgical tools, and other assistive data [9]. This integration of the physician with the data and sensors ensures the visualization of patient data in 3D and collects and analyses newly generated data. However, subjective assessment of ergonomics and functionalities from end-users was used for minor improvements on the interfaces. Besides this, the AR tool must be seamlessly integrated into the daily workflow of the clinical site. To accomplish this, further developments on the AR hardware are required.

In the context of geospatial experiments, research has been conducted on defining requirements for hologram positioning and display [8]. The presented work contributes to optimized experimental user testing in a real 3D spatial layout. It was concluded that as long as reliable and accurate tracking cannot be provided by the AR tool, the use of the device will be limited to spatially confined environments.

The construction industry faces the problem of determining the location of underground utilities before excavation work can be carried out [10]. AR can be useful for field workers for work planning and during excavations. Research showed that field workers want to implement a finished version of the AR prototype-tool of utility excavations [10]. Important functionalities for the operator were distance measuring, an estimate of leakage locations, and planning and coordination with other professionals.

Other new research describes innovative methods to integrate AR technologies with other technologies such as positioning sensors, tools for managing and visualizing geospatial data (GIS), and systems using high precision real-time kinematics (GNSS) [11]. The main criticism from the end-user was directed to the casing of the device. Additionally, work needs to be done in delivering a physical and cognitive ergonomic device which will facilitate the job of the operators and make their life easier during the field activities. Moreover, the maintenance industry is facing significant challenges nowadays in which costs, safety, availability, and reliability are demanding objectives [12]. In recent years, the evolution of digital technologies has given analog devices a digital footprint. This enables greater connectivity and provides possibilities to achieve higher levels of productivity and thus contributes to the objectives of the maintenance industry [13]. The integration of new digital technologies becomes possible and introduces the term "Industry 4.0". This requires a quick and efficient maintenance service to guarantee that companies implement an efficient production system [14]. An important characteristic of Industry 4.0 is the exploitation of data to evolve from scheduled, control-based processes and systems to smart processes and systems.

Opportunities arise to predict the behavior of operators, machines, and systems allowing faster decision-making and less downtime [15]. Predicting maintenance enables getting a holistic view of data sources, collection, and analysis to preserve asset reliability and management [13]. Integrating the Industry 4.0 paradigm in maintenance operations will have far-reaching consequences for the interactions between humans and technology [16]. The role of the human shifts from mainly being a spectator and machine operator towards a strategic decision-maker and a flexible problem-solver [17]. Due to the increasing complexity of production, humans need to be supported by assistance systems [18]. These systems need to aggregate and visualize information in an understandable way such that humans can make well-thought-out decisions and solve urgent problems on short notice. The focus in this research has turned to operators and the support given to them when performing a maintenance task. This study focuses on critical activities that take place on the shop floor of the maintenance facility.

AR can be useful for many situations in maintenance where users require real-time additional information tailored to the activity. Furthermore, if properly used and developed, AR visualization capabilities can transform maintenance processes [19]. Despite the visualization capabilities AR offers, the use of contextualized and customized information supply can be further explored. Not only will this contribute to the development of novel AR adaptive tool devices, but it will also convince users that they will forego traditional methods and opt for AR-assisted solutions.

New interactions between humans and AR support tools and the digital and physical world will directly influence the operator and the nature of work. Operator 4.0 is an experienced operator who can work cooperatively using human-machine interaction technologies to address complex problems [20]. However, not all operators have the same level of expertise, skills, preferences, expectations, and learning capabilities. An appropriate level of detailed instructions should be provided to the operator, tailored to their needs and expectations. Through AR, virtual information that is needed to support maintenance operators can directly be overlaid onto the real workspace. Novice operators can easily get real-life and real-time instructions, whereas off-site experts can collaborate remotely with them. AR guidance could significantly increase the efficiency and effectiveness of the maintenance operation, increase people and process safety, and minimize unplanned downtime [21]. Moreover, it is needed to support the operator to understand, map, and develop his/her competencies by developing an adaptive tool to provide tailored information to enhance the operator's task performance. To reach this goal, it is needed to provide a structured process that allows having a systematic approach for using the adaptive tool.

1.3. Scope

The purpose of this work is to propose an adaptive architectural framework for a structured procedure that enables a systematic approach to provide support to operators using an AR tool. This tool should support everyday practices and facilitate adaptive capabilities. This adaptive architectural framework can be used for everyday maintenance tasks by capturing the know-how and helpful tips of more experienced operators. Based on experience, expertise, external factors, and the current condition of systems, the operator will be able to get access to tailored information at the right time.

Figure 1 shows the main focus points that will contribute to the need for a dynamic tool. This research mainly focuses on the following aspects: (1) Operator 4.0, (2) AR capabilities, and (3) maintenance operations.

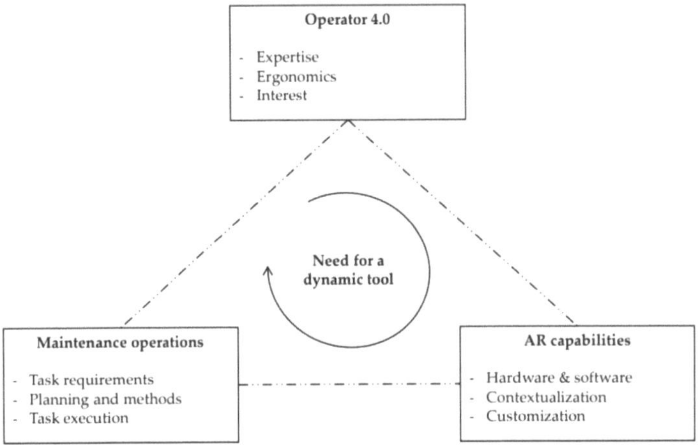

Figure 1. Focus points of the adaptive architectural framework.

2. Theoretical Background

The research objectives identified in the previous section indicate the need to review the existing literature. The description of the focus points is presented in Figure 1. Each step is explained in detail in the following subsections.

2.1. AR in Maintenance Operations

To ensure continuous system performance in their present operating context, efficient maintenance planning and task execution are required. A maintenance plan contains consolidated listings with descriptions of the condition monitoring, time or usage-based interventions and failure-finding tasks, the re-design decisions, and the run-to-failure decisions [22]. To systemize the process for determining the appropriate maintenance task requirements, the application of the AR tool will be explored using the adapted Reliability Centred Maintenance (RCM) process steps [22].

- *Step 1:* Select equipment. In the first step, the operator must decide what to analyze. Each system component has a unique combination of failure modes and failure rates [22]. When a failure occurs in a system, the operator should prioritize and analyze the impact each failure has on the process. High impact failures have high priority.
- *Step 2:* Determine the functions. The function of a system determines the action that it will perform. AR spatial mapping and tracking systems can be used for finding all major and less obvious failures in a system [23]. An operator can overlook less obvious failures, whereas the AR tool can capture and report all failures.
- *Step 3:* Describe failures. Overlapping virtual information to physical components, according to their real-world position, ensures identification of the failure. The operator can see that the virtual image and the real object are not in the same place.
- *Step 4:* Describe failure modes. A failure mode indicates how the system fails to perform its function [22]. Maintenance interventions such as checking, changing, and condition monitoring can be performed using AR.
- *Step 5:* Select maintenance action. Based on predefined actions and instructions, the operator can address the failure using tailored AR guidance and contextualization.
- *Step 6:* Document results. Technical manuals often recommend a maintenance method for certain equipment and systems. However, manuals or work descriptions are not often tailored to a particular operating environment and actual environmental conditions. The technology can capture the time needed for addressing the failure and what sequence of tasks has been performed. Hereafter, the technology provides a periodic intervention to eliminate failure to occur.

- *AR solution:* Select a flexible tool. This tool should assist the operator by systemizing the maintenance procedure. Besides this, the tool should contextualize and customize the information supply to the need and skills of the operator. This support tool should easily be embedded in everyday maintenance operations.

Figure 2 presents the use of AR within a maintenance problem-solving process. The focus of the AR support tool is put on sensing all failure causes, providing visual guidance in problem-solving, and alert the operator if a task is performed incorrectly. The tool also monitors and reports the time and sequence of steps required to perform the maintenance task. An accurate indication of the time needed to perform a task can be captured. Maintenance planning and schedules can be adapted to this information.

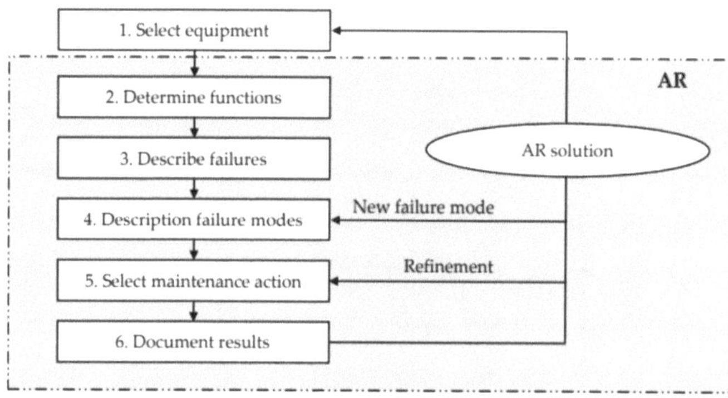

Figure 2. Added value augmented reality (AR) offers maintenance procedures adapted from Campbell [22].

2.2. Operator 4.0: Augmented Operator

Maintenance operations, in which the human workforce has a crucial role, need instruments able to manage complex systems. Nowadays most of the lower-skilled human jobs were eliminated and replaced by technology. This results in the remaining jobs becoming more complex, comprehensive, and increasing the importance of interdisciplinary cooperation [24]. Consequently, numerous complex, high-precision processes are and will be managed manually. However, many of them could collaboratively be executed by humans and interactive technology cooperation.

Several factors are important regarding work satisfaction and dissatisfaction, including the work environment, work organization, and whether the work is interesting [18]. The operator's cognitive ergonomics, interests, and expertise should be included carefully since interaction with the real and digital systems is established. During this process, the operator's physical and mental workloads are affected by both system and process features as well as by external factors. Operators produce a subjective experience depending on the surrounding environment, individual skills and characteristics, and task features [24].

In an augmented environment, the Operator 4.0 typology considered is the augmented operator [25]. More specifically, the augmented operator owns a superior knowledge of the working environment which is not only derived from the daily interactions related to the maintenance task or procedures, but it incorporates a variety of value-added contents that are suited to augment his/her skills and abilities to perceive and act within the working environment [26]. Operator 4.0 is expected to benefit from the information and guidance generated within the virtual context. Applicable resources that usually may not be available should be made available directly using interactive technology. Besides this, a remote consultation tool to get easy and fast information and advice, and a tool for safety and security enhancement has to become available. The AR technology ensures lowering the cognitive load of the operator, as the operator does not have to manually search for filter according to the current context, or interpret information on a screen, rather they can visualize it directly on the target object [27].

From a business perspective, maintenance KPIs can be analyzed that allow managers to have a proper overview of workstations and production lines in real-time monitoring [27]. It becomes possible to identify, analyze, diagnose, and resolve errors to keep maintenance processes moving towards operational efficiency.

2.3. AR Capabilities

Nowadays, the operator knows the current state of a component by consulting a paper or digital work description containing the right configuration and information. Instead of forcing the worker to waste time consulting paperwork and interpreting the information provided into the work instructions, the proposed technology projects directly on the component giving accurate maintenance instructions and relevant information.

AR is considered key for improving the transfer of information from the digital to the physical world of the smart operator [28]. Moreover, AR has incorporating capabilities to new human-machine interfaces to maintain IT applications and assets. It displays real-time feedback about the smart maintenance processes and assets to the smart operator to improve decision making. This technology supports the smart operator in real-time during manual maintenance procedures by becoming a digital assistance system [27]. Hereby, the operator reduces the time used for reading printed work instructions and documents, looking into computer screens or tablets, and following strict procedures. Operators no longer need to run the risk of using old or outdated paper documents and instructions. Digital information systems provide operators with the latest updated information. Other advantages the tool provides for the operator are its ability to offer intuitive information and combining operator intelligence and flexibility with error-proofing systems to increase the efficiency of manual steps [29]. AR offers a powerful tool for supplying the operator with contextualized and customized information [30]. Altogether, AR can improve the quality, reliability, maintenance time, and reduce the failure rate.

Operator 4.0 will have access to the data coming from the train components and sensors in the maintenance facility. Besides this information source, knowledge can be gathered from other operators in the maintenance facility or even outside the facility from professional (social) networks. All information needs to be delivered to the operator and adapted, contextualized, and transformed to make it understandable such that decisions can be made resolutely and thoroughly.

2.4. The Need for a Dynamic Tool

As the complexity of the maintenance operations grows, proper support tools and approaches are required for the operator [18]. Research suggests that Operator 4.0 is required to be highly flexible and should demonstrate adaptive capabilities in a very dynamic working environment [31]. Therefore, a need exists for an AR tool to support the operator in his/her daily work. Depending on the task environment, condition of the asset, failure description, level of expertise, and operator experience, the AR tool must provide tailored contextualized and understandable information into the operator's space in coexistence with real-world objects. To implement AR in the railway industry, the system has to be easy to maintain and modify. New content management tools are required as well as reconfigurability systems. The visualized information must be tailored to the operation and environment. Additionally, the way information is brought to the operator has to be studied. Future AR systems must be adaptive and able to systematically capture the operator's intentions in performing a maintenance task. Besides this, it should collect the data of any maintenance procedure. The information collected could be used for improving the training process of the tool or the maintenance procedure itself.

Before the tool can function as a support system for the operator, a structured approach should be proposed to systemize the adaption process. As can be seen in Figure 1, the realization of the tool requires a structured framework.

3. Methods

The developed adaptive architectural framework aims at accelerating the adaption of information supply assistance that AR provides to the augmented operator. The framework will help to understand what information should be captured, why this should be captured how it must be captured, and how the data is being reused for future AR experiences. In the following, the basics of the underlying decision support system, as well as boundary conditions and requirements are presented as they form the basis of this work.

3.1. Decision Support System

A human-centered approach to capture the knowledge of an operator is the decision support system. The AR tool is used to capture expert knowledge on maintenance task performance. The aim is to mimic how experienced operators make decisions based on using their experiences to form plausible approaches for new situations. Incremental learning for modeling complicated decisions support systems is required to quickly retrieve information by representing and organizing experts' knowledge [32]. Gathering expert knowledge is vital for the development of (1) specific domain knowledge, needed to generate example solutions, and (2) general domain knowledge, needed to develop the reasoning structure. Applying an adaptive tool to a decision support system could assist a novel operator in a similar way to how an expert would use their experience to solve a problem.

The distribution model of cognition has been adapted for this framework and focuses on developing an ensemble of distributed individuals and artifacts [33]. This model considers two indispensable parts: internal and external representations. Internal representations are the knowledge structures in the operator's head that can be retrieved from memory External representations are the knowledge structures coming from the environment. The environmental elements help to make sense of the dynamic working situation by providing information on physical symbols, objects, dimensions, constraints, and relations embedded in physical configurations [33]. Besides this, the environment provides information on what task is expected to be executed and who will participate in the procedure.

The task that needs to be performed, together with both external and internal representations, contributes to the mental representation of a task solution.

Opportunities arise for human and intelligent systems to collaborate, learn from each other, and work together to achieve common objectives. Effective collaboration can be established if the interactive technology is logical, explicable, and able to understand the human cognitive processes [34]. A cognitive and interactive tool can learn and improve with an operator acting as a mentor for the system, based on his/her experience and knowledge, whereas the system provides feedback to the human in return [35]. For the operator, the process of providing feedback and interactions ensures both increased efficiency and developed confidence in the system [33].

3.2. Boundary Conditions

Technical and technological issues related to the development and implementation of AR are debatable elements. Most studies that review technological solutions for visualization either use mobile devices with camera-overlay AR or head-mounted displays (HMDs) with see-through AR [36]. Mobile devices need to be held in the hand when used and therefore potentially hinder maintenance tasks. HMD solutions leave the hands free, allowing for a more natural and intuitive hand-based interaction with virtual objects. However, sometimes a limited view can be experienced by the operator using HMDs.

Initializing an operator's knowledge level must be established to verify the level of expertise an operator has. However, in real-life situations, it can be difficult to obtain an exact and reliable assessment of human competence. To estimate an operator's competence, different data collection methods could be explored, such as interviews, testing at the workplace, using empirical methods, and maintenance task simulations [31]. The collection of input data for competence analysis can be achieved by measuring the execution time of the maintenance task in the facility and evaluating the experiences of the technicians. For knowledge capturing, experienced operators must be recruited. To increase the effectiveness of the knowledge capturing method, a large number of operators should be involved which is important for providing decision support.

3.3. Architectural Requirements

The proposed architecture, apart from bringing together information from different digital databases, combines five major features by exploiting AR: (1) capturing the knowledge of the operator, (2) providing maintenance support, (3) performing failure analysis, (4) providing problem-solving strategy, and (5) providing learning capabilities. The adaptive architectural framework has been designed simultaneously while performing case studies. The studies are proving the value of the framework. The framework is based on the methodological decision support system and its main requirements include:

- Provide the augmented operator with real-time feedback and augmented reality content on tasks/procedures execution. Operators are guided by the supplying of visual and audible instructions to give tangible feedback.
- Based on expertise, experience, external factors, current conditions of the component, it is required to ensure the operator has a personal tailored digital knowledgeable assistant to interact with. Depending on the operator's ability, noncritical information can be supplied using subtle instructions in different visible frequencies.
- By capturing the knowledge of the operator and procedural steps, the system can learn from previous maintenance procedures. The time and sequence of steps used to perform a maintenance task can be captured and reported. This can indicate how much time is needed in the future to perform the task. Moreover, failure rates can be compared to this information, providing insight on the most sustainable procedure. Maintenance planning and schedules can be adapted to these accurate findings. Therefore, the efficiency of operation support will be increased.

Based on the architectural requirements, a functional analysis is presented to structure and identify potential solutions that exist for the adaptive tool. In Figure 3, the most viable solutions are proposed. The analysis explores potential solutions for a given function, the solutions are variable.

Based on the functional requirements from Figure 3, the architecture in Figure 4 has been designed and implemented in the maintenance process analysis of Figure 2.

- *Step 1:* Select equipment. The goal to be achieved is formulated. The operator will be guided by visual and audible instructions which also give tangible feedback.
- *Step 2:* Determine functions. The task that needs to be selected to reach the goal is stated. Besides this, the adaptive AR tool provides all failure causes and digital information on all potential solutions. It will let the operator be aware of the context to gather relevant information and/or services, relevancy depends on the operator's tasks [37]. Using context awareness systems, such as AR, accurate access to maintenance information is provided such that the operator's performance efficiency can be increased.
- *Step 3:* Describe failures. Initiation and evaluating the operator's expertise is currently based on the operator's or manager's perspective. The level of expertise varies from having no clue what is going on up to being an expert and able to train others. In this framework, initiating the level of expertise is performed manually but can become an automized process in the future. Operators can be equipped with sensors to activate psychomotor and cognitive responses that are beyond what operators can verbalize. Capturing gestures of experts can improve interactions with AR and ensures future knowledge capturing [33].
- *Step 4:* Describe failure modes. Dynamic behavior capturing is required to perform a successful fault diagnosis [33]. Based on time and process tracking, the operator should know what initiated the fault, what the current situation is, what is needed to solve the issue, and what time is required to solve the task. Varying business demands changes

in work routines, resource availability, and environmental conditions. Depending on the complexity and nature of the maintenance task, the operator adapts his/her maintenance concept.

- *Step 5:* Select maintenance action. The tool presents the task that aims to restore the functionality of a system. The actions that can be performed to restore the functionality of the product can be technical, administrative, and managerial [30]. Continuous assessment takes place of the operator's performance, task condition, and other external conditions. Besides this, the tool will send warning messages of improper maintenance operation execution. When the task or business demand increases, mental demand increases resulting in negative effects on physiological variables [38]. The likelihood that the operator fails in performing his/her task becomes subsequently larger, it is therefore needed to have a control or monitoring system that alarms the operator when tasks are not performed adequately.
- *Step 6:* Documents results. Documentation can support the detection of schedule derivations or the search for sources of defects and the responsibility of the operator [39]. Adequate process monitoring methods help managers and operators to document the current status of the maintenance work as well as to understand origins and defects. Some maintenance tasks require inactive input, for instance, to leave comments on specific objects. AR allows storing these annotations directly in relationship to the real environment.
- *AR solution:* Select a flexible tool. Incorporating different types of data, interfaces, visualization systems and sensors makes the adaptive tool applicable to multiple solutions.

Function \ Solution	Manual system	Sensor system	Interactive system	Augmented system	…
Capturing knowledge operator	Manual assessment by supervisor or operator him/herself	Experiments, interviews, simulation models	Tracking time and sequence of steps	Gesture tracking systems	…
Providing maintenance support	Static step-by-step guideline	User-tailored task description	Remote support	Augmented task illustration	…
Performing failure analysis	Static catalogue of typical failures	Reading sensor diagnostics	Displaying diagnostics automatically	Systemized experience and expertise tracking	…
Providing a problem-solving strategy	Manual task performance	Adaptive task performance	Contextualized task performance	Visualized task performance	…
Providing learning capabilities	Manual assessment of task performance	Stand-alone digital feedback system	A collaborative learning feedback system	Dynamic multiple feedback	…
Generating interactions	Human interactions	Programming, modelling, knowledge management	Adaptable system to transfer knowledge	Reconfigurability of systems	…
Assessing cognitive workload	Subjective questionnaires	Physiological parameters monitoring	Stress analysis	Motion capture system	…
Continuous assessment procedure	Verification by operator	Real-time condition monitoring	e-maintenance	Intelligent prognostics tools	…
Capturing results	Documentation on paper	Process monitoring	Object detection and instance segmentation	Capture contextualized and customized data	…
…	…	…	…	…	…

Figure 3. Functional analysis of adaptive capabilities and potential solutions.

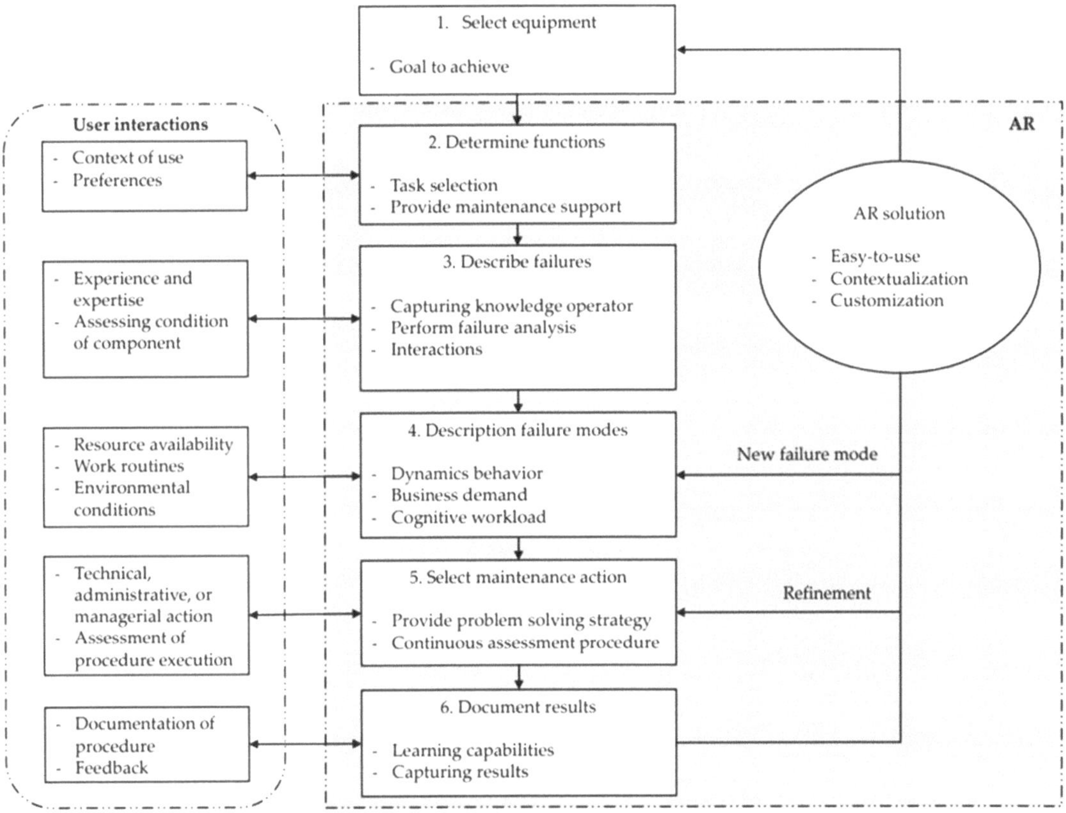

Figure 4. Schematic of the adaptive architectural framework.

4. Case Study

The purpose of a case study is to explore and depict a setting in which the researcher exploits data from direct observations, systematic interviewing from public or private archives [40]. This case study aims to increase understanding of the need for having an adaptive architectural framework to support Operator 4.0. The case study helps to identify technology adaption patterns, defining requirements needed to support the operator, and providing future steps needed for application in maintenance operations.

4.1. Case Study Characteristics

In this research, one single case study is examined for validation of the framework creating the opportunity for in-depth observation. This framework allows for a two-way learning system. More specifically, not only does the case study present the effects of the framework, the framework iterates after the case study is explored. An example of using one single case to examine reengineering service operations is performed by Narasimham and Jayaram [41]. Other research focused on motivating operators to take active decisions when transforming IT function profiles in healthcare organizations using one single case study [42]. More recent work examined the process of resource alteration underlying the digital manufacturing journey using a single case [43].

The Dutch Railway company (NS) is a Dutch state-owned company and the principal railway operator in the Netherlands. The Dutch rail network is Europe's busiest and will only become busier [44]. Among others, NS aims to achieve safe, sustainable, and reliable operations in which technological developments are key [45]. The company recognizes the added value Artificial Intelligence (AI), Internet of Things (IoT), and AR have to improve services and contribute to efficient operations. Additionally, they have noticed that the coronavirus is accelerating digitization and technological developments. In short, NS offers great opportunities to exploit a case study for framework verification.

Several information sources have been used through the research such as interviews, managerial presentations, student thesis on maintenance operations, public, and internal documents. Multiple interviews with managers and operators were held online. The participants were selected based on their knowledge of innovative technologies, data collection and management, and maintenance operations. In total, 28 participants were interviewed of which the majority are part of the NS technical department. This sample size represents 35% of the total sample size and is sufficient for this specific knowledge field [46]. The length of the interviews varied from 30 minutes to 1 hour. The interviews provided deep insight into actual focuses, potentials, and challenges that the case should include. Based on the information sources, a case description was prepared and verified by the company. To increase the validity of the framework, the completed version was discussed, completed, and improved together with the company.

4.2. Investigation of the Retractable Step

Previous research was conducted within the Dutch railway industry for adopting VR technologies for training and skilling of employees, and to assist train drivers during specific operations [47]. In this research, the focus was put on increasing operational and training efficiencies. Notwithstanding the positive results AR can achieve, there are still a few questions that remained unanswered about the employability of this emerging solution The verification of the usage of this adaptive architectural framework can be performed by exploring a case study for NS. For this case study, the Fast Light Innovative Regional Train (FLIRT) type is further examined. This train is a passenger electric multiple unit trainset [48]. All interviewees agreed on the validity of the framework using this case.

The railway company investigated the failure mechanism of the FLIRT electric door system in which the retractable step caused the system to fail repeatedly [49]. This system serves to bridge the gap between platform and vehicle. Within 50 operation days, 187 services were requested for the door system. Since the deployment of the FLIRT train series on the Dutch railways in 2016, 1099 service requests were already made in 2017 [47]. The door system failure accounts for 17.4% compared to other failing components and is only surpassed by the communication system of the train which represents 27.5% of all failures [47]. The failing door system has, therefore, a high priority since it has a direct and great impact on train operations.

4.3. Failure Description Retractable Step

Product description of the system is based on the technical documentation of the retractable step [50]. The entire system consists of the sliding step and the connection cable to the control unit. The extension unit basically consists of a frame in which the walking zone is stored. The walking zone is equipped with a step sensor and anti-slip coating. The drive takes place by means of a DC motor. The motor is mounted on a compact drive-bearing unit. The driving force is transmitted from the motor shaft via a hollow shaft to a toothed belt wheel. The toothed belt wheel drives the central drive belt. A carrier on the toothed belt established the connection with the sliding step and converts the motor drive power into the extension movement of the extension unit. The sliding step is guided by profile rollers on the extension unit. The upper and lower rails are both made of stainless steel. The profile below the lower rail is made of aluminum. If a vertical load of 150 N or more is applied to the sliding step during extension or retraction, the movement is stopped immediately. A schematic overview of the train door retractable step system is provided in Figure 5.

Number	Component
1	Base frame
2	Slide-out unit
3	E-motor
4	Toothed belt
5	Cove plate (bottom)
6	Water outlet with pipe flange
7	Cover plate
8	Safety switch
9	Cable box
10	Connection cable (socket)

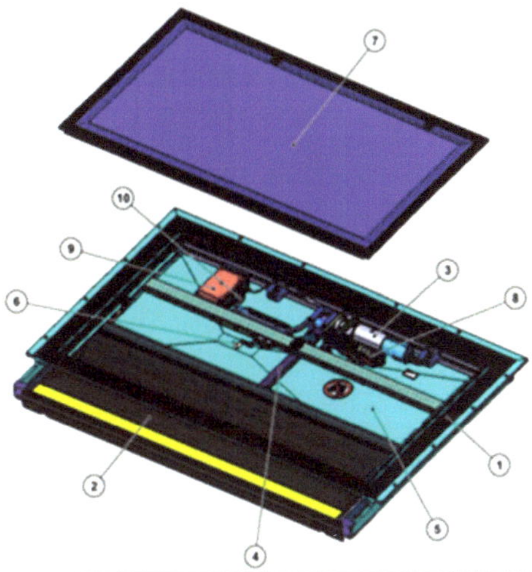

Figure 5. Component description FLIRT retractable step [49].

According to the research performed by NS, the FLIRT sliding step causes many problems [49]. The most critical problem is that the retractable step gets stuck due to clamping. In that research, clamping was caused by the fact that the left rail was raised about 2.4 mm. To find out why the bottom rail came up, a destructive test was performed. The part of the aluminium profile at the height of the elevation was cut out and the lower rail was removed. Details of the side profile with rollers are presented in Figure 6.

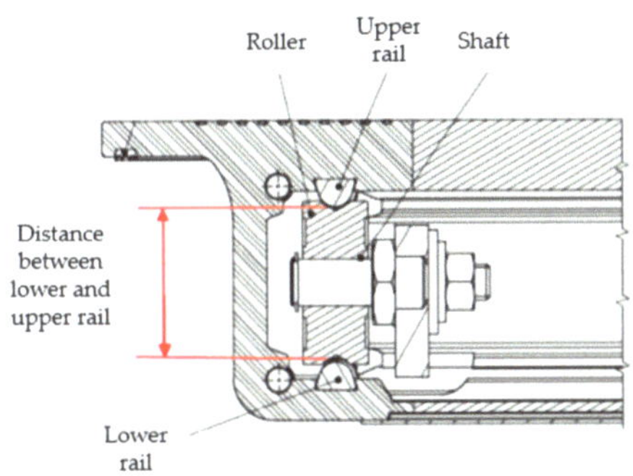

Figure 6. Details of side profile retractable step with rollers Fast Light Innovative Regional Train (FLIRT) [49].

The rise of the lower rail was caused by the formation of aluminum oxide between the stainless-steel rail and the aluminum profile. More specifically, it was caused by galvanic corrosion of the sliding step. The volume of aluminum oxide pushes the bottom rail up by 2.4 mm.

4.4. Application Adaptive Architectural Framework

The application of repairing the retractable step in the designed adaptive architectural framework is presented in Figure 7. General specifications for the application of the framework to this case are:

- Decisions are made based on the operator's perspective on his/her level of expertise. Let the amount of AR information and frequency of information supply be adapted to the specific user, task demand, and business demand.

- Knowledge is captured from expert operators to use their experience to assist a novel operator to solve a problem. General and specific domain knowledge should be gathered to provide an incremental learning method. The time needed and sequence of steps of the procedures involved can be derived from the task performance. Hereby, the operator and the company capture detailed knowledge on the procedure to become more efficient and adequate problem solvers.
- Safety and security measures should be taken into account more consciously. The framework ensures sending warning messages if procedures or tasks are not performed according to procedures or safety standards.

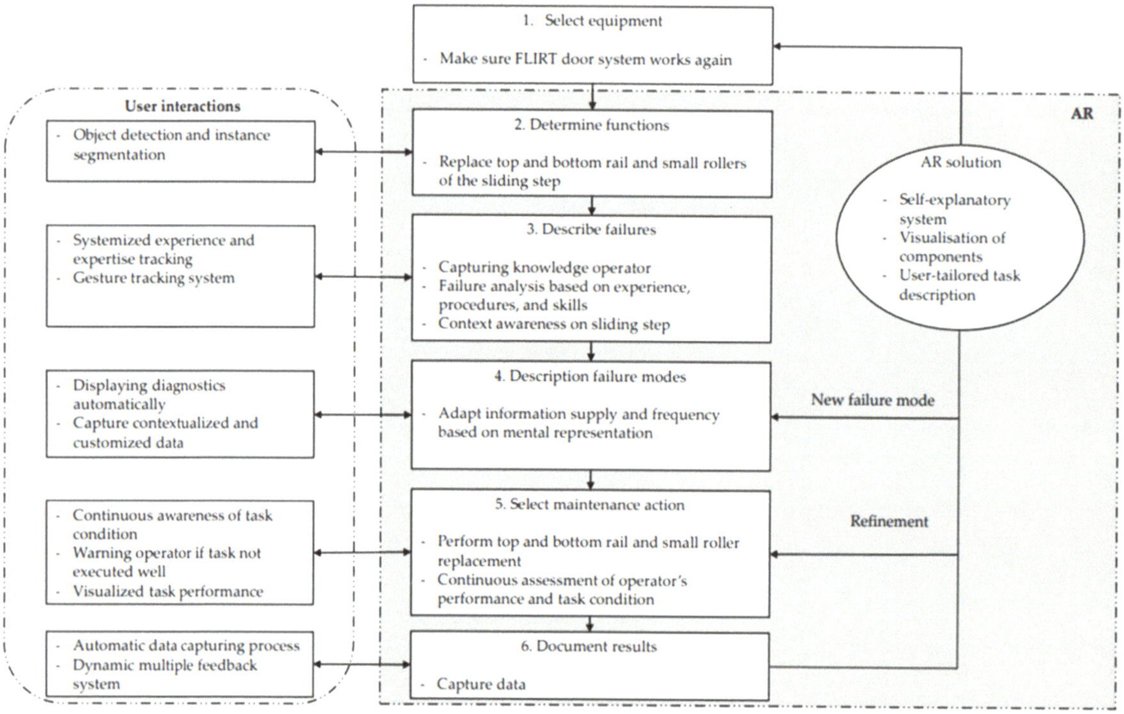

Figure 7. Applied case study for the designed adaptive architectural framework.

Some functional solutions from Figure 3 can be used for the case study by setting adaptive capability requirements. All solutions directly affect the user's interactions with the AR technology.

5. Discussion of the Main Results

The case study represents a unique endeavor about the importance of having a structured process to ensure support is provided to Operator 4.0 using AR. This adaptive architectural framework reveals what technology adaption patterns are identified. Besides this, requirements are provided to support Operator 4.0 in his day-to-day work. Finally, future trends are examined from the point of view of maintenance work.

5.1. Technology Adoption Patterns

Looking for technology adoption patterns, this research emphasizes the perceived ease of use of AR technologies, the perceived usefulness of the technology, and having self-explanatory systems. This contributes to better expectation management interaction and acceptance between the operator and the AR tool. Still, a lot of other technology adoption patterns can be found due to a lack of understanding of key challenges and success factors. Multiple hardware solutions can be proposed to enable adaptive instructions to operators. Research has been performed to assess AR tools to be implemented for practical everyday use [20]. This research investigates usability together with the achieved levels of productivity and quality. It was concluded that facilitating the operator with adapted displayed instructions is not only useful for novel operators but also experienced users. Not only hardware is important when it comes to technology adoption, ergonomics includes even more aspects of user acceptance. The study revealed that an inadequate design of the user interface can lead to distraction or disorientation [51]. Apart from the adoption patterns mentioned before, the case revealed the importance of having an analysis of mental and physical demands. A recent study supports the main findings of the case study and identifies success in adopting AR in the industry by achieving: user acceptance, visibility of information, ergonomics, and usability of the user interface [52].

5.2. Support Requirements for Operator 4.0

This research suggests information provided to the operator should be based on real-time, contextualized, and customized data. The information supply (and the related user interface) should be tailored to the expertise and experience of the operator, component condition, and other external factors such as organizational demand. The application of different visual computing technologies in industrial operator tasks was presented by Segura et al. [27]. They presented several cases to show how proper visual analytic systems can support the operator to better understand and easily detect wrong production situations. Their research emphasizes the need of adapting and balancing procedures with the experience and expertise of the operators. As suggested by earlier research, Operator 4.0 can be empowered by adapting the machine-user interface, machine behavior, and planning [18]. Based on the operator, user interfaces of the AR tool could be adapted to allow only showing functions that the worker understands. This will facilitate in identifying the role of the operator, AR capabilities, and maintenance operations.

5.3. Future Trends in Maintenance Operations

The case study depicts the ability to have an automatic data capturing process while also having a dynamic multiple feedback system. Capturing maintenance operation data and knowledge of the operator contributes to incremental learning capabilities. Consequently, maintenance planning and schedules can be adapted to this. Hence, future operations support increases. Digitized systems recognize changes in operations and continuously updates component performance [13]. Thus, ensuring optimum efficiency is always achieved. The maintenance operators can be supported with real-life instructions, diagnostic information, and remote assistance. However, the economic, environmental, and social challenges faced by the AR industry still require further investigation [21].

6. Conclusions

Although much progress in AR has been made in recent years, little attention has been paid to correct the integration of humans in the emerging context of AR in professional industrial (engineering) environments. A human-centered approach is necessary for the successful adaption of AR technology tools for the railway industries. Operator 4.0 will play an important role in facilitating the transition from traditional maintenance procedures to remotely, digitized, and autonomous operations. Few in-depth studies assess and evaluate human factors and interaction in (industrial) AR systems [27]. Attention has been drawn to operators as a key element to address new and unpredictable behaviors in AR. Since operators experience an increased complexity of their daily tasks, an adaptive tool is desired to support the operators. Based on the operator's competence, expertise, component condition, and external factors, the tool will provide contextualized and customized information. Looking at the spread of previous research, we conclude that they were all considering the need of (1) supporting Operator 4.0 interactive technology and (2) supplying tailored based information [18,25,53]. However, before tailored-based information can be supplied to support the operator, it is required to structure the process to enable systemizing this approach. Therefore, this research bridges the gap between the need to support operators using an AR tool and the approach of providing this.

The main research contribution is twofold: (1) proposing an adaptive architectural framework aimed at shaping and structuring the process that enables a systematic approach to provide a support tool to operators using an AR tool, and (2) a case study that implements the aforementioned framework. As a result, an adaptive architectural framework is suited to augment the operator's skills and abilities to perceive and act within the working environment. This digital assistant tool supports the operator with vocal and visual interaction capabilities. It is meant to provide quick, tailored, and efficient information on maintenance tasks.

The adaptive architectural framework can: (1) capture the knowledge of the operator, (2) support the operator in performing maintenance tasks, (3) conduct failure analysis to find all potential failure modes, (4) provide all problem-solving strategies, and (5) improve learning capabilities by documentation of procedural task performance. This framework can be adapted to be able to absorb and immerse the environment for preliminary training on new or complex procedures. To this end, the proposed adapted architectural framework is scalable and modular since the principle can be applied to different industries and infrastructures.

Many companies are considering implementing AR solutions to their maintenance operations and are willing to perform several experiments using the technology [52]. From the proposed framework, we suggest managers start exploiting opportunities for AR technology application fields. In a maintenance workshop, operators are directly linked to AR solutions. In our case company, we witnessed the relevance of structuring the process of providing customized and contextualized data to operators. Using this framework, operators will find all potential failure modes of a component and define all problem-solving strategies needed to solve the issue. Therefore this framework increases the safety, efficiency, and availability of the operators. However, managers should be aware that the decision of this trajectory is costly. Besides the costs, managers should bear in mind that using AR strategies is only an intermediate element of the digitization of the company structure. Furthermore, the development of AR technologies continues and, therefore, organizations should be ready for frequent iteration and adjustments in application strategies.

Industry 4.0 is still an open research field where already much has been done but there is still more to do to accomplish its vision. This research proposed a forward-looking adaptive framework for Operator 4.0. From a technical point of view, additional work can be done to improve and optimize the technical performance of the framework in terms of capturing the operator's knowledge and transforming this expert knowledge into a database for the development of a decision support technology [32]. More specifically, future work is required in the specification of the expert-level of the operators. Capturing knowledge is currently based on the perceived perception of the operator or manager. However, knowledge capturing should become autonomous [54]. In addition, verification and validation of the framework can be performed in a simulation and experimentation setting. A methodological limitation of this research is the use of one case study. Adding to this, the case was based on a limited number of interviews. The case provides useful information on an emerging topic like AR. But one case is still not enough to draw generalisable conclusions. The external validity of the research could be enhanced by examining other companies in a similar situation. In terms of research perspectives, future work will be needed toward the development of prognostic capabilities. Integration of sophisticated algorithms for real-time monitoring and process control will support maintenance operations.

Finally, the close relationship with the case study enables us to follow up its journey in the future. Efforts and the involvement of other managers bear the opportunity to continue the current case.

Author Contributions: Conceptualization and methodology, S.S., R.D., A.M., and S.T.; data collection and analysis, S.S., R.D., A.M., and S.T.; writing—original draft preparation, S.S.; writing—review and editing, S.S., R.D., A.M., S.T., and L.v.D.; supervision, R.D., A.M., S.T. and L.v.D.; project administration and funding acquisition, A.M., L.v.D. All authors have read and agreed to the published version of the manuscript.

Funding: This research was funded by PPP surcharge for Research and Innovation from the Dutch Ministry of Economic Affairs and Climate, from the Netherlands Railways, and the University of Twente.

Institutional Review Board Statement: Not applicable.

Informed Consent Statement: Informed consent was obtained from all subject involved in the study.

Data Availability Statement: Not applicable.

Acknowledgments: The authors acknowledge with gratitude the support of the Netherlands Railways.

Conflicts of Interest: The authors declare no conflict of interest. The funders had no role in the design of the study; in the collection, analyses, or interpretation of data; in the writing of the manuscript, or in the decision to publish the results.

References

1. Lima, E.S.; McMahon, P.; Costa, A.P.C.S. Establishing the relationship between asset management and business performance. *Int. J. Prod. Econ.* **2021**, *232*, 107937. [CrossRef]
2. Maletič, D.; Maletič, M.; Al-Najjar, B.; Gomišček, B. An analysis of physical asset management core practices and their influence on operational performance. *Sustain. J. Rec.* **2020**, *12*, 9097. [CrossRef]
3. Gavrikova, E.; Volkova, I.; Burda, Y. Strategic aspects of asset management: An overview of current research. *Sustain. J. Rec.* **2020**, *12*, 5955. [CrossRef]
4. Infoholic Research LLP. Augmented Reality for MRO (Maintenance, Repair and Overhaul) Market—Global Forecast to 2024. Available online: https://www.infoholicresearch.com/report/augmented-reality-for-mro-market/ (accessed on 14 December 2020).
5. Nee, A.; Ong, S.; Chryssolouris, G.; Mourtzis, D. Augmented reality applications in design and manufacturing. *CIRP Ann.* **2012**, *61*, 657–679. [CrossRef]
6. Ratnayake, D.; Lohit, P.; Singh, B.; Mishra, V.P. Analysis of machine learning algorithms in smart manufacturing. In *Proceedings of the 8th International Conference on Reliability, Infocom Technologies and Optimization (Trends and Future Directions) (ICRITO), Noida, India, 4–5 June 2020*; IEEE: New York, NY, USA, 2020; pp. 707–712.
7. Mekni, M.; Lemieux, A. Augmented reality: Applications, challenges and future trends. *Appl. Comput. Sci.* **2014**, *20*, 205–214.
8. Keil, J.; Edler, D.; Dickmann, F. Preparing the HoloLens for user studies: An augmented reality interface for the spatial adjustment of holographic objects in 3D indoor environments. *KN J. Cartogr. Geogr. Inf.* **2019**, *69*, 205–215. [CrossRef]
9. Velazco-Garcia, J.D.; Shah, D.J.; Leiss, E.L.; Tsekos, N.V. A modular and scalable computational framework for interactive immersion into imaging data with a holographic augmented reality interface. *Comput. Methods Programs Biomed.* **2021**, *198*, 105779. [CrossRef]
10. Hansen, L.H.; Wyke, S.S.; Kjems, E. Combining reality capture and augmented reality to visualise subsurface utilities in the field combining reality capture and augmented reality to visualise subsurface utilities in the field. In Proceedings of the 37th International Symposium on Automation and Robotics in Construction (ISARC 2020), Kitakyushu, Japan, 26–30 October 2020.
11. Stylianidis, E.; Valari, E.; Pagani, A.; Carrillo, I.; Kounoudes, A.; Michail, K.; Smagas, K. Augmented reality Geovisualisation for underground utilities. *PFG J. Photogramm. Remote Sens. Geoinf. Sci.* **2020**, *88*, 173–185. [CrossRef]
12. Del Amo, F.I.; Erkoyuncu, J.A.; Roy, R.; Palmarini, R.; Onoufriou, D. A systematic review of Augmented Reality content-related techniques for knowledge transfer in maintenance applications. *Comput. Ind.* **2018**, *103*, 47–71. [CrossRef]
13. Saidy, C.; Valappil, S.P.; Matthews, R.M.; Bayoumi, A. *Development of a Predictive Maintenance 4.0 Platform: Enhancing Product Design and Manufacturing*; Springer International Publishing: Cham, Switzerland, 2020; Volume 166, pp. 1039–1049.
14. Masoni, R.; Ferrise, F.; Bordegoni, M.; Gattullo, M.; Uva, A.E.; Fiorentino, M.; Carrabba, E.; Di Donato, M. Supporting remote maintenance in industry 4.0 through Augmented Reality. *Procedia Manuf.* **2017**, *11*, 1296–1302. [CrossRef]
15. Ruiz-Sarmiento, J.-R.; Monroy, J.; Moreno, F.-A.; Galindo, C.; Bonelo, J.-M.; Gonzalez-Jimenez, J. A predictive model for the maintenance of industrial machinery in the context of industry 4. *Eng. Appl. Artif. Intell.* **2020**, *87*, 103289. [CrossRef]
16. Nardo, M.; Forino, D.; Murino, T. The evolution of man–machine interaction: The role of human in Industry 4.0 paradigm. *Prod. Manuf. Res.* **2020**, *8*, 20–34. [CrossRef]
17. Hermann, M.; Pentek, T.; Otto, B. Design Principles for Industrie 4.0 Scenarios. In *Proceedings of the 2016 49th Hawaii International Conference on System Sciences (HICSS), Koloa, HI, USA, 5–8 January 2016*; Institute of Electrical and Electronics Engineers (IEEE: USA): New York, NY, USA, 2016; Volume 2016, pp. 3928–3937.
18. Kaasinen, E.; Schmalfuß, F.; Özturk, C.; Aromaa, S.; Boubekeur, M.; Heilala, J.; Heikkilä, P.; Kuula, T.; Liinasuo, M.; Mach, S.; et al. Empowering and engaging industrial workers with Operator 4.0 solutions. *Comput. Ind. Eng.* **2020**, *139*, 105678. [CrossRef]

19. del Amo, I.F.; Erkoyuncu, J.A.; Roy, R.; Wilding, S. Augmented Reality in maintenance: An information-centred design framework *Procedia Manuf.* **2018**, *19*, 148–155. [CrossRef]
20. Holm, M.; Danielsson, O.; Syberfeldt, A.; Moore, P.; Wang, L. Adaptive instructions to novice shop-floor operators using Augmented Reality. *J. Ind. Prod. Eng.* **2017**, *34*, 362–374. [CrossRef]
21. Kaczmarek, M.J.; Gola, A. Maintenance 4.0 Technologies for Sustainable Manufacturing—An Overview. *IFAC PapersOnLine* **2019**, *52*, 91–96. [CrossRef]
22. Campbell, J. The Reliability Handbook, from Downtime to Uptime—In No Time! *Ed. Glob. Lead. Phys. Asset Manag. Price Waterhouse Coopers LLP* **1999**, *23*, 23–36.
23. Bellucci, A.; Ruíz, A.; Díaz, P.; Aedo, I. Investigating augmented reality support for novice users in circuit prototyping. In *Proceedings of the 2018 International Conference on Big Data and Computing, New York, NY, USA, 25–30 June 2018*; Association for Computing Machinery (ACM): New York, NY, USA, 2018; p. 35.
24. Peruzzini, M.; Grandi, F.; Pellicciari, M. Exploring the potential of Operator 4.0 interface and monitoring. *Comput. Ind. Eng.* **2020**, *139*, 105600. [CrossRef]
25. Romero, D.; Stahre, J.; Wuest, T.; Noran, O. Towards an operator 4.0 typology: A human-centric perspective on the fourth industrial revolution technologies. In Proceedings of the International Conference on Computers and Industrial Engineering, Tianjin, China, 29–31 October 2016; pp. 1–11.
26. Longo, F.; Nicoletti, L.; Padovano, A. Smart operators in industry 4.0: A human-centered approach to enhance operators' capabilities and competencies within the new smart factory context. *Comput. Ind. Eng.* **2017**, *113*, 144–159. [CrossRef]
27. Segura, Á.; Diez, H.V.; Barandiaran, I.; Arbelaiz, A.; Álvarez, H.; Simões, B.; Posada, J.; García-Alonso, A.; Ugarte, R. Visual computing technologies to support the Operator 4. *Comput. Ind. Eng.* **2020**, *139*, 105550. [CrossRef]
28. Bigliardi, B.; Bottani, E.; Casella, G. Enabling technologies, application areas and impact of industry 4.0: A bibliographic analysis. *Procedia Manuf.* **2020**, *42*, 322–326. [CrossRef]
29. Egger, J.; Masood, T. Augmented reality in support of intelligent manufacturing—A systematic literature review. *Comput. Ind. Eng.* **2020**, *140*, 106195. [CrossRef]
30. Palmarini, R.; Erkoyuncu, J.A.; Roy, R.; Torabmostaedi, H. A systematic review of augmented reality applications in maintenance. *Robot. Comput. Manuf.* **2018**, *49*, 215–228. [CrossRef]
31. Małachowski, B.; Korytkowski, P. Competence-based performance model of multi-skilled workers. *Comput. Ind. Eng.* **2016**, *91*, 165–177. [CrossRef]
32. Smith, J.; Yazdanpanah, F.; Thistle, R.; Musharraf, M.; Veitch, B. Capturing expert knowledge to inform decision support technology for marine operations. *J. Mar. Sci. Eng.* **2020**, *8*, 689. [CrossRef]
33. Illankoon, P.; Tretten, P. Collaborating AI and human experts in the maintenance domain. *AI Soc.* **2020**, 1–12. [CrossRef]
34. Kambhampati, S. *Explicable Planning and Replanning for Human-in-the-Loop Decision Support*; NASA: Washington, DC, USA, 2017.
35. Crowder, J.A.; Carbone, J.N. *Collaborative Shared Awareness: Human-AI Collaboration*; Springer International Publishing: New York, NY, USA, 2004.
36. Bottani, E.; Vignali, G. Augmented reality technology in the manufacturing industry: A review of the last decade. *IISE Trans.* **2019**, *51*, 284–310. [CrossRef]
37. Erkoyuncu, J.A.; del Amo, I.F.; Mura, M.D.; Roy, R.; Dini, G. Manufacturing technology improving efficiency of industrial maintenance with context aware adaptive authoring in augmented reality. *CIRP Ann. Manuf. Technol.* **2017**, *66*, 465–468. [CrossRef]
38. Huang, T.-L.; Liao, S. A model of acceptance of augmented-reality interactive technology: The moderating role of cognitive innovativeness. *Electron. Commer. Res.* **2015**, *15*, 269–295. [CrossRef]
39. Gattullo, M.; Scurati, G.W.; Fiorentino, M.; Uva, A.E.; Ferrise, F.; Bordegoni, M. Towards augmented reality manuals for industry 4.0: A methodology. *Robot. Comput. Manuf.* **2019**, *56*, 276–286. [CrossRef]
40. Cousin, G. Case study research. *J. Geogr. High. Educ.* **2005**, *29*, 421–427. [CrossRef]
41. Narasimhan, R.; Jayaram, J. Reengineering service operations: A longitudinal case study. *J. Oper. Manag.* **1998**, *17*, 7–22. [CrossRef]
42. Guillemette, M.G.; Mignerat, M.; Paré, G. The role of institutional work in the transformation of the IT function: A longitudinal case study in the healthcare sector. *Inf. Manag.* **2017**, *54*, 349–363. [CrossRef]
43. Demeter, K.; Losonci, D.; Nagy, J. Road to digital manufacturing—A longitudinal case-based analysis. *J. Manuf. Technol. Manag.* **2020**. [CrossRef]
44. International Union of Railways (UIC). *Railway Statistics—Synopsis*. Available online: https://uic.org/support-activities/statistics/ (accessed on 10 February 2021).
45. NS. Expected Future Developments. Available online: https://www.nsjaarverslag.nl/jaarverslag-2020/inleiding/onze-strategie/verwachte-ontwikkelingen-op-lange-termijn (accessed on 26 February 2021).
46. Malterud, K.; Siersma, V.D.; Guassora, A.D. *Sample Size in Qualitative Interview Studies: Guided by Information Power*; Qualitative Health Research: Salt Lake City, UT, USA, 2016.
47. Martinetti, A.; Hart, K.; Damgrave, R.; van Dongen, L.A.M.; Turkenburg, R.; Nouwens, A. There is no spoon: Applying virtual reality for maintenance training of rolling stock technicians. *Int. J. Syst. Syst. Eng.* **2018**, *8*, 398–415. [CrossRef]
48. Stadler. *Electric Low-Floor Multiple-Unit*; Stadler: Pierce Brendan Brosnan, Switzerland, 2016; Volume 41.
49. Mertens, P.; van Hal, B. *Rapport Onderzoek Schuiftrede FLIRT HRN*; Nederlandse Spoorwegen: Utrecht, The Netherlands, 2020.

50. Bode. *Technische Documentatie Schuiftrede met Elektrische Aandrijving*; Bode: Hamburg, Germany, 2016.
51. Funk, M.; Bachler, A.; Bachler, L.; Kosch, T.; Heidenreich, T.; Schmidt, A. Working with augmented reality? A long-term analysis of in-situ instructions at the assembly workplace. *ACM Int. Conf. Proc. Ser.* **2017**, *F1285*, 222–229.
52. Masood, T.; Egger, J. Adopting augmented reality in the age of industrial digitalisation. *Comput. Ind.* **2020**, *115*, 103112. [CrossRef]
53. Romero, D.; Stahre, J.; Taisch, M. The Operator 4.0: Towards socially sustainable factories of the future. *Comput. Ind. Eng.* **2020**, *139*, 106128. [CrossRef]
54. Bortolini, M.; Faccio, M.; Gamberi, M.; Pilati, F. Motion Analysis System (MAS) for production and ergonomics assessment in the manufacturing processes. *Comput. Ind. Eng.* **2020**, *139*, 105485. [CrossRef]

applied sciences

MDPI

Article

WARM: Wearable AR and Tablet-Based Assistant Systems for Bus Maintenance

Diego Borro [1,2,*], Ángel Suescun [1,2], Alfonso Brazález [1,2], José Manuel González [1,2], Eloy Ortega [3] and Eduardo González [4]

1 CEIT-Basque Research and Technology Alliance (BRTA), Manuel Lardizábal 15, 20018 San Sebastián, Spain; asuescun@ceit.es (Á.S.); abrazalez@ceit.es (A.B.); jmgonzalez@ceit.es (J.M.G.)
2 Tecnun, Universidad de Navarra, Manuel Lardizábal 13, 20018 San Sebastián, Spain
3 TCMAN, Passeig de Maragall 120, 08027 Barcelona, Spain; eortega@tcman.com
4 DBUS—Compañía del Tranvía de San Sebastián, Fernando Sasiain 7, 20015 San Sebastián, Spain; egonzalez@dbus.es
* Correspondence: dborro@ceit.es

Featured Application: Comparison two digital solutions (tablet based and Augmented Reality based) for bus maintenance against the traditional solution based on paper.

Abstract: This paper shows two developed digital systems as an example of intelligent garage and maintenance that targets the applicability of augmented reality and wearable devices technologies to the maintenance of bus fleets. Both solutions are designed to improve the maintenance process based on verification of tasks checklist. The main contribution of the paper focuses on the implementation of the prototypes in the company's facilities in an operational environment with real users and address the difficulties inherent in the transfer of a technology to a real work environment, such as a mechanical workshop. The experiments have been conducted in real operation thanks to the involvement of the public transport operator DBUS, which operates public transport buses in the city of Donostia—San Sebastian (Spain). Two solutions have been developed and compared against the traditional process: one based on Tablet and another one based on Microsoft HoloLens. The results show objective metrics (Key Performance Indicators, KPI) as well as subjective metrics based on questionnaires comparing the two technological approaches against the traditional one based on manual work and paper.

Keywords: augmented reality; intelligent garage; bus maintenance; public transportation

Citation: Borro, D.; Suescun, Á.; Brazález, A.; González, J.M.; Ortega, E.; González, E. WARM: Wearable AR and Tablet-Based Assistant Systems for Bus Maintenance. *Appl. Sci.* **2021**, *11*, 1443. https://doi.org/10.3390/app11041443

Academic Editor: Jasiulewicz-Kaczmarek Małgorzata

Received: 22 December 2020
Accepted: 1 February 2021
Published: 5 February 2021

1. Introduction

This work targets the applicability of wearable/mobile and augmented reality (AR) technologies for maintenance of bus fleets. These technologies have already shown its ability to enrich the human perception, increasing operators' performance and reducing their error rate. In addition, the actual emergence of new augmented reality (AR) headset provides a better way of integration of such technologies for non-experts. Despite the advantages of AR, which have been proven in many previous publications, few companies have been quick to adopt this technology for industrial manufacturing applications due to the hardware for AR available in the market [1]. As the review shows, this area has only been sparsely covered by research. This might be due to the fact that devices powerful enough to support high-quality AR are only recently available at reasonable prices, and enabling technologies, such as ARCore and ARKit, have been released as free SDKs to developers only starting in 2017 [2].

Maintenance mechanics spent a valuable time reporting their activities, especially when unexpected reparations are accomplished. These reports have the form of handwritten annotations that have to be manually typed into the maintenance system by another maintenance worker. This manual data processing is error prone and often lacks completeness. In addition, only text information is reported; pictures, videos, and audio cannot be included.

Assistive systems could help workers to maintain or even increase their productivity [3]. These systems could be in the form of tablet-assisted systems or more sophisticated AR-based systems. Augmented reality is capable of projecting assistive instructions in-view of the user or directly in situ at the object of interest. The proposed assistant WARM helps the mechanics to report their activity, including rich multimedia data if so requested. The WARM assistant connects with the back office and with the vehicle through the IT standard EN13149 parts7/8/9 (https://www.en-standard.eu/pd-cen-ts-13149-7-2020-public-transport-road-vehicle-scheduling-and-control-systems-system-and-network-architecture/) for data exchange.

WARM has been demonstrated in DBUS garage, under real operational conditions. To this end, four maintenance actuations such as M1, M2, M3 (standard maintenance actuations) and PRE_ITV (prior to the official vehicle inspection) have been tested.

The main goal of the WARM assistant is, on one side, the reduction of the time spent by the maintenance workers in paperwork activities and, on the other side, the augmentation, enrichment and completion of the information reported by the maintenance operators at the end of the maintenance task. More in details, the system is committed to:

1. Reduce maintenance reporting time. The maintenance activity requires not only for the "reparation task" but also for the "reporting task". This paperwork usually comprises a checklist of 1–2 complete pages that has to be handwritten by the worker at the end of the task. In addition, the worker has to annotate relevant aspects of the task, particularly when an unexpected damage shows up and additional spares are requested.
2. Allow maintenance operators to easily keep track and collect verbal comments, pictures, audio tracks and video clips concerning the particularities of the maintenance tasks accomplished, especially in the case of unexpected repairs. At present, the reporting consists of completing a checklist form.
3. To demonstrate the feasibility of the standard EN13149 to exchange data between the maintenance assistance system and the vehicle.

In this paper, the authors have tested two different technologies for the WARM assistant, one based on tablet and another one based on AR headset, that make feasible the concept of "intelligent garage and maintenance", targeting the applicability of AR technologies to the maintenance of bus fleets. The rest of the paper is structured as follows. The state of the art is presented in Section 2. The proposed solutions and their architectures are described in the Section 3. Sections 4 and 5 show the description of the experiments and the results, respectively. Finally, Section 6 discusses and summarizes the main contributions of this work.

2. State of the Art

The main challenge of an AR system is to obtain perfect alignment between real and virtual objects in order to create the illusion that both worlds coexist. To that end, the position and orientation of the observer (i.e., the localization of the human with respect to the environment) has to be determined in order to configure a virtual camera that displays the virtual objects in their corresponding position. This problem is known as tracking and, although there are many alternatives to address it by using different sensors, tracking based on optical sensors is the most popular solution.

In the last years, markerless monocular solutions have gained in popularity due to their simplicity and low cost (one single camera) and by avoiding having to position markers in the scene. They take advantage of the visual cues that are naturally in the scene. Depending on whether the scene geometry is known or not, the markerless tracking is divided into two groups [4]. In Structure From Motion (SFM) approaches the camera movement is estimated while the 3D reconstruction of the scene is performed [5]. On the other hand, model-based techniques store the knowledge about the scene in a 3D model, which is available before the camera tracking begins. The 3D model could be represented by its simple 3D geometry [6], or by a more detailed description, that includes the geometry and the texture of its surface [7].

However, despite many years of research, optical tracking of industrial objects is not a solved problem (metallic objects with reflections, poor textures, dirty environment, etc.) and many studies have to be conducted yet [8,9] to provide reliable solutions that can be used in a factory. Errors in the estimation of the position of the virtual objects added to a scene greatly reduces the realism and quality of the integration of the information in the real context. This can severely limit the relevance of the information received by the user.

Focusing on industrial environments, AR solutions can also be found but, even markers-based solutions, most of them are in an experimental/lab prototype status [1–3]. Several works [8,10–14] demonstrate the benefits of AR based approaches applied to the assistance in maintenance, assembly and repair operations in terms of operation efficiency. These works argue that AR enriches the way in which users understand the real world, i.e., AR let users understand clearly what to do at any time. Compared to VR, AR offers information that is integrated into the real world, while VR manipulates virtual objects. For this reason, AR is used for both training as VR, but it is mainly used for guidance.

Some works [15] studied the long-term usage of high-quality AR technology in industrial training and determined that it reduces stress for the user compared to traditional training procedures, and therefore, improves worker satisfaction.

Current approaches use AR to present the worker with an "on-line" virtual manual of the task. In many cases, this is simply a list of the different steps that the user has to complete in order to finish a task. More advanced approaches present a virtual model integrated into the real world [16]. These works describe algorithms that recognize the 3D objects and track them meanwhile disassembly instructions are overlapping in the real scene guiding the user along the task. In this case, the disassembly sequence is computed in a pre-process following automatic methods based on the object geometry [17,18].

A very systematic review papers can be seen in [19–21] works. In all of these works, the common points are that AR has a lot of previous works showing tracking and registering applications in most of the market sectors. However, they agree that AR is still not mature for complying with industrial requirements of robustness and reliability. In fact, looking the experiments, we have the same conclusions in our work.

AR technology tries to improve the traditional concept of digital maintenance. Concerning to this, the maintenance management software market is highly fragmented with a poorly differentiated product, low-skilled support service, and very high availability. IBM has a comprehensive business asset management package for asset maintenance and lifecycle management called Maximo Asset Management [22]. The big drawback of this solution is its very high price. PRISMA 3 EAKM (Enterprise Asset Knowledge Management) [23] of Sisteplant is a 100% web application that provides intelligence to maintenance management and visibility aimed at all types of users from the perspective of life cycle management. MMS solutions (Maintenance Management Solutions) of Idasa [24] allow to quickly and efficiently manage equipment and facilities, assets and real estate, the available stock and its supply needs, maintenance services and service requests, own or contracted service personnel, as well as controlling the associated costs in detail and at all times. One of the largest provider of business applications and services in the world, Infor, has an EAM (Entreprise Asset Management Solutions) that improves maintenance schedules, increases manufacturing cell uptime, improves reliability and risk management policies, and provides deep insight within the company for more precise strategic planning [25]. Primavera's solution allows planning, scheduling and managing maintenance according to the availability of technical means and the operational condition of the assets from anywhere with an Internet connection, thanks to its 100% Web platform [26]. As a summary, the following Table 1 shows the characteristics of the main systems and a comparison with WARM proposal.

Table 1. Feature comparison among different maintenance management software.

Name	Asset Management	Mobile Devices	Voice Recognition	Vehicle Identification	Augmented Reality
IBM	Yes	Yes	No	Yes	No
Sisteplant	Yes	Yes	No	No	No
Idasa	Yes	Yes	No	No	No
Infor	Yes	Yes	No	No	No
Primavera	Yes	Yes	No	No	No
WARM	Yes	Yes	Yes	Yes	Yes

3. Solution Approach

In this work, the use of wearable devices (tablet and AR device (e.g., smart headset)) is proposed. The devices do not need to be worn during the whole operation but just when required, to support the operator to complete the checklist and collect relevant multimedia data of the extraordinary facts of the activity.

The maintenance system currently deployed in DBUS is the system GIM (Gestión Integral del Mantenimiento, Integral Maintenance Management in English), which is a Computerized maintenance management system (CMMS) developed and commercialized by the company TCMAN. Recently, TCMAN has introduced the software GIM Android App to ease the collection of maintenance data by nomadic users. This application could benefit from some of the innovations pursued by WARM, as it lacks a friendly and efficient graphic user interface (GUI) well adapted to mechanic's needs.

Figure 1 shows a schematic of the WARM system. Once the maintenance worker receives the work order sheet from the maintenance system (back office) using Web Services, the wearable device of the WARM system connects to the bus on-board computer through the WiFi network on-board. Using this connection, the wearable device identifies the bus and reads the relevant data for the maintenance task. This connection uses a proprietary interface. Additionally, to complete the data read from the bus, a direct connection to the CAN is done. At this point, the wearable device starts guiding the work to be completed, accordingly to the work order sheet and collects all relevant information. This guidance is adapted to the "work style" and preferences of each maintenance worker participating in the study. The worker will use the wearable device for different purposes:

- For dictating "done/undone" at the corresponding entries of the checklist form.
- For dictating annotations concerning the work done.
- For dictating the list of spares needed.
- For taking photographs and recording audio or video clips concerning the particularities of the reparation.

Once the work is finished and the report complete, this document is sent to the back office. All communications are compliant with the standard EN13149 via WiFi.

WARM system has been developed upon the application program interface (API) of the GIM system. Data collected with the WARM system are naturally uploaded and integrated in the DBUS maintenance servers running GIM. Besides, GIM API implements a communication layer to exchange information with GIM servers by means of Web Services. This layer has been adapted to comply with the standard EN13149. Finally, a new development was necessary to connect GIM Android APP with the bus (i.e., the CAN/FMS interface) using again the standard EN13149.

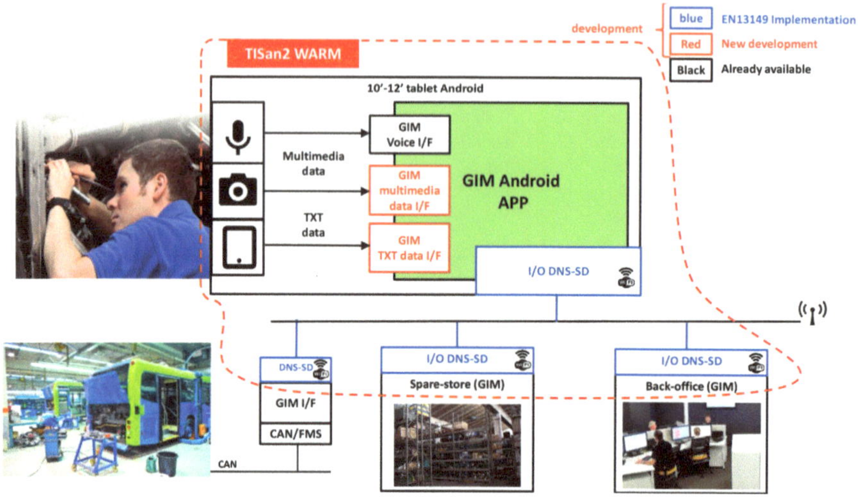

Figure 1. Description of the Wearable Augmented Reality (AR) Maintenance Assistant System (WARM) featuring its integration with the GIM system, already deployed in DBUS company. Red boxes refer to new developments; blue boxes and lines refer to the implementation of the standard EN13149 (new development as well); black boxes refer to previously available features.

3.1. Tablet-Based Solution

A new graphic user interface has been developed to gather all the controls and information required by the WARM system. The participation of the mechanics in the design of the interface showed to be fundamental for the success of the subsequent tests. The whole flowchart associated with the Tablet application is shown in Figure 2.

The navigation through WARM has been done in a simple and intuitive way so that the operators do not spend too much time to fill in the maintenance lists. Once the user has logged into the system (Figure 3a), the next screen (Figure 3b) offers the operator the following options are shown:

- Capture multimedia, to take photographs and record videos to describe new incidences found during reparations or prior to begin with them.
- Select a bus to begin/continue the work orders (WO) associated with the user logged in, which opens the corresponding checklist.

The "Maintenance sheet" screen (Figure 4) will load the list of the checks required for the type of maintenance that will be performed on the vehicle (ITV, M1, M2 . . .). The user can see the fields that have not yet been evaluated with a red background and clicking on them they will be verified and change their background color to white with a green verification symbol. The user must mark the field to record that the revised component works correctly or leave it unchecked to warn otherwise.

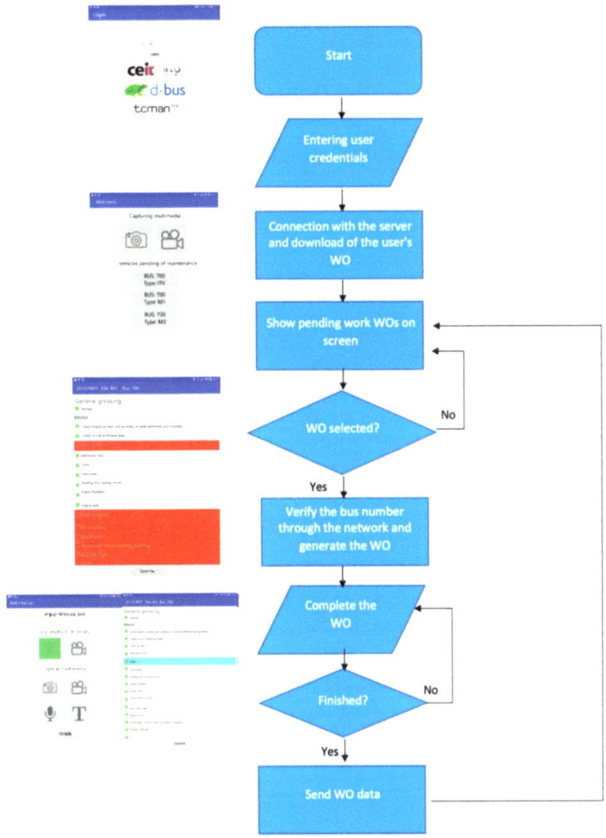

Figure 2. Flowchart for the Tablet application.

(a) (b)

Figure 3. WARM system (Android mobile app). Images corresponding to the login (**a**) and the main menu (**b**).

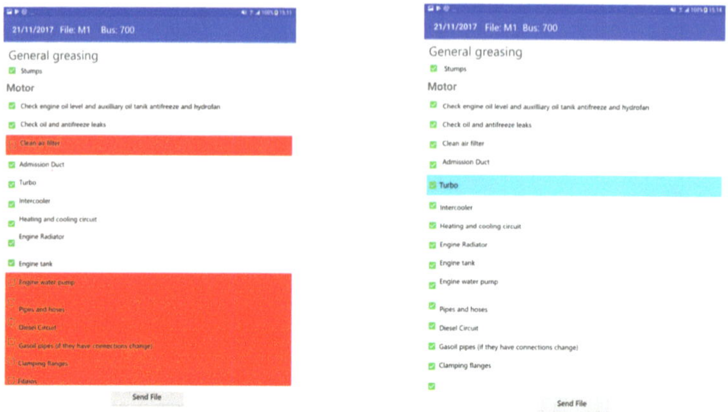

(a) (b)

Figure 4. Maintenance sheet M1 with unedited field (**a**) and maintenance sheet M1 with incident notification (**b**). Red background marks not yet started tasks and blue background marks reported tasks.

In the event that an item is not functional, and additional information needs to be attached, the user has to press and hold the item in the list and the "Notify Incident" screen (Figure 5) will show up. This last screen allows you to upload an image or video previously taken, to capture a new photo, video or audio, or to annotate a comment to enrich the description of the incident. A field that has received an incident notification will appear on the screen with a blue background.

(a) (b)

Figure 5. Notify incident screen (**a**) and notify incident screen with attached image (**b**).

Clicking on the "Send file" button on the "Maintenance sheet" will mean that the maintenance of the vehicle has been completed and it will cause the application to generate a PDF. The complete maintenance information (Figures 6b and 7b) that will be attached to an email, also generated by the application, that will be sent to the server and the person in charge of the vehicle maintenance.

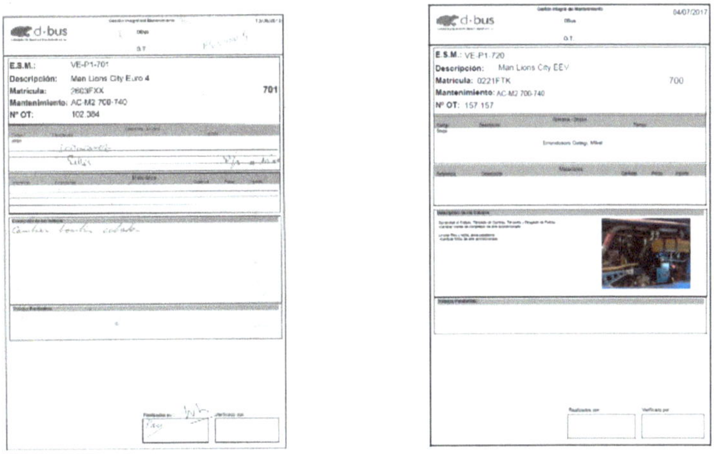

(a) (b)

Figure 6. Maintenance report (1st page). Handwritten report (**a**). Digital report automatically produced and transmitted to the back office by the WARM system (**b**).

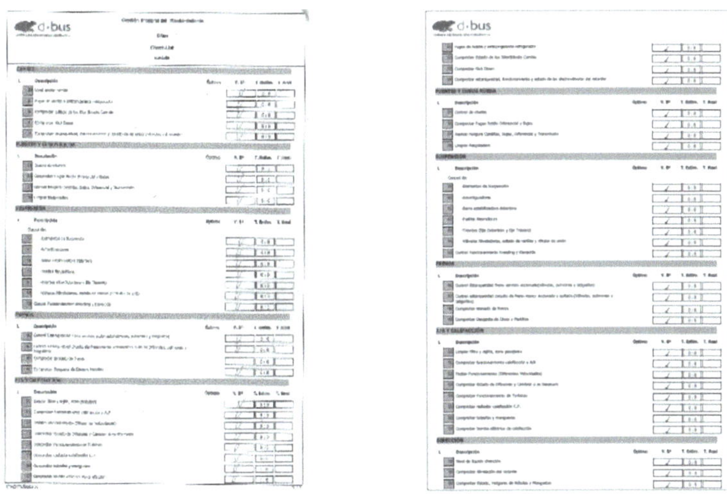

(a) (b)

Figure 7. Maintenance report (2nd page). Handwritten report (**a**). Digital report automatically produced and transmitted to the back office by the WARM system (**b**).

Figures 6 and 7 compare the current reports produced by DBUS mechanics (handwritten document) with the digital reports produced by WARM (shown in pdf format). Notice the picture complementing the annotations of the mechanic.

3.2. AR-Based Solution

To develop the AR solution, an application based on AR with smart headset (such as Microsoft HoloLens, Figure 8) is proposed. These types of AR headset are a new type of hardware that integrates a video camera, a screen and an audio system. The appearance on the market of this type of headset provides a better way of integrating these technologies for non-experts.

Figure 8. Microsoft AR HoloLens.

AR specific algorithms have been implemented whose objective is to solve the problem of detection and tracking of objects through artificial vision. Due to the fact that objects in industrial environments are characterized by the absence of visual characteristics that can be used in the detection (the usual surfaces are metallic, without patterns or other visual characteristics), two different tracking methods were used to detect the bus and to locate the user respected to the work environment. The geometry and related coordinates that describe our problem are illustrated in Figure 9.

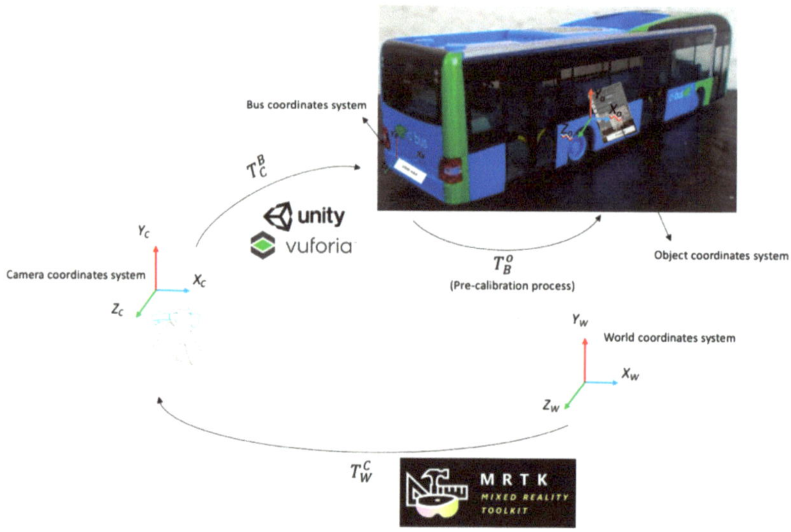

Figure 9. All the proposed system coordinates and transforms.

A brief description of the coordinate systems will be given below:

- W corresponds to the world coordinates system;
- C corresponds to the camera coordinates system
- B corresponds to the bus coordinates system with the origin in the license plate;
- O corresponds to the object coordinates system, i.e., the virtual plane with checklist showed to the user through the headset.

The tracking problem consists in finding a rigid transform function T_A^B which maps points from coordinates system A to coordinates system B. This transform is composed of a rotation and an offset. Rigid transforms are invertible by definition. Therefore, if a transform that goes from system A to B is known, the inverse transform that goes from B to A can be obtained directly by inverting the original matrix. In addition, if three coordinate systems are present (A, B and C) and the transforms from A to B and from B to C are known, it is possible to obtain the transform that goes from A to C by left-multiplying the transform matrices:

$$T_A^C = T_B^C T_A^B \quad (1)$$

The two tracking systems implemented in our approach are the following:

1. Detection of the environment and positioning of the operator (spatial mapping): using the 4 infra-red cameras and inertial sensors of HoloLens device, the detection of the

most relevant geometric characteristics of the objects of the environment is performed. Through the information obtained with this algorithm, the transformation matrix T_W^C of the device's camera is estimated and thus, it is able to track the operator in the environment. The MixedRealityToolkit-Unity toolkit (MRTK) has been used as an AR support for this spatial mapping and user interaction. It is a collection of scripts and components designed to accelerate the development of applications aimed at the HoloLens device and Windows Mixed Reality.

2. Vehicle detection: the detection of objects must be much more accurate and rigorous than the environment since it is the core of the activity and this depends on the exact positioning of all virtual objects for the operator's guidance. Due to the large size and poor homogeneity of bus surfaces, it is difficult to position them accurately through a direct detection of the vehicle or its textures. Therefore, it has been decided to develop an algorithm that analyze in real time the images captured by the HoloLens in search of certain fixed elements in all vehicles (license plates). Once the fixed element has been detected, its position can be computed respected to the camera (T_C^B). For the image processing, Unity 3D native Vuforia libraries have been used. Vuforia is an AR software development kit (SDK) for mobile devices. It recognizes and tracks flat images (image objectives) and simple 3D objects, in real time.

For the license plate recognition, Vuforia library offers a tool (Image Target Behavior) to recognize and position images in the environment. Through this tool an analysis of the images captured by the HoloLens camera is performed. The algorithm implemented transforms the images of the camera to images with high contrast to apply edge and vertex detection techniques, which it is called "feature points" (Figure 10). Through these feature points, the algorithm makes comparisons to detect known shapes defined by the user like the license plate.

Figure 10. Image of a license plate and the image processing performed detecting feature points.

The position of the rest of elements of the vehicle can be estimated knowing its relative position respected to the license plate (T_B^O). This can be done in a pre-process calibration stage. Thanks to all of this, it is possible to compute in every single frame the global position of the virtual objects that the system should project to the user.

$$T_W^O = T_B^O T_C^B T_W^C \tag{2}$$

Once all the data of the work order are obtained, the algorithms for the spatial mapping and vehicle detection are executed in real time. When all the elements to be verified are located, AR objects are placed in points close to them to guide the operator and assist him in maintenance work. The positioned objects are panels that contain the elements to be verified in maintenance. The panels group elements that have characteristics in common (motor, batteries, control panel, wheels, etc.) and are positioned in an area close to the place where the elements it contains must be verified. Finally, once the entire scene is complete, the operator can interact with AR objects to verify each of the elements of the vehicle that are required. The flowchart associated with the HoloLens application is shown in the following Figure 11:

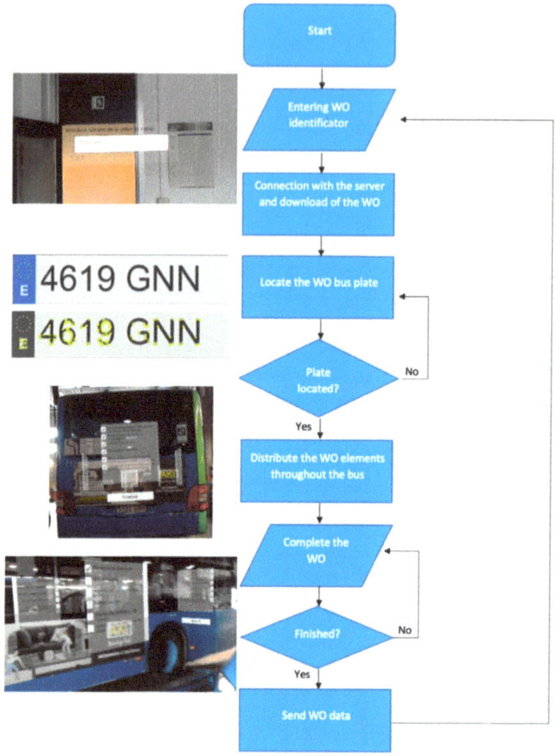

Figure 11. Flowchart for the HoloLens application.

Each of the panel elements contains a checkbox to inform that this maintenance section has been reviewed and verified (Figure 12). To verify an item, the operator must look at it and by a certain gesture it will be verified. During the element verification process, there are a number of tools in order to attach additional information to the work order. These tools are activated by voice commands for notes dictation and image/video capture.

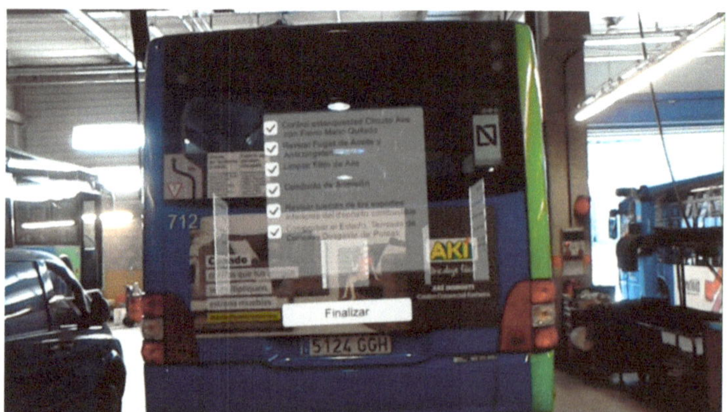

Figure 12. Example of an AR panel showing it through the HoloLens device.

We would like to emphasize that the contribution of the paper focuses on the implementation of the prototypes of the system in the company's facilities in an operational environment with real users (TRL7 target) and address the difficulties inherent in the transfer of a technology developed in laboratory (TRL 4) to a real work environment, such as a mechanical workshop. In this scenario, we have tested that workers do not welcome an application when it is too complex. For this reason, we have developed an easier AR application with simple AR objects projections (planes with checklists). As in the tablet-based solution, the designed GUI has been developed to gather all the controls and information required by the WARM system taking into account all suggestions provided by the mechanics.

4. Description of the Experiments

The evaluation of the WARM system has followed the EBSF2 EU project methodology [27], which assesses the performance variation in situations of "do nothing" and "do something" by measuring a sufficient set of Key Performance Indicators (KPIs) [28]. This evaluation framework has been applied to all the case studies of the EBSF2 project and other projects involving multisite demonstrators [29,30]. The methodology defines Validation Objectives (VO) that state general aspects in which the innovation tested is expected to be beneficial. The expected benefits of each VO are quantified through one or more performance targets (PT) that are actually measured through some KPIs. The evaluation method consists of analyzing whether the variation of the KPI values from "do nothing" to "do something" situations meet the expectations for each performance target. To assess whether a technological innovation is worth to retain or not, quantitative target values for each KPI have been set to serve as reference during the evaluation of test results. These improvement thresholds have been decided by the authors to assess the maturity of the innovations, i.e., to measure if the actual performance improvement is sufficient to continue the innovation development from current TRL 6/7 to TRL 9.

For the evaluation of the WARM system, the "do nothing" situation corresponds to the current procedures, which is essentially paper-based handwritten data collection, and it is explained in Section 4.1. For the "do something" situation, two scenarios have been developed: one for the tablet-based solution, explained in Section 4.2, and another for the AR headset-based solution, detailed in Section 4.3.

DBUS operations require that every morning at 6:00 am, a minimum of 103 vehicles of which 2218 m buses and 7312 m buses must be ready to run. The total fleet is composed of 130 buses. Maintenance tasks are organized in three periodical levels (M1 each 10,000 km, M2 each 30,000 km and M3 each 60,000 km), each level contains the previous one. In addition to these three actuations, the PRE_ITV assessment prepares the bus to succeed in the official technical inspection of the vehicle (known as ITV in Spain). In order to reduce the number of days that buses are out of service, the PRE_ITV assessment is made to coincide with one of the three periodical actuations. The garage (Figure 13) is able to service 10 buses simultaneously.

Predictive maintenance work is planned with 7 to 10 days in advance. In addition to maintenance tasks, the daily work at the garage comprises eventually other reparations and actuations due to accidents, breakdowns and incidents occurred during daylight and nightly operations. To cope with this unpredictable workload, maintenance planning is revised every afternoon to guarantee that the needed vehicles to start operations the day after will be available.

Figure 13. DBUS garage: two of the ten maintenance posts available.

These four preventive maintenance tasks (M1, M2, M3 and PRE_ITV) were selected for testing the WARM system. The main purpose of the test was to compare the traditional reporting activity, i.e., handwritten reports, with the WARM-assisted reporting process using the tablet-based solution and the Holoens-based solution. In order to have a homogeneous data volume, the test plan has been oriented to the maintenance of MAN Lion's City bus models with numbers between 700 and 780. This bus model has the largest number of units in the fleet of DBUS. To compare the WARM system with the system currently used, a series of quantitative and qualitative Key Performance Indicators (KPIs) that represent the most relevant aspects of both maintenance management systems have been defined:

- Dedicated effort in data management: time spent in planning and organizing work orders.
- Dedicated effort in data processing: time spent completing the documentation of work orders.
- Operator perception of workload: subjective perception of different methodologies by maintenance technicians.
- Maintenance costs per vehicle.

A test plan has been developed for each of the defined cases capable of collecting the necessary data for the calculation of the indicators for the evaluation.

4.1. Paper-Based Handwritten Data Collection Test Plan (the Current Procedure)

The collection of quantitative data of the system currently used in the maintenance workshop, has been done manually accompanying the maintenance operator in his working day. For each test, the following data have been taken to calculate the indicators:

- Maintenance date.
- Type of maintenance (M1, M2, M3 or PRE_ITV).
- Bus number.
- Operator in charge of the work order.
- Time to complete the work order.
- Verified items of the work order.
- Total elements to verify.
- Times the checklist has been used.
- Maintenance start and end maintenance time.
- Number of comments added to the work order.
- If maintenance starts on the day or is the continuation of a previous one.

For qualitative data, after performing a maintenance the operators complete an assessment sheet of aspects of their work. Through a series of questions, we can assess the following aspects:

- Workload acceptance.
- Valuation of work orders.
- Valuation notes and attachments.
- General opinion.

4.2. Tablet-Based Solution Test Plan

The data to be collected will be similar to the previous case with the difference that in this case, all will be collected automatically by the device and stored in an internal file. In order to perform the tests with this new device, it is necessary to instruct the operator before starting the tests, for this it is necessary to train worker in the following points:

- Navigation through the application interface.
- Options to activate the different editing functions of the work order (verify item, add multimedia, change report types, quick verification by sections, etc.).
- Attach multimedia files to a work order (audio, video and photos).
- Add comments to document errors or repairs.
- Sending of the finished work order.

Like the previous case, at the end of a work order the operator will be required to complete a similar survey about the application and the device.

4.3. AR Headset-Based Solution Test Plan

Like the previous case, the data to be collected will be similar and will be collected by the HoloLens device and stored in an internal file. Similarly, when introducing a new device in maintenance management, it will be necessary to instruct operators in advance. Being a completely unknown device by the workshop technicians, the instruction should be more accurate. Therefore, it is necessary to explain the following points:

- General description of the test to be performed.
- Comment on what the device and AR is.
- How it fits on the head.
- A reduced simulation of the work to be done is done to explain and practice the basic gestures for handling the device.
- Once the basics have been learned, the maintenance application is executed and all navigation and use through it is explained step by step.
- Finally, the worker is asked if he needs additional information or any extra explanation.

During the whole process, the maintenance technician is accompanied, trying to interfere as little as possible, in order to be able to resolve any questions or problems that arise during the completion of the work order with the new device.

The way to qualitatively evaluate the AR headset system will be similar to the previous cases but adapting some of the questions to the new device.

4.4. Tests Execution and Data Collection

WARM have been demonstrated in DBUS garage, under real operational conditions. Four preventive maintenance tasks (M1, M2, M3 and PRE_ITV) have been the test cases. All systems have been tested in the DBUS maintenance garage. As defined above, the data has been collected manually for the case of paperwork and automatically for the cases of applications for Android tablet and HoloLens AR headset solutions. Informed consents were performed before conducting the study and all data was treated anonymously. Data collection has involved:

- 25 buses of the fleet;
- 23 maintenance operators;
- 3 maintenance managers.

where the evaluation was within-subject and with randomized order.

The data corresponding to the subjective part have been made manually through surveys for each of the cases raised. The surveys are similar for all cases (Figure 14). Two experimental work scenarios are proposed in which tests will be carried out:

- WITHOUT_WARM: in this scenario, the maintenance operator will perform the work as before, this is without the assistance of the WARM system. This will be the control group.
- WITH_WARM: in this scenario, the operator will use the WARM assistance system with Tablet and HoloLens. This will be the test group.

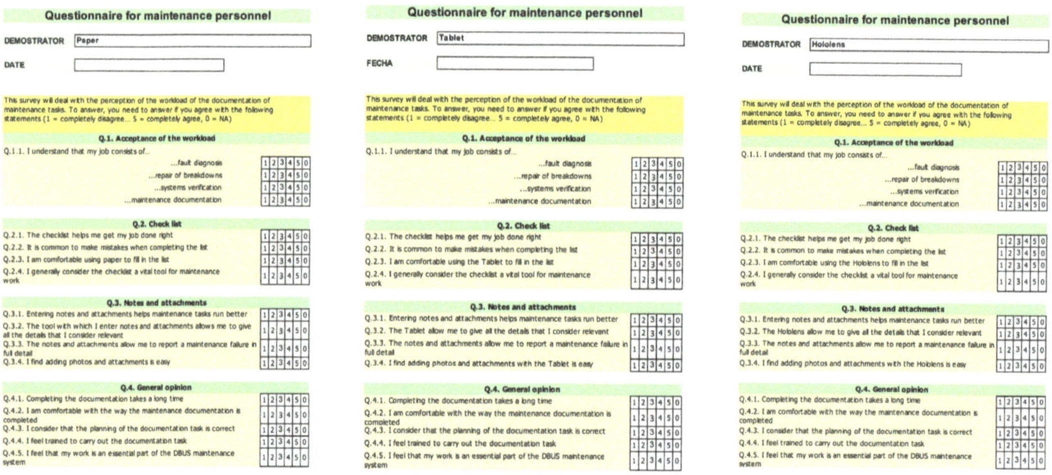

(a) (b) (c)

Figure 14. Survey on usability: questionnaires for the mechanics. Current system scenario (**a**), tablet-based scenario (**b**), and AR-based scenario (**c**).

In both scenarios, the same KPIs will be collected during the execution of maintenance work of the same type. The following table shows the structure of the experimental work to be carried out, specifying the validation objectives and the KPIs finally selected for evaluation. Besides, once the bulk of the test data are obtained, it is necessary to debug them in order to obtain the defined indicators and to be able to make a reliable comparison between the proposed steps. Table 2 summarizes the KPIs to be measured, including the formulae used to calculate them through the data collected and their corresponding improvement objectives.

Table 2. Selected KPI in the test and control scenarios to evaluate and validate the system.

# KPI	Validation Objective VO	Performance Target PT	Improvement Target (Threshold)	KPI Description	Units	Formula	Inputs
1.a	Speed up data management	Time improving of data management	10%	Effort dedicated to data management	% workers full time	$\frac{1}{N_d}\sum_{i=1}^{N_d}\left(W_i\frac{T_i}{t}\right)$	W_i: Workers dedicated to data management on the day *i* T_i: Total time dedicated to data management on the day *i* N_d: Number of days *t*: Duration of the workday
2.a	Reduction of workload	Time reduction of data processing	25%	Effort dedicated to data processing	% workers full time	$\frac{1}{N_d}\sum_{i=1}^{N_d}\left(W_i\frac{T_i}{t}\right)$	W_i: Workers dedicated to data processing on the day *i* T_i: Total time dedicated to data processing on the day *i* N_d: Number of days *t*: Duration of the workday
2.b			10%	Workload perception from questionnaires	[1–10] being 1 high workload perception and 10 low workload perception	$\frac{1}{N_W}\sum_{i=1}^{N_Q}\frac{1}{N_A}\sum_{j=1}^{N_A}R_{ij}$	R_{ij}: Answer to question *j* of worker *i* N_W: Number of workers N_A: Number of questions
3.a	Minimizing operating and maintenance cost	Reduce maintenance costs per vehicle	10%	Cost of maintenance employees per vehicle every 10,000 km	€/(vehicle × 10,000 km)	$\frac{1}{N_b}\sum_{i=1}^{N_b}C_i\,T_i$	N_b: Number of buses C_i: Cost of maintenance employees for bus *i* T_i: Maintenance time spent for bus *i*

Some comments about the selected KPIs:

- Effort for data management: for this KPI, the data management is the assignment of work orders to employees. This task is done by managers.
- Effort for data processing: for this KPI, the data processing is the filling of the work orders. This task is done by technicians.
- Staff's perception of workload: The questionnaire evaluates the comfort to fill the maintenance sheets and report problems. A high value means that the technician finds it easier to work and therefore less workload. All the questions are similar for all scenarios, for this reason we can compare it directly. This task is done by technicians.
- Cost of maintenance staff per vehicle: The data for the cost staff has been provided by DBUS.
- Improvement targets are the threshold values that test results must pass to assess whether a technological innovation is worth retaining.

To evaluate the traditional handwritten reporting method (i.e., without using the WARM system) maintenance mechanics have been accompanied by personnel of CEIT during their activity, who have collected all the necessary data (e.g., times) to produce the requested KPI. Conversely, to evaluate the WARM-assisted reporting process, the WARM device was in charge of collecting data concerning times and other aspects. In this scenario, maintenance workers were also accompanied by personnel of CEIT during their activity, to support them in the use of the tool. After every maintenance task (traditional and WARM-assisted) the survey was conducted to measure subjective perceptions of the workers. At the end of each maintenance work, the maintenance mechanic completed a questionnaire.

5. Results of Experimental Case-Studies

The following Table 3 shows the summary of the experiments.

Table 3. Summary of results.

KPI	Units	Test Cases			
		Paper	Tablet	AR Headset	
1.a—Effort for data management	% workers full time	0.77	0.35	1.67	Less is better
2.a—Efforts for data processing	% workers full time	1.03	0.73	0.73	Less is better
2.b—Staff's perception of workload	Scale 1–10	7.76	8.44	7.22	More is better
3.a—Maintenance staff costs per vehicle	EUR/(vehicle × 10,000 km)	24.78	24.7	25.09	Less is better

As it can be deduced from the KPIs, clearly the case of a mobile device with Android (Tablet) system is superior to the rest in all aspects. The tablet solution scores the best result in KPI 1.a (effort for data management). This is a consequence of the wide introduction of the tablet technology nowadays. The assignment of a work order to the concerned mechanic is done directly thorough the GIM system, which transmits it to the tablet app seamless. This is much more efficient than the current paper-based procedure that requires to print out a paper copy of every work order and to hand them to the corresponding mechanics. With the tablet solution, all the mechanics get the work orders assigned to them immediately. The mechanics only have to pick-up the tablet from the maintenance office. Once the work is done, the tablet transmits the information to the back office without the need of further participation of maintenance managers. Although similar arguments would also play in favor of the AR headset solution, its result shows poorer performance even compared to the current paper-based procedure. This discrepancy is due to the fact that the AR headset take longer to turn on and the maintenance officer needs more time to verify that the AR headset are ready for the mechanic to use. This fact also penalizes the KPI 3.a, since maintenance workers finally need to spend more time with the AR Headset solution than with the other two.

Both AR headset and tablet solutions perform better on KPI 2.a (data processing efforts), demonstrating the effectiveness and speed of the checklist completion methods implemented by the two technologies.

The tablet solution also scores the best result in KP 2.b, which measures the subjective perception of the workload of the mechanics. AR headset solution scores the lower value, slightly smaller than the current paper-based procedure. This is due to the fact that, today, all the workers in the maintenance shop are familiar with the use of this type of device and do not require any training to use the application. An important issue, apart from the improvement of time and costs, is that the operator perceives a lower workload because the process of documenting work orders is automated and can be detailed with multimedia files unlike the current paper-based system.

On the contrary, in the case of the AR headset solution with AR, a good acceptance by the operators has not been achieved. The device is quite heavy and not very ergonomic for the work environment in which they work. In addition, the process of documenting work orders has been slowed down because this device does not allow to complete them as quickly as the others.

The use of an Android tablet has been beneficial for the good acceptance of this solution, since many mechanics are familiar with Android devices, since they own Android smartphones. Furthermore, mechanics already use their smartphones in their workplace, so they are used to using them safely and cleanly in a very dirty environment. The tablet's large screen has contributed to its good readability and interaction performance.

Table 4 presents the relative improvement or worsening in the KPIs of the tablet-based and AR headset solutions with respect to the current paper-based procedure. The table also includes the expected target for each KPI. It shows that neither the tablet solution nor the AR Headset solution exceeds the target thresholds of all KPIs. However, the tablet-based solution produces much higher improvements in the effort for data management and processing and gets close to the threshold in the staff's perception of workload. The maintenance cost per vehicle (KPI 3.a) seems to be unaffected by the tested innovations, probably because the worktime reduction (KPI 1.a and 2.a) is very small when compared with worktime required by the reparation tasks.

Table 4. Final assessment: relative improvements (in %). Negative values correspond to worse values than those obtained with the current procedure (Paper).

KPI	Target	Tablet vs. Paper	AR H. vs. Paper
1.a—Effort for data management	10%	54.5%	−116.9%
2.a—Efforts for data processing	25%	29.1%	29.1%
2.b—Staff's perception of workload	10%	8.8%	−7.0%
3.a—Maintenance staff cost per vehicle	10%	0.3%	−1.3%

6. Conclusions

The result of this work has been the development of two solutions for improving the maintenance of bus fleets: one based on Android-based tablet and another one based on Microsoft HoloLens AR headset. The developed solutions are designed as a front-end for interaction with the GIM system. This development is a simplified application of Android GIM that helps fill in the maintenance sheets. This application connects directly through Web Services to the database server.

As a summary of the AR task, different tracking methods (spatial mapping and object detection) have been used to analyze the environment and locate the user respected to the environment itself and the bus. Once the bus is detected and located, it has been possible to include virtual objects in it.

The two solutions have been tested in real operation condition against the current paper-based procedure. Tests have taken place in the garage at the premises of DBUS, which operates the public transport buses in the city of San Sebastian (Spain). Experimental results prove the superior performance of the tablet-based solution in terms of reduction of effort for data management and processing and staff's perception of workload. Other conclusions are that workers are ready to adopt these tools (tablet-based solutions) that make their work easier. Regarding the additional information collected (videos, photos, etc.), it has been well received by the workshop manager verifying that the multimedia data enriches the information system.

Finally, the analysis of the relative improvement of the proven innovations, presented in Table 4, suggests that the tablet-based maintenance assistance system is worth retaining and therefore seems suitable for further developments to achieve TRL 9.

Taking into account the results and the company feedback, as for a quick and comfortable implementation for operators, future steps will be made towards the implementation of kiosks, instead of tablets. Workers have shown good acceptance of this type of device because it leaves them hands free and they do not need much more to carry out their daily work.

As for further research in AR, future lines should go towards more comfortable devices (it depends on the market) and GUIs that really give real benefits to the user in the form of more functionalities and/or user experience than tablet-based solutions.

Author Contributions: Conceptualization, D.B. and Á.S.; methodology, D.B. and Á.S.; software, J.M.G. and E.O.; validation, J.M.G. and E.G.; investigation, D.B., A.B., Á.S., J.M.G., and E.O.; resources, D.B., A.B. and Á.S.; writing—original draft preparation, D.B.; writing—review and editing, D.B., A.B. and Á.S.; supervision, D.B., A.B. and Á.S.; project administration, D.B., A.B. and Á.S.; funding acquisition, D.B. and Á.S. All authors have read and agreed to the published version of the manuscript.

Funding: This research was funded by the European Commission in the frame of EBSF 2 project ("European Bus System for the Future 2", grant number 636300) and by Economy and Competitiveness Ministry, RETOS-COLABORACIÓN program, in the frame of MANBUS project ("System for assistance Bus Maintenance", grant number RTC-2015-3419-4).

Institutional Review Board Statement: Ethical review and approval were waived for this study, due to the following reasons:

- Volunteers received complete information about the test procedure
- No personal data was collected during the tests

Furthermore, the study did not involve:

- The usage of any non-CE marked device
- The collection or analysis of data that could be used to identify participants (including email addresses or other contact details)
- Any physical contact with participants
- Any risk of discomfort or inconvenience to participants
- Any risk of psychological distress to participants or their families
- Any participant recruited from vulnerable groups

Informed Consent Statement: Informed consent was obtained from all subjects involved in the study

Conflicts of Interest: The authors declare no conflict of interest.

References

1. Blattgerste, J.; Renner, P.; Pfeiffer, T. Augmented reality action assistance and learning for cognitively im-paired people: A systematic literature review. In Proceedings of the 12th ACM International Conference on PErvasive Technologies Related to Assistive Environments, Rhodes, Greece, 5–7 June 2019; ACM: New York, NY, USA, 2019; pp. 270–279.
2. Blattgerste, J.; Renner, P.; Strenge, B.; Pfeiffer, T. In-situ instructions exceed side-by-side instructions in augmented reality assisted assembly. In Proceedings of the 11th Pervasive Technologies Related to Assistive Environments Conference, Corfu, Greece, 26–29 June 2018; ACM: New York, NY, USA, 2018; pp. 133–140.
3. Hoover, M. *An Evaluation of the Microsoft HoloLens for a Manufacturing-Guided Assembly Task*; Iowa State University Digital Press: Ames, IA, USA, 2018.
4. Teichrieb, V.; Lima, M.; Lourenc, E.; Bueno, S.; Kelner, J.; Santos, I.H.F. A survey of online monocular marker-less augmented reality. *Int. J. Modeling Simul. Pet. Ind.* **2007**, *1*, 1–7.
5. Longuet-Higgins, H.C. A computer algorithm for reconstructing a scene from two projections. *Nat. Cell Biol.* **1981**, *293*, 133–135. [CrossRef]
6. Drummond, T.; Cipolla, R. Real-time visual tracking of complex structures. *IEEE Trans. Pattern Anal. Mach. Intell.* **2002**, *24*, 932–946. [CrossRef]
7. Álvarez, H.; Leizea, I.; Borro, D. A new marker design for a robust marker tracking system against occlusions. *Comput. Animat. Virtual Worlds* **2012**, *23*, 503–518. [CrossRef]
8. Serván, J.; Mas, F.; Menendez, J.L.; Rios, J. Using augmented reality in airbus A400M shopfloor assembly work instructions. *AIP Conf. Proc.* **2012**, *1431*, 633. [CrossRef]
9. Álvarez, H.; Borro, D. Junction assisted 3D pose retrieval of untextured 3D models in monocular images. *Comput. Vis. Image Underst.* **2013**, *117*, 1204–1214. [CrossRef]
10. Matsas, E.; Vosniakos, G.-C. Design of a virtual reality training system for human–robot collaboration in manu-facturing tasks. *Int. J. Interact. Des. Manuf.* **2017**, *11*, 139–153. [CrossRef]
11. Canetta, L.; Redaelli, C.; Flores, M. *Digital Factory for Human-Oriented Production Systems*; Springer: Berlin/Heidelberg, Germany, 2011.
12. Webel, S.; Bockholt, U.; Engelke, T.; Gavish, N.; Olbrich, M.; Preusche, C. An augmented reality training plat-form for assembly and maintenance skills. *Rob. Auton. Syst.* **2013**, *61*, 398–403. [CrossRef]
13. Raghavan, V.; Molineros, J.; Sharma, R. Interactive evaluation of assembly sequences using augmented reality. *IEEE Trans. Robot. Autom.* **1999**, *15*, 435–449. [CrossRef]
14. Serván, J.; Rubio, L.M.; Mas, F.; Gómez, A.; Rios, J. Augmented Reality using Laser Projection for the Airbus A400M Wing Assembly. In Proceedings of the 29 International Manufacturing Congress, Belfast, UK, 29–30 August 2012.
15. Henderson, S.; Feiner, S. Exploring the benefits of augmented reality documentation for maintenance and re-pair. *IEEE Trans. Vis. Comput. Graph.* **2011**, *17*, 1355–1368. [CrossRef] [PubMed]
16. Álvarez, H.; Aguinaga, I.; Borro, D. Providing Guidance for Maintenance Operations Using Automatic Marker-less Augmented Reality System. In Proceedings of the 2011 10th IEEE International Symposium on Mixed and Augmented Reality, ISMAR 2011, Basel, Switzerland, 26–29 October 2011; pp. 181–190.
17. Aguinaga, I.; Borro, D.; Matey, L. Path-planning techniques for the simulation of disassembly tasks. *Assem. Autom.* **2007**, *27*, 207–214. [CrossRef]
18. Aguinaga, I.; Borro, D.; Matey, L. Parallel RRT-based path planning for selective disassembly planning. *Int. J. Adv. Manuf. Technol.* **2007**, *36*, 1221–1233. [CrossRef]
19. Palmarini, R.; Erkoyuncu, J.A.; Roy, R.; Torabmostaedi, H. A systematic review of augmented reality applications in maintenance. *Robot. Comput. Manuf.* **2018**, *49*, 215–228. [CrossRef]

20. Fernández del Amo, I.; Erkoyuncu, J.A.; Roy, R.; Palmarini, R.; Onoufriou, D. A systematic review of Augmented Reality content-related techniques for knowledge transfer in maintenance applications. *Comput. Ind.* **2018**, *103*, 47–71. [CrossRef]
21. Boboc, R.; Gîrbacia, F.; Butilă, E.V. The application of augmented reality in the automotive industry: A systematic literature review. *Appl. Sci.* **2020**, *10*, 4259. [CrossRef]
22. Available online: http://www-03.ibm.com/software/products/es/maximoassetmanagement (accessed on 1 September 2020).
23. Available online: http://www.sisteplant.com/es/soluciones/innovative-technologies/smart-maintenance-prisma-cmms/ (accessed on 1 September 2020).
24. Available online: http://idasa.com/rosmiman/ (accessed on 1 September 2020).
25. Available online: http://es.infor.com/solutions/eam/ (accessed on 1 September 2020).
26. Available online: https://es.primaverabss.com/es/software/soluciones-especializadas/mantenimiento/eam/eam/ (accessed on 1 September 2020).
27. Cascajo, R.; Monzon, A.; Román, A.; López, J.M. EBSF_2 Deliverable 2.3 Report on Evaluation Framework, Restricted Document. 2016.
28. Tozzi, M.; Corazza, M.; Guida, U.; Musso, A. Testing innovations for increased energy efficiency of electric buses: Evidence from the EBSF_2 project. *Transp. Res. Procedia* **2020**, *48*, 2166–2175. [CrossRef]
29. Corazza, M.V.; Guida, U.; Musso, A.; Tozzi, M. A new generation of buses to support more sustainable urban transport policies: A path towards "greener" awareness among bus stakeholders in Europe. *Res. Transp. Econ.* **2016**, *55*, 20–29. [CrossRef]
30. Corazza, M.V.; Guida, U.; Musso, A.; Tozzi, M. A European vision for more environmentally friendly buses. *Transp. Res. Part D Transp. Environ.* **2016**, *45*, 48–63. [CrossRef]

applied sciences

MDPI

Article

A Combined Anomaly and Trend Detection System for Industrial Robot Gear Condition Monitoring

Corbinian Nentwich * and Gunther Reinhart

Institute for Machine Tools and Industrial Management, Technical University Munich, 85747 Garching, Germany; emeritus.reinhart@tum.de
* Correspondence: corbinian.nentwich@iwb.tum.de

Abstract: Conditions monitoring of industrial robot gears has the potential to increase the productivity of highly automated production systems. The huge amount of health indicators needed to monitor multiple gears of multiple robots requires an automated system for anomaly and trend detection. In this publication, such a system is presented and suitable anomaly detection and trend detection methods for the system are selected based on synthetic and real world industrial application data. A statistical test, namely the Cox-Stuart test, appears to be the most suitable approach for trend detection and the local outlier factor algorithm or the long short-term neural network performs best for anomaly detection in the application of industrial robot gear condition monitoring in the presented experiments.

Keywords: condition monitoring; industrial robots; anomaly detection; trend detection

Citation: Nentwich, C.; Reinhart, G. A Combined Anomaly and Trend Detection System for Industrial Robot Gear Condition Monitoring. *Appl. Sci.* **2021**, *11*, 10403. https://doi.org/10.3390/app112110403

Academic Editor: Jasiulewicz-Kaczmarek Małgorzata

Received: 27 September 2021
Accepted: 3 November 2021
Published: 5 November 2021

1. Introduction

Currently, industrial robots are the workhorses of highly automated production systems [1]. A challenge to the productivity of such systems remain faults of industrial robot gears as they can cause extended downtimes. Condition monitoring (CM) of the gears can be a measure for countering this issue. CM describes a maintenance strategy in which sensor data is used to determine the health state of a robot gear. For this, sensor data is transformed into health indicators that correlate with the gear's health state. Critical monitored values within the time series of the health indicators form the decision criterion for a maintenance action [2]. Usually, there are many industrial robots operating in a production system and the health state of each of the axes must be monitored. Hence, manual monitoring is not feasible and an automated system is required. Such a system must be able to detect anomalies and trends in the health indicator data reliably. Anomalies in the data can be related to faults that occur abruptly (e.g., breaking of a gear tooth) and trends can be an indicator for increasing wear [3]. The occurrence of such events should be presented to the maintenance crew while showing only few false alarms. To the best of our knowledge, such a combined system does not yet exist for industrial robot gear condition monitoring. Hence, the contribution of our publication is threefold. Firstly, a combined anomaly and trend detection system (CATS) for industrial robot gear CM and secondly a method for selecting suitable anomaly detection (AD) and trend detection (TD) models for this defined application are presented. Thirdly, the suitability of different AD and TD models for the defined use case is evaluated by applying the method. Thus, the remainder of this publication is structured as follows: in Sections 1.1 and 1.2 an overview of industrial robot CM systems, AD and TD models is given and the addressed research gap is refined. In Section 2, CATS and the AD and TD model evaluation method is described. In Section 3, the method is applied to state-of-the-art AD and TD models and suitable models for CATS are selected. In Section 4, the limitations of the presented approach are discussed. In doing so, the outlook discussed in Section 5 is derived, which also includes a summary of our contribution. Through the remainder of this publication the term application refers to the condition monitoring of industrial robot gears.

1.1. State of the Art

In this section, first supervised and unsupervised approaches for robot condition monitoring are presented. As this research area does not present the fields of anomaly detection and trend detection models completely, a broader overview of these research fields is given subsequently. Finally, the state of the art is summarised and the research gap is presented that we are addressing.

1.1.1. Industrial Robot Condition Monitoring

Different approaches for the CM of industrial robots exist in the literature. These can be classified by the type of model used, i.e., supervised or unsupervised machine learning models or the raw data used, which are mainly acceleration sensor data or robot controller data.

In the field of supervised models and robot controller data, several models such as XGBoost and different neural networks based on both joint specific data such as speed and torque and operational specific data (e.g., number of emergency stops) were compared from a fleet of 6000 robots. A maximum AUC value (area under the curve) of 0.87 could be achieved for a neural network model for fault detection in axis 2 [4]. A similar model comparison for logistic regression, support vector machines, random forests and ensemble stacking was performed in [5]. Here, angle, angle speed, acceleration and torque data were used from 26 robots to classify gear faults. The best AUC value of 0.77 was reached by the random forest classifier. Fault detection for loose gear belts was performed with a decision tree, a gradient booster and a random forest and statistical features derived from current data. Here, the random forest performed best with F1-scores around 0.9 [6].

In the section of unsupervised models and robot controller data, a kernel density estimator was used to detect faults based on motor angle, angle velocity and torque in combination with the Kullbach-Leibler divergence. Data from accelerated wear tests show a clear increase in the health indicator [7]. In another publication, the transferability of models was investigated for a combination of principle component analysis and Q-residuals. Anomalies were assumed if the distance measure was above a set threshold. The study shows that the use of the differences between measured and set quantities such as torques as raw data perform best in terms of transferability. In this context, transferability describes the training of the model based on the data of only one robot and then also using this model for other robots [8]. A model based on the deviations of a dynamic equation of a robot relative to actual measurements of the robot is combined with Hotelling's T^2 test statistic to determine robot faults [9]. A sliding-window convolutional variational autoencoder was used to detect anomalies in pick-and-place operations of a robot simulated by little strikes on the robot. The method outperforms benchmark models with an F1-score of 0.89 [10]. A long short-term memory neural network was successfully used to detect anomalies within the grinding process of an industrial robot based on speed, position and torque data. Anomalies were generated by applying a force to the robot hand during the process [11].

Turning to supervised learning approaches based on acceleration sensor data, multiple methods are worthy of note. A sparse autoencoder was trained with data from an attitude sensor (collecting acceleration and velocity signals at 100 Hz) attached to the tool centre point of the robot. The sensor collected data from normal behaviour and different fault conditions such as pitting and broken teeth of a gear. The classification results showed accuracy values of 90 percent [12]. Wavelet-based features in combination with a neural network were used to classify backlash faults for a six axis industrial robot [13]. Multiple supervised models such as a support vector machine, neural networks, gaussian processes and random forests were combined with different dimensionality reduction methods based on data from acceleration sensors attached to the gear caps for gear fault classification. The SVM and GP showed the best performance with accuracy values over 91 percent [14].

In the area of unsupervised models and acceleration sensor data, a gaussian mixture model was used based on health indicators derived from time and the time-frequency domain to differentiate measurements from a degreased robot from normal measurements of the robot. Classification performances over 94 percent for recall and precision values were achieved [15]. Time domain and frequency domain features derived from a residual signal were used in combination with thresholding for gear fault detection for different test trajectories [16]. A one-class generative adversarial autoencoder was used for the detection of artificially introduced faults in a robot gear in [17]. Classification accuracies of 97 percent were achieved for the identification of different faults.

1.1.2. Anomaly Detection Models

The state of the art provides various anomaly detection models for point, collective and contextual anomalies of uni- and multivariate time series and spatial data. One possibility for clustering such models is presented in [18]. Here, anomaly or novelty detection methods are structured in probabilistic, distance-based, reconstruction-based, domain-based and information theoretic approaches. For a detailed review of anomaly detection methods, refer to [18] or more recently to [19]. Below, only those approaches that are considered in the method evaluation of our publication are presented. Different approaches from the above mentioned classification scheme are compared. From the field of probabilistic models, a kernel density estimator (KDE) based on the values of the time series [20] is used. This model fits a non-parametric probability density function on the data. By calculating the probability that a sample (one step of a time series) belongs to this density and by comparing this value with a threshold, anomalies can be determined. Furthermore, a gaussian process (GP) for one-class classification is used, which works based on a similar principle [21]. From the field of distance based approaches, the local outlier factor (LOF) [22], the isolation forest (IF) [23] and the DBSCAN algorithm [24] are used. LOF is based on determining the density of data points and detects anomalies as data points with few close neighbors. IF is based on multiple tree classifiers for one-class classifcation. DBSCAN is a clustering algorithm that determines anomalies based on their distance to reachable points from cluster core points. Multiple representatives from the reconstruction-based model class are used. An autoregressive (AR) [25] and autoregressive moving average model (ARMA) [26] are applied and compared with a convolutional and a long short-term neural network [27,28]. All four models are used as regression models between the past time steps of the signals and a time step of the signal in the future. The deviations between these predictions and the actual progress of the signal are then compared with a threshold. If the deviation exceeds the threshold, an anomaly can be assumed. Furthermore, the one class support vector machines (OCSVM) [29] as a domain-based model is included for the comparison. This model builds a domain of inliers based on support vectors and the border data points of this domain. Data points outside this border line are classified as anomalies. As a simplistic baseline model, an approach is considered where a data point is compared to a multiple of the standard deviation of the reference data (abbreviated STD). If this distance exceeds a defined threshold, an anomaly is assumed.

1.1.3. Trend Detection Models

In the context of this publication a trend is defined as the gradual change in future events from past data in a time series [30]. Trend detection can be differentiated from remaining useful life (RUL) estimation by several aspects. In contrast to RUL estimation, trend detection methods do not extrapolate existing time series into the future. Furthermore, no thresholds for the extrapolated time series are defined which describe the end of lifetime of an asset. Trend detection methods have different purposes. It is possible to differentiate between models for change point detection, trend description and identification of trend presence in a time series. For the considered application, a model is required that answers the question of whether a trend is present. This is why the remainder of this subsection focuses on the field of trend presence identification. Here, various statistical tests exist. The Mann-Kendall test (MK) is a sign test based on pairs of all samples of a time series and their predecessors [31] to detect trends. The Cox-Stuart (CS) test uses a reduced amount of data pairs for a sign test [32] to achieve the same objective. The Wilcoxon-Mann-Whitney trend test builds a test statistic based on the signs of the slopes between samples and the rank sums of the samples with an increasing and decreasing slope [33] for this purpose. The Durbin-Watson test checks for auto-correlation in the residuals of a regression fit. If the residuals do not show autocorrelation, a trend can be assumed [34]. Furthermore, slope based approaches in combination with thresholds exist. The most simple approach from this field is to fit a linear or quadratic function to the time series data, calculate the slope of this function and compare it with a threshold. This model will be named linear regression model, short LR, for the rest of the publication. If the slope exceeds the threshold value, a trend can be assumed. A more complex approach for trend detection is based on the clustering of a time series. In a first step, a clustering algorithm (e.g., Fuzzy-K-Means) is used to detect clusters within the time series. Then, the slope between the cluster centres is determined. Finally, the slope values of the cluster centres are compared with a threshold to decide, whether a trend exists [35]. The last approach for trend detection presented in this section is based on the comparison of the time series' moving average with its overall mean (moving average model, short MA). In a first step, these two quantities are calculated. Afterwards, the time series' standard deviation multiplied by a factor is added to the overall mean to determine a threshold. Then, it is determined, whether the moving average of the signal rises above this threshold for a defined time window. If this is the case, it can be assumed that a trend is present in the signal. The principle behind this method is also illustrated in Figure 1.

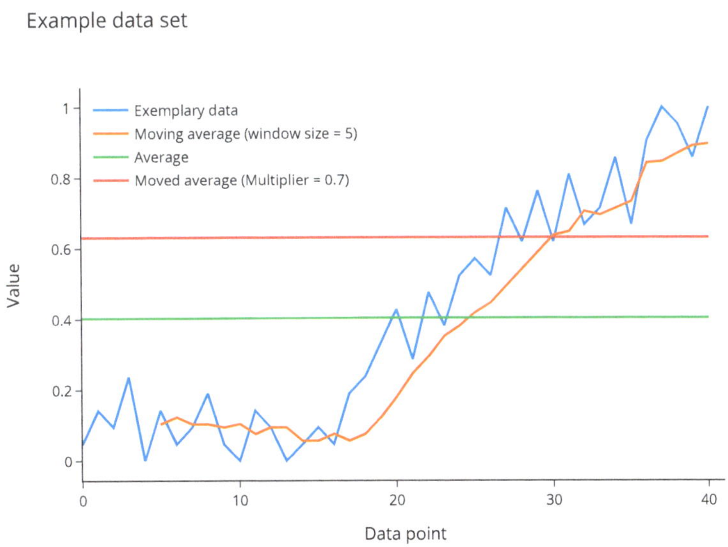

Figure 1. Example for the trend detection method using moving averages.

1.2. Considered Research Gap

In the field of industrial robot gear condition monitoring no combined AD and TD model has been presented up to now to the best of our knowledge. Therefore, the research objective of this publication is to present such a system. For the detailed design of this system, a suitable AD and TD model must be chosen. As no comparison of AD and TD models for univariate time series of HIs derived from acceleration sensors has been performed up to date, a method to select suitable AD and TD models for the application of industrial robot gear condition monitoring is formulated. Afterwards, it is applied to choose models for the presented combined system. In the context of the framework presented in [36], we address the question of algorithm selection for the inference task. By doing so we support the transfer of state of the art AI models into practice and reduce the effort of model selection for practitioners. The identification of suitable data acquisition systems or the selection of features is not considered in this publication. This is e.g., considered in [3]. Therefore, the presented work builds up on assumptions derived from this publication. These assumptions are summarized in Section 2.1.1. Furthermore, we limit our research frame to the field of six-axis articulated robots as we can not provide comprehensive experiments for other asset classes and hence validate our approach for such assets.

2. Materials and Methods

In this section, firstly CATS is described. Subsequently, the method for selecting suitable AD and TD models for CATS is described.

2.1. Combined Anomaly and Trend Detection Model

The objective of CATS is the reliable detection of trends and anomalies in industrial robot gear health indicator data. In the following, the assumptions that the system is based on, are defined. Then, the system itself is presented.

2.1.1. System Assumptions

The presented model builds upon certain assumptions. Data ingested in the system must be collected from a setup with a constant robot trajectory and load. The system analyses only univariate time series data of one health indicator per axis derived from acceleration sensor data. A suitable HI is described for example in [3]. The HI exhibits stationary behaviour when the robot axis is in a healthy state. The considered time series can be subject to trends $x_{trend}(t)$, seasonality $x_{seasonality}(t)$, noise $x_{noise}(t)$ and anomalies $x_{anomaly}(t)$. Noise can be caused by changing environmental conditions or sensor effects. Trends can occur due to wear. Trends due to sensor drifts are prevented by the sensor setup or suitable data preprocessing (e.g., high pass filtering of the raw data). Seasonality can occur due to changing temperatures of the gears. These temperature changes lead to variations in the HI (for example, see [37]). These temperature changes result from varying utilisation in the production system. They could be caused for instance by a three shift working model with reduced utilisation during night shift. Summarising, this time series can be expressed as in Equation (1).

$$x(t) = x_{trend}(t) + x_{seasonality}(t) + x_{noise}(t) + x_{anomaly}(t) \quad (1)$$

2.1.2. System Design

The objective of the presented system is to evaluate whether $x_{anomaly}(t) \neq 0$ or $x_{trend}(t) \neq 0$. For this, an anomaly detection model and a trend detection model are deployed in parallel. The detection of an anomaly in a defined number of sequential measurements leads to the recommendation of immediate maintenance actions. The detection of trends in the data of a defined number of a sequential measurements leads to the proposal of maintenance actions in the near future. The working principle of the system is summarized in Figure 2. The design of the system addresses different aspects of the industrial robot gear condition monitoring use case. Faults, whose manifestation but not the underlying fault mechanism progress (e.g., tracking of the growth of a crack in a gear tooth) can be tracked with HIs, will cause point or collective anomalies. The AD model will be used for the detection of such faults. Other faults, whose progress can be tracked (e.g., increasing wear), will cause trends in the HI. These trends will be detected by the trend detection model.

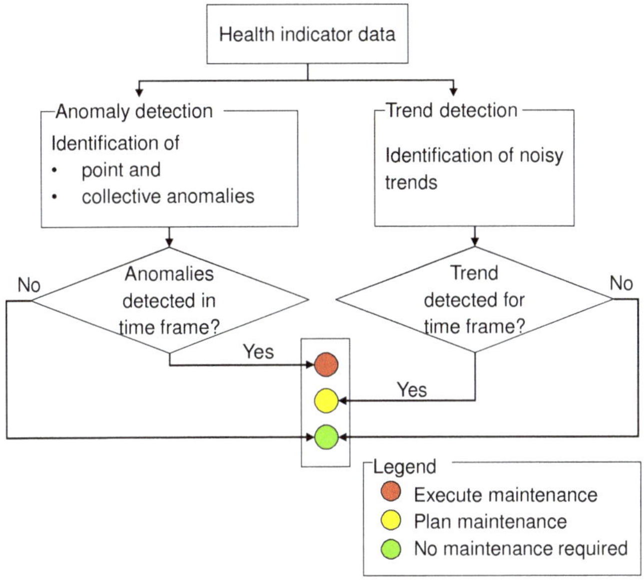

Figure 2. Overview of the condition monitoring system.

2.2. *Method for Anomaly and Trend Detection Model Selection*

In this section, the overall model evaluation method is proposed. Then, more detailed information is given about the generation of synthetic data and the model evaluation criteria.

2.2.1. Overall Method and Selected Models

To select suitable AD and TD models for the presented system a three step approach was followed to ensure that the most suitable models are chosen. Firstly, potential models were identified in the literature. Secondly, these models were applied on synthetic data meeting defined characteristics of the considered application and evaluated in respect of different quality criteria to reduce the solution space. Thirdly, the best performing models were evaluated using real world data taken from accelerated wear tests of industrial robots. The overall selection process is summarised in Figure 3. In the following, these steps are explained in detail.

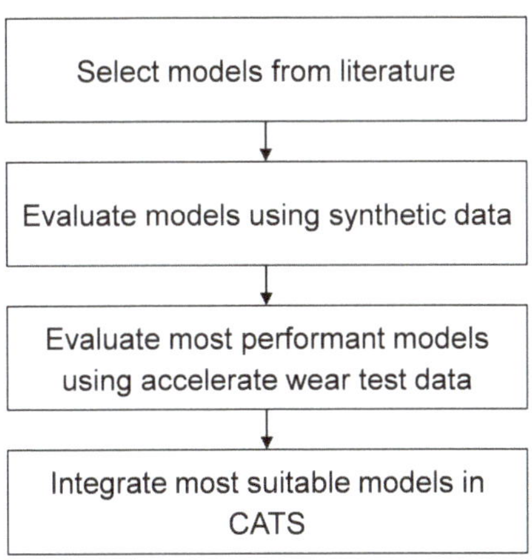

Figure 3. Overview of the model evaluation method.

As described in Section 1.1, a large number of AD and TD models exist. Hence a holistic comparison of existing approaches is not feasible. Therefore, models from the classes as described in [18] were chosen for the AD model comparison. In detail, the models listed in Table 1 were used. The models are explained in detail above in Section 1.1.2. For TD model comparison, the MK test, the CS test as well as the LR and MA based approaches described in Section 1.1.3 were chosen. The implementation of the models is described in an open source repository [38].

Table 1. Models considered.

Anomaly Detection Model	Anomaly Detection Model Type	Reference
CNN	Reconstruction based	[28]
LSTM	Reconstruction based	[27]
AR	Reconstruction based	[25]
ARMA	Reconstruction based	[26]
KDE	Probabilistic	[20]
GP	Probabilistic	[21]
OCSVM	Domain based	[29]
IF	Distance based	[23]
DBSCAN	Distance based	[24]
LOF	Distance based	[22]
STD	Distance based	[-]
Trend detection model	**Trend detection model type**	**Reference**
MK	Statistical test	[31]
CS	Statistical test	[32]
LR	Slope based	[-]
MA	Slope based	[-]

2.2.2. Synthetic Data Generation

For the model comparison based on synthetic data, a data generator was implemented to create time series as described in Equation (1). Different trend, noise, seasonality and anomaly functions were considered. In detail, linear and quadratic trend functions were implemented. White noise and uniform noise with different variances or ranges were used as noise functions. Sine functions and a hand crafted function as described in Equation (2) were applied for seasonality. Here, t is the current time step, which would relate to the length of one hour of the time series and a is the magnifier factor, which is further described in Table 2. An example of this function is depicted in Figure 4 on the upper right side.

$$f(x) = \begin{cases} (t\%24) \times a/4 & \text{if } (t\%24) \leq 4 \\ 1, & \text{if } 4 < (t\%24) \leq 20 \\ \frac{24-(t\%24)}{4} \times a & \text{otherwise} \end{cases} \tag{2}$$

For the anomaly function, a uniform distribution was used to define the anomaly positions. Different lengths for collective and different amplitudes for both collective and point anomalies were applied. To derive reasonable parameter ranges, certain realistic assumptions were made. A time series consists of 8736 samples representing 24 measurements per day for one year. The range of the trend functions' slopes should allow a doubling of the HI value in no less than one week and no more than half a year. Noise and seasonality should as a minimum result in a deviation of the time series by the factor 0.3 and as a maximum by the factor 9 from the mean of the signal. These assumptions were based on collected HI data from industrial robots in a car manufacturing plant. Due to confidentiality reasons, this data can not be published. The different functions, their parameters, the range of the parameters used and underlying assumptions for the parameter range choice are specified in Table 2. In the first three months of the time series no anomaly or trend occurs. In the last nine months anomalies may occur. Figure 4 shows a typical synthetic time series.

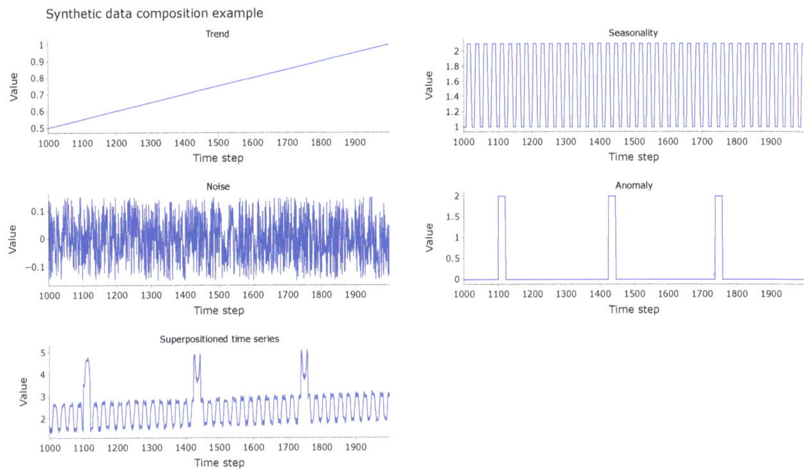

Figure 4. Example of a synthetic time series.

Table 2. Overview of used parameter ranges for the synthetic time series.

Signal	Parameter Type	Parameter Values	
		Synthetic Data Set 1	**Synthetic Data Set 2**
$x_{trend}(t)$	Trend type	Linear	
		Quadratic	
	Trend slope	Linear: 0.012	Linear: 4.58×10^{-4}
		Quadratic: 7.09×10^{-5}	Quadratic: 1.05×10^{-7}
$x_{seasonality}(t)$	Seasonality type	Sine	
		Production cycle (Formula 2)	
	Amplitudes a	Sine: 0.15	Sine: 3
		Production cycle: 1.1	Production cycle: 2
$x_{noise}(t)$	Noise type	Uniform noise	
		White noise	
	Noise parameters	Uniform noise range: 0.15	Uniform noise range: 1
		White noise mean: 0 White noise standard deviation: 0.03	White noise mean: 0 White noise standard deviation: 0.8
$x_{anomaly}(t)$	Anomaly types	Point anomaly	
		Collective anomaly	
	Anomaly parameters	Amplitude: 2	Amplitude: 1.1
		Collective anomaly lengths: 20 measurements	Collective anomaly lengths: 5 measurements

Based on this parameter range, over 26 million unique time series could be modeled. To reduce the computational effort, two reduced data sets were created. The first data set (synthetic data set 1) was used for an initial screening of the models' performance.

It consisted of time series with low noise, trends with a high slope, and large anomaly magnitude values and lengths. Furthermore, a second data set (synthetic data set 2) with more difficult conditions for the detection of trends and anomalies was generated. Here time series with high noise, low trend slopes, and low anomaly magnitudes and lengths were calculated. In each time series 40 anomalies were present. Each created time series was analysed by each model to detect trends and anomalies. In total, 16 unique time series were analysed per data set.

2.2.3. Model Evaluation

To measure the models' performance, the ROC curves (receiver operating characteristic curves) for different parameter choices of the models were determined. This means that different model parameters were varied and the True Positive Rate (TPR) and False Positive Rates (FPR) of the models for the synthetic data were determined. More precisely the models were presented with slices of the time series and had to determine, whether trends or anomalies were present in the time series. For the trend detection task, these slices were increased in size per time series with a window size of 1008 samples and an initial size of 2016 samples. This is equivalent to 24 measurements per day for a length of 12 weeks for the initial window. For the anomaly detection, the first 168 values were used to train the models. This is equivalent to 24 measurements per day for one week. The models were then tested on time series with a length of 6720 samples. The parameters that were varied for the different models are summarized in Table A1. The most robust models with high TPR and low FPR and high average AUC values (area under the curve) were then applied to data sets from accelerated robot gear wear tests. A data set, which is based on an accelerated wear test with an ABB IRB 6600-255/2.55, was used to test the trend detection models (Accelerated wear test 1). The experiment caused different faults in the robot gear of the second axis. In total, 2425 measurements over a time span of roughly one year were used from the experiment; these were acquired with an acceleration sensor at the robot gear cap. From this data the HI described in [3] was derived. For more information regarding the experiment, see [39,40]. The same data set and another data set, which was acquired during another accelerated wear test with an ABB IRB 7600-340/2.8, to test the anomaly detection models (Accelerated wear test 2). Here, 920 measurements were acquired over three months at the second axis gear cap with an acceleration sensor, and the same HI was calculated and various gear faults were subsequently detected in the second axis gear. As no obvious trend could be seen in this data set, it was just used for the AD model evaluation.

More information regarding this experiment is given in [3]. Figure 5 presents the various faults of both accelerated wear tests. For analyzing these data sets, the models parameters were chosen that yielded the best compromise in TPR and FPR during the experiments with the synthetic data. In a real world setup, other parameter sets could be more reasonable in respect of the trade-off between false alarms and undetected faults. A method of how to choose the best parameters given the maintenance circumstances of an individual robot is discussed in Section 4. Based on the results of the accelerated wear test experiments, a suggestion of which models to use for trend and anomaly detection in the CM system is made. The detailed model evaluation method based on synthetic data is depicted in Figure 6.

Figure 5. Overview of the faults of the accelerated wear tests following [3,39].

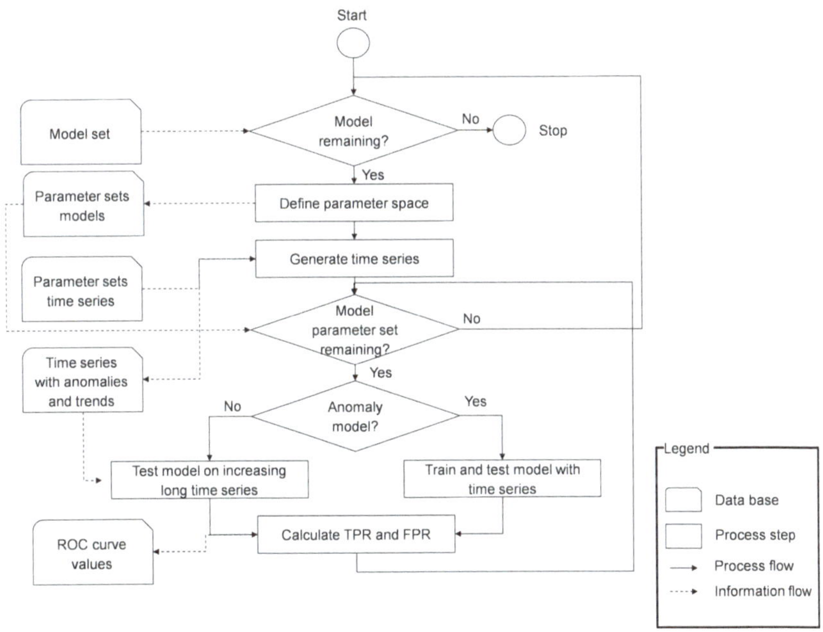

Figure 6. Overview of the model evaluation method.

3. Results

In the following, the presented method from the last section is applied to the AD and TD models listed in Table 1. First, the results for the TD models are shown, then the results of the AD models.

3.1. Trend Detection Model Comparison

Here, first the evaluation of the TD models based on synthetic data are presented. Subsequently, the results based on the accelerated wear test are analysed.

3.1.1. Evaluation Based on Synthetic Data

Figure 7 shows the ROC curve derived from the synthetic data set 1 and the model parameters described in Table A1. Ideally, the plots would show a dot in the upper left corner for a model. Such a dot would refer to a perfect classifier. This means that the model has a TPR of 1 and FPR of 0. Such a model would detect all trends and trigger no false alarms. The LR model and the MA model achieve these perfect classification results. The variation of parameters of the CS model does not influence the model performance and the MK model shows high TPR values only at the expense of an increased false positive rate. The results of synthetic data set 2 with the same model parameters are shown in Figure 8. Here, the CS model shows the best performance as a parameter combination exists where no false alarms are triggered and all trends are detected. It is followed by the MK model, which also yields a performance where all trends are detected and the FPR is small. The LR and the MA models achieve high TPR values only at the expense of increased FPR. The AUC values of the models for both data sets are presented in Table A2. Based on these results, it was decided to apply the CS and the MK model to the accelerated wear tests as they performed best on the more difficult data (synthetic data set 2) and based on their average AUC values.

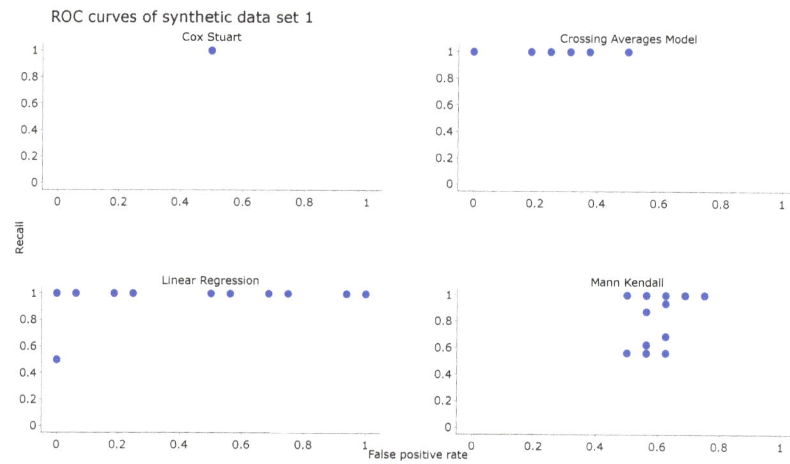

Figure 7. Trend detection model comparison based on the synthetic data set 1.

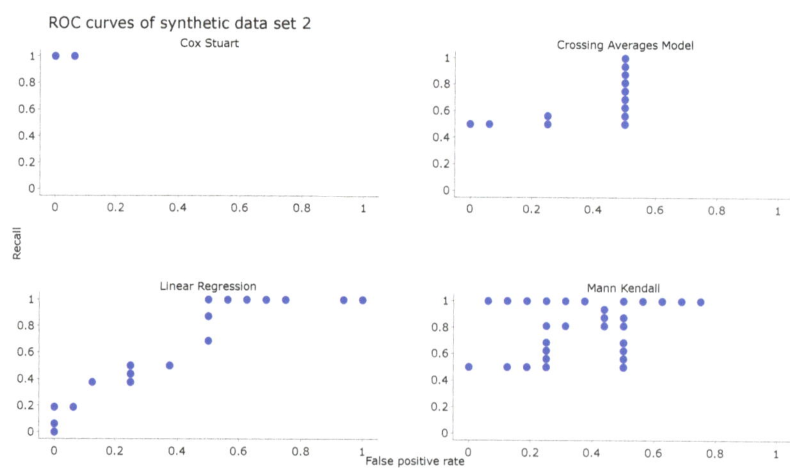

Figure 8. Trend detection model comparison based on the synthetic data set 2.

3.1.2. Evaluation on Accelerated Wear Test Data

The data from the accelerated wear test was analysed using the two chosen models. The results are depicted in Figure 9. The blue line shows the health indicator values, the dots indicate the models' decision of whether a trend is present in the time window of the last 504 samples (which equals a time frame of 2.5 months) while the horizontal yellow line shows, when more then 50 percent of the last 504 decisions were positive.

In such a case, a maintenance action should be planned. It can be seen that both models show similar behaviour for the beginning of the data set where they both detect a trend in the data after the initialisation phase of the first 504 measurements. The outlier at measurement 1000 leads to the rejection of the hypothesis that a trend is present for the following measurements in the MK model. It can be assumed that the CS model interprets the outlier correctly so that even for the following measurements a trend is detected. Both models detect the more stationary behaviour of the time series at its end. As the CS model handles the outlier around measurement 1000 better compared to the MK model, it is suggested to use the CS model in CATS. In this experiment, the confidence level parameters from the ROC curve of synthetic data set 2 were chosen for the models that yielded the highest TPR values with the lowest FPR at the same time.

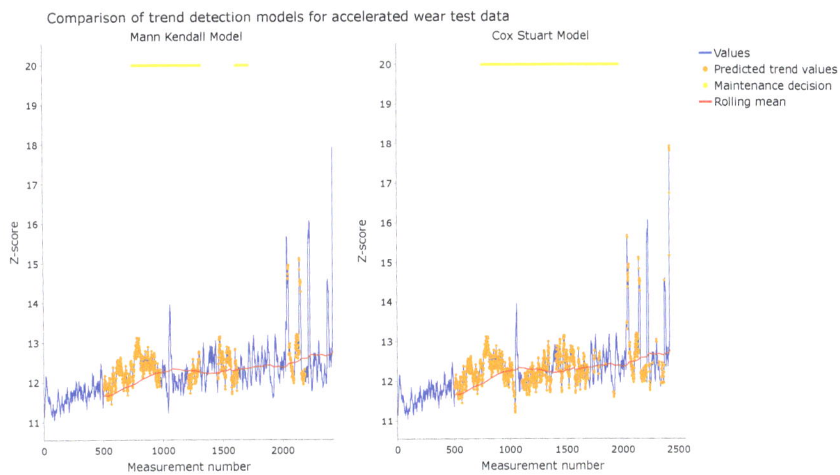

Figure 9. Results of the trend detection models based on accelerated wear test data.

3.2. Anomaly Detection Model Comparison

The presentation of the results of the AD model comparison follows the same scheme as Section 3.1.

3.2.1. Evaluation Based on Synthetic Data

The ROC curves of different models for the synthetic data set 1 are shown in Figure 10. Again, as described in Section 3.1.1 the plot would ideally show dots for the models at the upper left corner. Most of the models show good results except the OCSVM for which parameter combinations exist that yield poor classification performance. This means that all models are capable of identifying anomalies reliably and with a low false alarm rate in the case of high anomaly amplitudes and low noise level. In contrast, the models' overall performance regarding the synthetic data set 2 is rather poor. Figure 11 summarises the ROC curves for this data set. No perfect classifier was found for all models and the distance of the models' ROC curves to the upper left corner is large. Here, it can be concluded that the models struggle to detect anomalies at high noise levels and low anomaly amplitudes. This fact will also be discussed in Section 4. The AUC values for all models and both data sets are provided in Table A3. The individual ROC curves of all models for bothd data sets are presented in Figures A1 and A2. The best overall performance show the LSTM, STD and LOF models based on their average AUC values. Hence, it was decided to use the LSTM, STD and the LOF model on the accelerated wear test data.

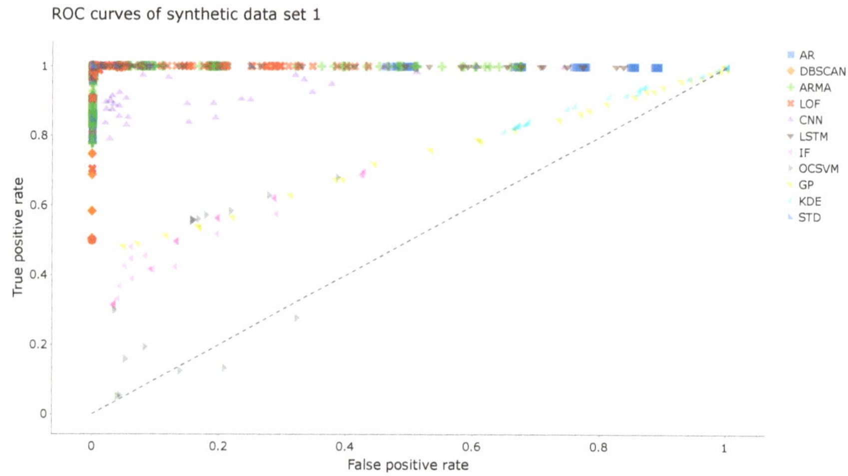

Figure 10. Results of the anomaly detection models based on synthetic data set 1.

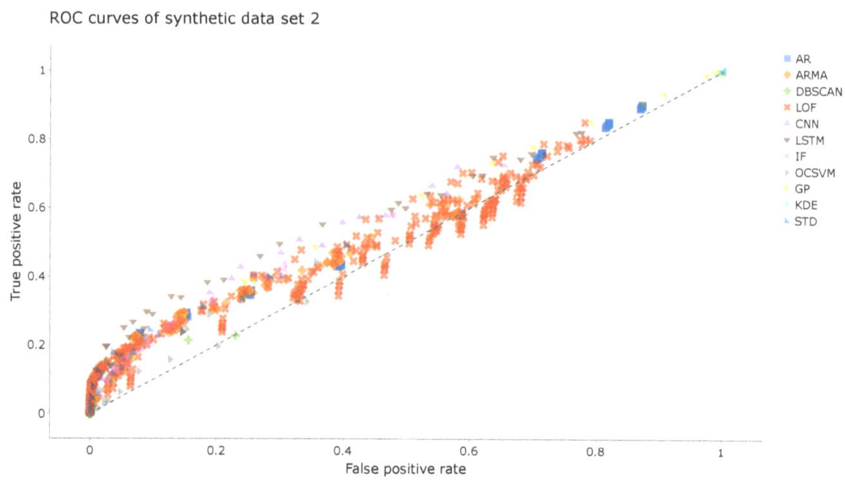

Figure 11. Results of the anomaly detection models based on synthetic data set 2.

3.2.2. Evaluation on Accelerated Wear Test Data

The results of applying the LSTM, STD and LOF models to the data from the accelerated wear test 1 are depicted in Figure 12. For this, all models were trained based on the first 500 measurements with model parameters of the ROC curves that yielded the best compromise between high TPR and low FPR values. It can be seen that all models correctly identify the anomalies at the end of the time series. The LOF model detects the outlier around measurement 1000 as an anomaly. Given a maintenance action decision criterion of 10 detected anomalies in the last 24 measurements, maintenance actions would have been triggered at the end of the data set for all models and a false alarm would have been triggered around measurement 1000 for the LOF model and for many more time ranges for the STD model. The AD models' behaviour on the second data set are summarized in a similar manner in Figure 13. In this scenario, the models were trained using the first 200 measurements with the same model parameters. It can be seen that the LSTM model and the STD model detect more anomalies than the LOF model along the time series. The apparent anomaly at the end of the time series is detected by all models.The LSTM triggers two false alarms around measurement 300. The STD model triggers many false alarms. Summarising, the STD shows more false alarms compared to other models. The LOF and LSTM model detect only the apparent anomalies with a low false alarm rate. Hence, it is suggested that either the LOF model or LSTM model is used in CATS as the AD model.

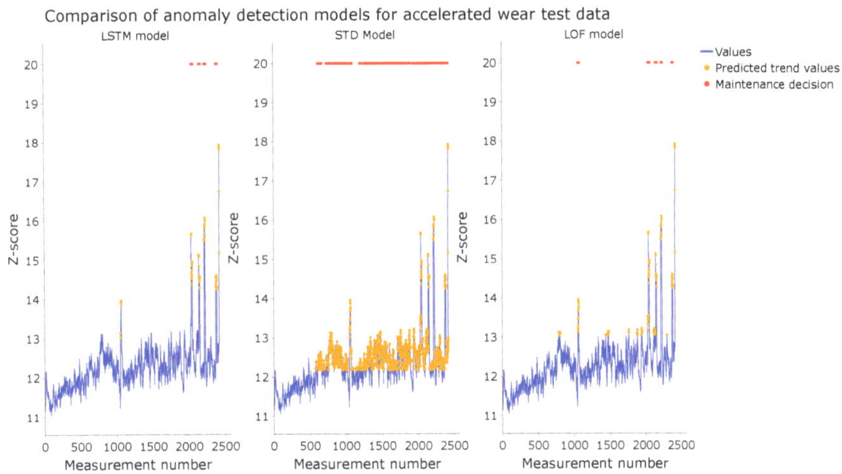

Figure 12. Results of selected anomaly detection models for accelerated wear test 1.

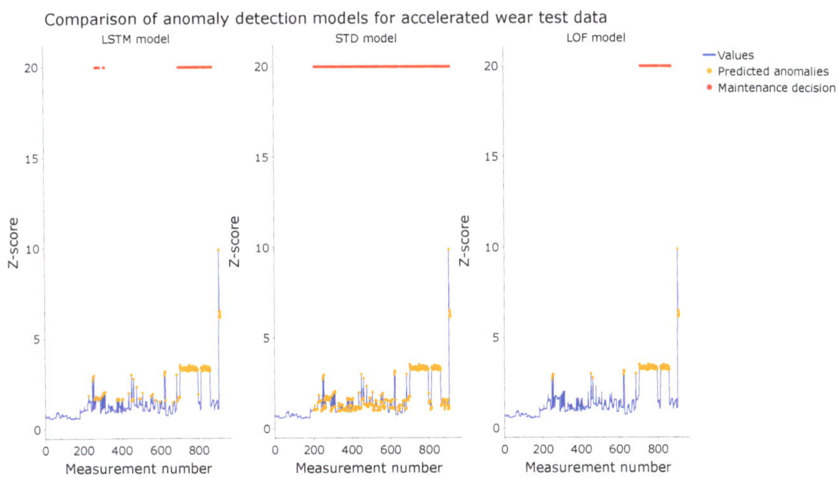

Figure 13. Results of selected anomaly detection models for accelerated wear test 2.

4. Discussion

The presented results highlight some interesting aspects that will be discussed in this section. We will justify our initial choice of models and highlight some aspects of the models' performance on the synthetic data. Then, we will explain the models' parameter choice and end with organisational thoughts regarding the integration of CATS in a real world production site.

As emphasised in Section 1.1.2, a comprehensive comparison of AD and TD models is not feasible due to the high variety of existing models. Our motivation for selecting models from different categories as presented in [18] was to test how their underlying detection mechanisms cope with the different characteristics of time series. The fact that AD and TD models were found that detect the trends and anomalies in the accelerated wear test data reliably, strengthens the argument that the comparison of the selected models is sufficient for the application. From our point of view, the results of the AD model comparison based on synthetic data set 2 clearly highlights the limitations of anomaly detection models in general. High noise levels in the data make it difficult for such models to detect anomalies. Figure 14 shows a typical time series of this data set. Even as a human operator, it is difficult to identify the anomalies. However, from our experience, such extreme noise does not appear in the HI time series as shown in Figure 9 or Figure 13 for the accelerated wear tests. When deploying AD or TD models in real world applications, suitable model parameters must be chosen. For this, from our point of view, the parameters have to be configured for the individual robot considering the common trade-off between false alarms (higher FPR) and undetected faults (lower TPR). If no ideal anomaly or trend detection model can be used considering the ROC curves, this trade-off can be tackled by considering a maintenance score for an individual robot. This maintenance score can be influenced for example by the position of the robot in the production systems in respect of the distance to buffers or the effort required to exchange the robot. Other criteria could be the required calibration effort after the replacement or the response time of the maintenance team if a replacement is required. For robots with a higher maintenance score, model parameters with high TPR and higher FPR should be chosen. For robots with a lower maintenance score, model parameters with lower TPR and low FPR should be selected. This principle is also depicted in Figure 15. The reconfiguration of such models might also be required if the FPR or TPR do not meet the expected behaviour over time. Finally, the implications that the formulated assumptions in Section 2.1.1 yield must be discussed. To meet these assumptions, two aspects must be considered in a real world application. First of all, a measurement trajectory must be used for data acquisition so that the HI data is comparable and has a low noise level. Secondly, CATS must be extended by mechanisms to ensure that anomalies or trends in the HI data are only due to wear and not changing environmental conditions, new robot programs or faulty data acquisition systems.

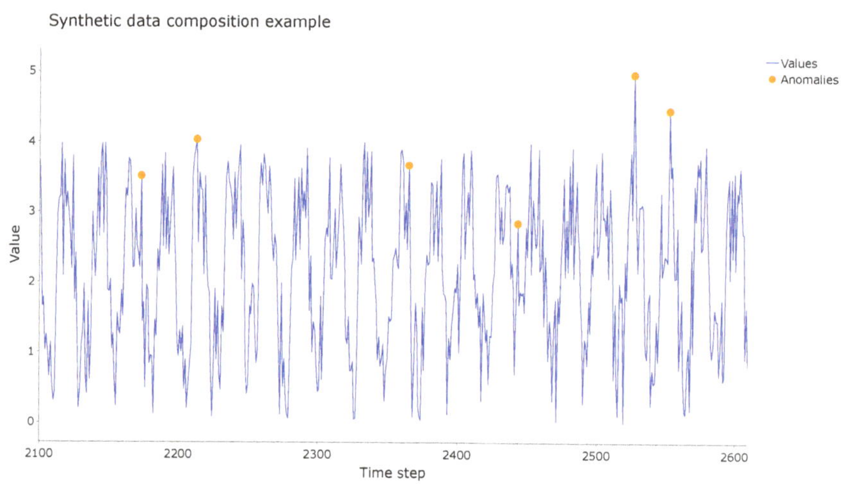

Figure 14. Example of a noisy time series from synthetic data set 2.

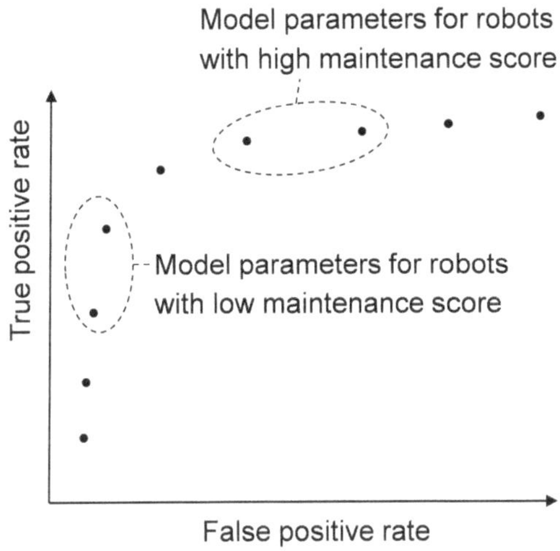

Figure 15. Selection of model parameters based on a maintenance score.

5. Conclusions

A combined anomaly detection and trend detection system for the condition monitoring of industrial robot gears has been presented. To select suitable models for these tasks, a method in which models are evaluated based on synthetic and accelerated wear test data was formulated. The synthetic data consists of time series with noise, cyclic behaviour, trends and anomalies based on realistic assumptions that were gathered from industry data. The accelerated wear test data was collected during two experiments with six-axis industrial robots, which provoked multiple gear faults and exhibited both trends and anomalies. By applying the presented method, it was found that the Cox-Stuart test is most suitable for trend detection and the local outlier factor algorithm or the long short-term neural network are capable of detecting the anomalies in the accelerated wear test data. For future research, we believe that the considerations in Section 4 regarding the extensions of CATS with functionalities to detect reasons for false alarms such as robot program changes or the change of the robot tool and the automatic reconfiguration of models in case of too many false alarms are the most important topics for enabling the automated condition monitoring of industrial robot gears in industry.

Author Contributions: Conceptualization, C.N.; methodology, C.N.; software, C.N.; validation, C.N.; formal analysis, C.N.; investigation, C.N.; resources, G.R.; data curation, C.N.; writing—original draft preparation, C.N.; writing—review and editing, G.R.; visualization, C.N.; supervision, G.R.; project administration, C.N.; funding acquisition, G.R. All authors have read and agreed to the published version of the manuscript.

Funding: We express our gratitude to the Bavarian Ministry of Economic Affairs, Regional Development, and Energy for the funding of our research. The formulated outlook will be investigated as part of the research project "KIVI" (grant number IUK-1809-0008 IUK597/003) and will be further developed and implemented.

Data Availability Statement: The data presented in this study are available on request from the corresponding author. The data are not publicly available due to confidentiality reasons.

Conflicts of Interest: The authors declare no conflict of interest. The funders had no role in the design of the study; in the collection, analyses, or interpretation of data; in the writing of the manuscript, or in the decision to publish the results.

Acknowledgments: We express our gratitude to our project partners BMW, Fluke and KUKA, who participate in the KIVI project for the fruitful discussions and creative ideas.

Appendix A. Model Parameters for ROC Curves

Table A1. Considered Models.

Model	Parameter	Range
MK	Confidence interval	0.9–0.999 in variablestep sizes
CS	Confidence interval	0.9–0.999 in variable step sizes
LR	Slope threshold	0–1 in variable step sizes
MA	Amplifier	min: 0.5, max: 0.8, step: 0.1
	Length above threshold	min: 24, max: 72, step: 2
	Moving Average Window	0.04×(dataset length)–0.18×(dataset length)
ARMA	Autoregression lags	min: 1, max: 9, step: 2
	Moving average lags	min: 0.2, max: 1, step: 0.2
	Anomaly threshold	min: 0.01, max: 0.1, step: 0.02
AR	Autoregression lags	min: 0.2, max: 1, step: 0.2
	Anomaly threshold	min: 0.01, max: 0.1, step: 0.02
CNN	Training epochs	10, 20, 50
	Anomaly threshold	0.1, 0.2, 0.3, 0.4, 0.5, 0.9, 0.95, 0.98, 0.99, 0.999
LSTM	Training epochs	10, 20, 50
	Anomaly threshold	0.1, 0.2, 0.3, 0.4, 0.5, 0.9, 0.95, 0.98, 0.99, 0.999
DBSCAN	Epsilon	min: 0.1, max: 1.3, step: 0.2
	Minimal number of samples	13, 21, 34, 55, 89, 144, 233, 377
GP	Anomaly threshold	0.7, 0.8, 0.9, 0.95
	Kernel upper bound	0.0001, 0.0005, 0.001, 0.002, 0.003, 0.005, 0.008, 0.013, 0.021, 0.034, 0.055, 0.089, 0.144, 0.233, 0.377, 0.61, 0.987
IF	Number of estimators	50, 100, 200
	Contamination	0.01, 0.02, 0.03, 0.05, 0.08, 0.13, 0.21, 0.34
LOF	Number of neighbors	5, 10, 20, 30, 50, 80
	Contamination	0.001, 0.01, 0.02, 0.03, 0.05, 0.08, 0.13, 0.21, 0.34, 0.5
OCSVM	Kernel	'rbf, sigmoid
	Nu	0.01, 0.02, 0.03, 0.05, 0.08, 0.13, 0.21, 0.34
KDE	Bandwidth	0.2, 0.3, 0.5, 0.8, 1.3, 2.1, 3.4, 5.5
	Anomaly threshold	0.75, 0.9, 0.95, 0.99

Appendix B. AUC Tables

Table A2. Overview of the AUC values of the trend detection models.

	Synthetic Data Set 1	Synthetic Data Set 2
Cox Stuart	0.750000	0.968750
Crossing Averages Model	1.000000	0.679688
Linear Regression	0.984375	0.707031
Mann Kendall	0.732422	0.861328

Table A3. Overview of the AUC values of the anomaly detection models.

	Synthetic Data Set 1	Synthetic Data Set 2
AR	0.998641	0.550152
ARMA	0.999713	0.553447
CNN	0.955253	0.596032
DBSCAN	1.000000	0.553332
GP	0.716030	0.591000
IF	0.706679	0.580334
KDE	0.584487	0.499234
LOF	0.999880	0.550426
LSTM	0.995874	0.612189
OCSVM	0.656618	0.542888
STD	0.999951	0.576333

Appendix C. Individual ROC Curves

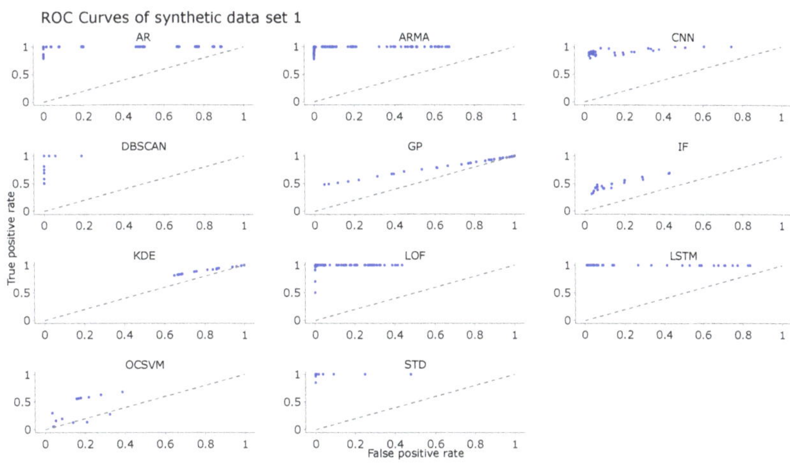

Figure A1. Results of the anomaly detection models based on synthetic data set 1.

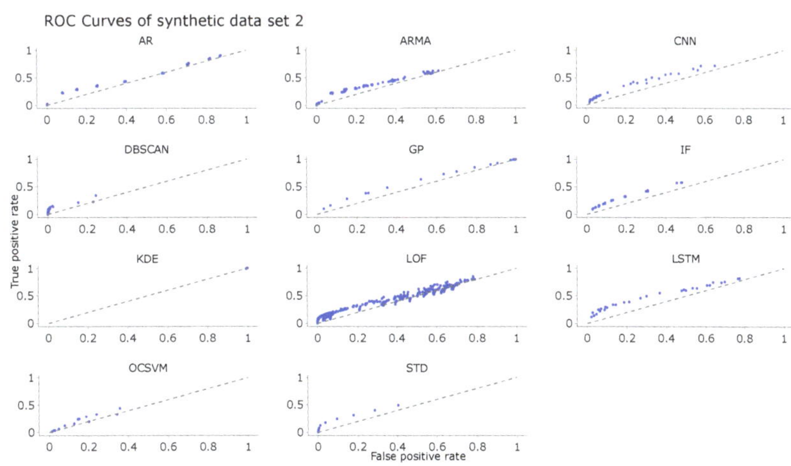

Figure A2. Results of the anomaly detection models based on synthetic data set 2.

References

1. Krockenberger, O. Industrial Robots for the Automotive Industry. In *SAE Technical Paper Series*; SAE International: Warrendale, PA, USA, 1996. [CrossRef]
2. 17359: 2018-05, Zustandsüberwachung und-Diagnostik von Maschinen—Allgemeine Anleitungen (ISO_17359: 2018). 2018 Available online: https://www.iso.org/standard/71194.html (accessed on 4 November 2021).
3. Nentwich, C.; Reinhart, G. A Method for Health Indicator Evaluation for Condition Monitoring of Industrial Robot Gears *Robotics* **2021**, *10*, 80. [CrossRef]
4. Costa, M.A.; Wullt, B.; Norrlöf, M.; Gunnarsson, S. Failure detection in robotic arms using statistical modeling, machine learning and hybrid gradient boosting. *Measurement* **2019**, *146*, 425–436. [CrossRef]
5. Sathish, V.; Orkisz, M.; Norrlof, M.; Butail, S. Data-driven gearbox failure detection in industrial robots. *IEEE Trans. Ind. Inform.* **2019**, *16*, 193–201. [CrossRef]
6. Cerquitelli, T.; Nikolakis, N.; O'Mahony, N.; Macii, E.; Ippolito, M.; Makris, S. (Eds.) *Predictive Maintenance in Smart Factories*, Information Fusion and Data Science; Springer: Singapore, 2021. [CrossRef]
7. Bittencourt, A.C.; Saarinen, K.; Sander-Tavallaey, S. A Data-driven Method for Monitoring Systems that Operate Repetitively-Applications to Wear Monitoring in an Industrial Robot Joint1. *IFAC Proc. Vol.* **2012**, *45*, 198–203. [CrossRef]
8. Sathish, V.; Ramaswamy, S.; Butail, S. Training data selection criteria for detecting failures in industrial robots. *IFAC-PapersOnLine* **2016**, *49*, 385–390. [CrossRef]
9. Trung, C.T.; Son, H.M.; Nam, D.P.; Long, T.N.; Toi, D.T.; Viet, P.A. Fault detection and isolation for robot manipulator using statistics. In Proceedings of the 2017 International Conference on System Science and Engineering, Ho Chi Minh City, Vietnam, 21–23 July 2017; IEEE: Piscataway, NJ, USA, 2017; pp. 340–343. [CrossRef]
10. Chen, T.; Liu, X.; Xia, B.; Wang, W.; Lai, Y. Unsupervised Anomaly Detection of Industrial Robots Using Sliding-Window Convolutional Variational Autoencoder. *IEEE Access* **2020**, *8*, 47072–47081. [CrossRef]
11. Wen, X.; Chen, H. Heterogeneous Connection and Process Anomaly Detection of Industrial Robot in Intelligent Factory. *Int. J. Pattern Recognit. Artif. Intell.* **2020**, *34*, 2059041. [CrossRef]
12. Hong, Y.; Sun, Z.; Zou, X.; Long, J. Multi-joint Industrial Robot Fault Identification using Deep Sparse Auto-Encoder Network with Attitude Data. In Proceedings of the 2020 Prognostics and Health Management Conference (PHM-Besançon), Besancon, France, 4–7 May 2020; IEEE: Piscataway, NJ, USA, 2020; pp. 176–179. [CrossRef]
13. Jaber, A.A.; Bicker, R. Industrial Robot Backlash Fault Diagnosis Based on Discrete Wavelet Transform and Artificial Neural Network. *Am. J. Mech. Eng.* **2016**, *4*, 21–31.
14. Nentwich, C.; Junker, S.; Reinhart, G. Data-driven Models for Fault Classification and Prediction of Industrial Robots. *Procedia CIRP* **2020**, *93*, 1055–1060. [CrossRef]
15. Cheng, F.; Raghavan, A.; Jung, D.; Sasaki, Y.; Tajika, Y. High-Accuracy Unsupervised Fault Detection of Industrial Robots Using Current Signal Analysis. In Proceedings of the IEEE International Conference on Prognostics and Health Management (ICPHM), San Francisco, CA, USA, 17–20 June 2019.
16. Kim, Y.; Park, J.; Na, K.; Yuan, H.; Youn, B.D.; Kang, C.S. Phase-based time domain averaging (PTDA) for fault detection of a gearbox in an industrial robot using vibration signals. *Mech. Syst. Signal Process.* **2020**, *138*, 106544. [CrossRef]
17. Pu, Z.; Cabrera, D.; Bai, Y.; Li, C. A one-class generative adversarial detection framework for multifunctional fault diagnoses. *IEEE Trans. Ind. Electron.* **2021**. [CrossRef]
18. Pimentel, M.A.; Clifton, D.A.; Clifton, L.; Tarassenko, L. A review of novelty detection. *Signal Process* **2014**, *99*, 215–249. [CrossRef]
19. Braei, M.; Wagner, S. Anomaly Detection in Univariate Time-series: A Survey on the State-of-the-Art. *arXiv* **2020**, arXiv:2004.00433.
20. Bishop, C.M. *Pattern Recognition and Machine Learning*, corrected at 8th printing 2009 ed.; Springer: New York, NY, USA, 2009.
21. Kemmler, M.; Rodner, E.; Wacker, E.S.; Denzler, J. One-class classification with Gaussian processes. *Pattern Recognit.* **2013**, *46*, 3507–3518. [CrossRef]
22. Breunig, M.M.; Kriegel, H.P.; Ng, R.T.; Sander, J. LOF. *ACM SIGMOD Rec.* **2000**, *29*, 93–104. [CrossRef]
23. Liu, F.T.; Ting, K.M.; Zhou, Z.H. Isolation Forest. In Proceedings of the 2008 Eighth IEEE International Conference on Data Mining, Pisa, Italy, 15–19 December 2008; IEEE: Piscataway, NJ, USA, 2008; pp. 413–422. [CrossRef]
24. Ester, M.; Kriegel, H.P.; Sander, J.; Xu, X. A Density-Based Algorithm for Discovering Clusters in Large Spatial Databases with Noise. In Proceedings of the Second International Conference on Knowledge Discovery and Data Mining, KDD'96, Portland, OR, USA, 2–4 August 1996; AAAI Press: Palo Alto, CA, USA 1996; pp. 226–231.
25. Kumar, V.; Banerjee, A.; Chandola, V. Anomaly Detection for Symbolic Sequences and Time Series Data. 2009. Available online: https://conservancy.umn.edu/bitstream/handle/11299/56597/Chandola_umn_0130E_10747.pdf;jsessionid=A026AFB0208E56D91DD811BA4A134056?sequence=1 (accessed on 4 November 2021).
26. Aggarwal, C.C. *Outlier Analysis*, 2nd ed.; Springer International Publishing: Cham, Switzerland, 2017. [CrossRef]
27. Chauhan, S.; Vig, L. Anomaly detection in ECG time signals via deep long short-term memory networks. In Proceedings of the 2015 IEEE International Conference on Data Science and Advanced Analytics (DSAA), Paris, France, 19–21 October 2015; IEEE: Piscataway, NJ, USA, 2015; pp. 1–7. [CrossRef]
28. Munir, M.; Siddiqui, S.A.; Dengel, A.; Ahmed, S. DeepAnT: A Deep Learning Approach for Unsupervised Anomaly Detection in Time Series. *IEEE Access* **2019**, *7*, 1991–2005. [CrossRef]

29. Zhang, R.; Zhang, S.; Muthuraman, S.; Jiang, J. One Class Support Vector Machine for Anomaly Detection in the Communication Network Performance Data. 2007. Available online: http://www.wseas.us/e-library/conferences/2007tenerife/papers/572-618.pdf (accessed on 4 November 2021).
30. Sharma, S.; Swayne, D.A.; Obimbo, C. Trend analysis and change point techniques: A survey. *Energy Ecol. Environ.* **2016**, *1*, 123–130. [CrossRef]
31. Kendall, M.G.; Gibbons, J.D. *Rank Correlation Methods*, 5th ed.; Arnold: London, UK, 1990.
32. Cox, D.R.; Stuart, A. Some Quick Sign Tests for Trend in Location and Dispersion. *Biometrika* **1955**, *42*, 80. [CrossRef]
33. Crawford, C.G.; Slack, J.R.; Hirsch, R.M. *Nonparametric Tests for Trends in Water-Quality Data Using the Statistical Analysis System*; USGS Numbered Series; U.S. Geological Survey: Reston, VA, USA, 1983. [CrossRef]
34. DURBIN, J.; WATSON, G.S. Testing for Serial Correlation in Least Squares Regression. I. *Biometrika* **1950**, *37*, 409–428. [CrossRef]
35. Melek, W.W.; Lu, Z.; Kapps, A.; Fraser, W.D. Comparison of trend detection algorithms in the analysis of physiological time-series data. *IEEE Trans. Bio-Med Eng.* **2005**, *52*, 639–651. [CrossRef]
36. Modi, S.; Lin, Y.; Cheng, L.; Yang, G.; Liu, L.; Zhang, W.J. A Socially Inspired Framework for Human State Inference Using Expert Opinion Integration. *IEEE/ASME Trans. Mechatron.* **2011**, *16*, 874–878. [CrossRef]
37. Carvalho Bittencourt, A. *Modeling and Diagnosis of Friction and Wear in Industrial Robots*; Linköping University Electronic Press: Linköping, Sweden, 2014; Volume 1617. [CrossRef]
38. Nentwich, C. CATS—A Combined Anomaly Detection and Trend Detection Model. Available online: https://github.com/xorbey/CATS_public (accessed on 4 November 2021).
39. Danielson, H.; Schmuck, B. Robot Condition Monitoring: A First Step in Condition Monitoring for Robotic Applications. Master's Thesis, Lulea University of Technology, Lulea, Sweden, 2017.
40. Karlsson, M.; Hörnqvist, F. Robot Condition Monitoring and Production Simulation. Master's Thesis, Lulea University of Technology, Lulea, Sweden, 2018.

applied sciences

MDPI

Article

Bi-Objective Optimization for Industrial Robotics Workflow Resource Allocation in an Edge–Cloud Environment

Xingju Xie [1,2], Xiaojun Wu [2] and Qiao Hu [1,3,*]

1 School of Mechanical Engineering, Xi'an Jiaotong University, Xi'an 710049, China; xiexingjv@foxmail.com
2 School of Software Engineering, Xi'an Jiaotong University, Xi'an 710049, China; xiaojunw@xjtu.edu.cn
3 Shaanxi Key Laboratory of Intelligent Robots, Xi'an Jiaotong University, Xi'an 710049, China
* Correspondence: hqxjtu@xjtu.edu.cn

Abstract: The application scenarios and market shares of industrial robots have been increasing in recent years, and with them comes a huge market and technical demand for industrial robot-monitoring system (IRMS). With the development of IoT and cloud computing technologies, industrial robot monitoring has entered the cloud computing era. However, the data of industrial robot-monitoring tasks have characteristics of large data volume and high information redundancy, and need to occupy a large amount of communication bandwidth in cloud computing architecture, so cloud-based IRMS has gradually become unable to meet its performance and cost requirements. Therefore, this work constructs edge–cloud architecture for the IRMS. The industrial robot-monitoring task will be executed in the form of workflow and the local monitor will allocate computing resources for the subtasks of the workflow by analyzing the current situation of the edge–cloud network. In this work, the allocation problem of industrial robot-monitoring workflow is modeled as a latency and cost bi-objective optimization problem, and its solution is based on the evolutionary algorithm of the heuristic improvement NSGA-II. The experimental results demonstrate that the proposed algorithm can find non-dominated solutions faster and be closer to the Pareto frontier of the problem. The monitor can select an effective solution in the Pareto frontier to meet the needs of the monitoring task

Keywords: industrial robot-monitoring system; industrial robot-monitoring workflow; workflow resource allocation; edge–cloud collaboration; bi-objective genetic algorithm

Citation: Xie, X.; Wu, X.; Hu, Q. Bi-Objective Optimization for Industrial Robotics Workflow Resource Allocation in an Edge–Cloud Environment. *Appl. Sci.* **2021**, *11*, 10066. https://doi.org/10.3390/app112110066

Academic Editor: Jasiulewicz-Kaczmarek Małgorzata

Received: 3 October 2021
Accepted: 24 October 2021
Published: 27 October 2021

1. Introduction

The use of industrial robots in manufacturing industry is increasing rapidly [1], and industrial robot-monitoring systems (IRMS) have played an important role in maintaining the normal operation of industrial robots and even the whole factory, most of the IRMSs are based on B/S or C/S architecture remote monitoring by the Internet [2]. With the development of IoT and cloud computing technology, IRMSs based on cloud computing architecture have emerged. For example, Hanbo Yang et al. [3] implemented a cloud manufacturing monitoring platform based on 5G and SIM (Standard Information Model), Rachmad Andri Atmoko et al. [4] implemented a cloud monitoring industrial arm robot based on MQTT protocol. In addition, cloud robotics has become an emerging area of robotics research [5], where the technological key is computational offloading, when the robot controller generates compute-intensive tasks to the cloud in order to reduce the requirements for controller performance and the computational energy consumption of the robot. However, since the robot cannot rely excessively on cloud resources due to its physical bandwidth limitation, computational offloading strategies for cloud robots have become a hot research topic. For example, G. Hu [6], B. Kehoe [7], and J. Wan [8] offload computationally intensive tasks such as robot grasping, simultaneous localization and mapping (SLAM), and navigation of cloud robots to the cloud.

In fact, most tasks in IRMSs, such as fault diagnosis, environmental monitoring, object recognition, and posture awareness, require IRMSs to continuously collect and process large amounts of environmental data in real time to maintain the accuracy of monitoring results; however, the collected data, especially internal and external sensor data, are semi-structured and unstructured [9], with high information redundancy and low value density, if all of them are offloaded to the cloud computing will occupy a large amount of communication bandwidth, and the congestion of data channels may also lead to an increase in latency, then the benefits brought by cloud computing will be greatly reduced. In this context, edge computing [10] as an emerging computing architecture, can place some computing tasks on edge servers close to the devices rather than in distant cloud centers, which can effectively reduce the pressure on the communication bandwidth and reduce the communication latency.

In this paper, IRMS is targeted at monitoring fixed-position multi-degree-of-freedom industrial robots which can perform tasks such as handling, palletizing, painting, assembly and welding, and common faulty units include bearings, gearboxes, and motors [11]. cloud-based architecture IRMSs are combined with cloud robotics and edge computing architecture. Edge computing resources can be considered to be an effective complement to cloud resources, reducing both the computational burden of the local monitor and the communication network burden of using cloud computing. The monitor can obtain real-time operation data of a set of industrial robots through internal and external sensors, process the monitoring data and obtain the corresponding monitoring results by executing workflow-based monitoring computation tasks. The monitor can obtain real-time monitoring data of a set of industrial robots through internal and external sensors, process the monitoring data and obtain the corresponding monitoring results by executing workflow-based monitoring computation tasks. To solve the computing resource allocation problem of monitoring workflows, a bi-objective optimization problem of time and monetary cost is modeled, and it can be solved by the proposed genetic algorithm. In experiments on this work, optimal solution of the two optimization objectives obtained by mixed integer quadratic programming (MIQP) technique are used a reference for comparison. The performance of the proposed algorithm and benchmark NSGA-II [12] are compared in various aspects such as different types and amounts of computing resources, different types and amounts of tasks, and evolutionary generations. The main contributions of this paper are listed as follows:

1. The edge–cloud architecture IRMS is architected to allow industrial robot-monitoring tasks to perform as workflows and can be allocated to computing resources in the edge–cloud environment.
2. The Industrial Robot-Monitoring Workflow Assignment Problem (IRMWAP) is defined in terms of the characteristics of industrial robot-monitoring workflow task execution as an NP-hard bi-objective (latency and cost) optimization problem.
3. The Improved NSGA2 based on Transcription Gene and Heuristic Recombination (INSGA2-TGHR) algorithm is proposed by means of improved genetic factors and recombination operators to provide a set of Pareto frontiers for the monitor with computing resource allocation schemes.

The rest of this paper is organized as follows. Section 2, relevant research works are reviewed. Section 3, describes an IRMS Architecture in Edge–Cloud Environment. In Section 4, the industrial robot-monitoring workflow computation allocation problem is modeled, and an improved algorithm of this work is proposed. In Section 5, experiments are conducted on the problem described and the algorithm proposed in Section 4. Section 6 concludes the whole paper and presents future plans.

2. Related Work

At present, most of the IRMSs are implemented in the mode of remote monitoring [2], and the monitoring of the equipment is realized through the upper computer client or web, for example, XuHong Mei [13] proposed a C/S (client and server) for remote monitoring of industrial robots, and Hongli Yin [14] proposed an Internet-based and sensor-driven architecture that combines remote monitoring and control. With the development of

IoT and cloud computing technology there are also industrial robots based on cloud computing architecture, for example, Hanbo Yang [3] and others have implemented a cloud manufacturing monitoring platform based on 5G and SIM (Standard Information Model), Rachmad Andri Atmoko [4] and others have implemented a cloud monitoring industrial arm robot based on MQTT protocol.

Table 1 summarizes the research related to cloud robotics and computing resource allocation in recent years. The computing resource allocation algorithm research mainly focuses on genetic algorithm and integer linear programming, and the optimization objectives mainly focus on latency, price, etc. Recently, many scholars have brought the edge computing in cloud robotics system to reduce latency and robot energy consumption Chen Wuhui [15] defined robotic streaming workflow (RSW) and networked cloud robotics (NCR) as the basic data structures for studying the allocation of workflows and computing resource topology, and by defining data flow graph (DFG) for the problem of allocating computing resources to workflows, the three optimization objectives of latency, price, and energy consumption are weighted linearized, and the above problems are solved by heuristic graph partitioning and MILP techniques, but their study only considers multi-robot and cloud-centric resource allocation, and metrics and units of three optimization objectives are not uniform, so simply weighted linearization cannot be used to represent the complete problem. Mahbuba Afrin [16] redesigned NSGA-II by pre-sorted initial population and minimum distant selection of chromosome that gives balanced solution for all objectives and obtained effectively performance improvement, but this study did not consider the constraints of communication resources in task assignment, so it still has some distance from practical applications.

Table 1. Summary of relevant works in cloud robotics and computing resource allocation.

Work	Solution Approach	Target Application	Workflow Scale Reduction	Communication Restrictions	Multi-Objective Optimization	Resource Type		
						Robot (Local)	Edge (Fog)	Cloud
[15]	Heuristic MILP	Cloud robotic	YES	YES	NO	YES	NO	YES
[16]	Augmented NSGA-II	Smart factory	No	NO	YES	YES	NO	YES
[17]	Benchmark NSGA-II	Mobile Computing Resource Allocation	No	No	YES	YES	NO	YES
[18]	Benchmark NSGA-II	Application placement	NO	NO	YES	YES	YES	YES
[19]	GA	Smart city	NO	NO	NO	YES	NO	YES
[20]	genetic-based ICA	Cloud robotic	NO	NO	NO	YES	YES	YES
[21]	Heuristic ILP	Cloud robotic	YES	YES	NO	YES	NO	YES
[22]	Heuristic devices sorting	Task scheduling in Dew	NO	YES	NO	YES	YES	YES
[23]	Heuristic scheduling algorithm	Task scheduling in Dew	NO	YES	NO	YES	YES	YES
This	INSGA2-TGHR	Industrial Robot Monitoring	YES	YES	YES	YES	YES	YES

In this paper, the cloud-based architecture IRMS has been further upgraded to edge–cloud collaboration architecture with the advantages of both cloud robotics and edge computing. Industrial Robot-Monitoring Workflow Allocation Problem (IRMWAP), with latency and price as the optimization objectives, is modeled and its solution is proposed as an improved NSGA-II based on the Transcription Gene and the Heuristic Recombination (INSGA2-TGHR) algorithm, which provides a set of computing resource allocation solutions for the monitor so that it can make decisions according to the actual environment.

3. Edge–Cloud Collaborative Architecture

3.1. System Model and Assumptions

In this paper, the architecture for IRMSs is designed as a local edge–cloud three-layer as shown in Figure 1 in which industrial robot-monitoring workflows can run, and the tasks in the workflow can be allocated computing resources in all three layers. The monitor in this system can connect and monitor multiple industrial robots, the monitor itself also has few computing resources, multiple monitors form a local monitoring network, the monitor cloud share its computing resources in the local monitoring network, the network shares the occupation of local computing resources, the monitor have smaller computing capacity, but its computing price can be disregarded. The monitor connects to the edge server through a fabric network. The edge server, which is typically a metropolitan-area-level computing service provider, has medium computing capacity and is more expensive, but has low latency for data transmission. The monitor connects to the cloud center. The cloud center, which is a wide-area-level computing service provider, has higher computing capacity and lower cost than the edge server, but higher latency for data transmission. The edge servers are also connected to the cloud center. The latency of data transfer within the local, edge, and cloud centers is low.

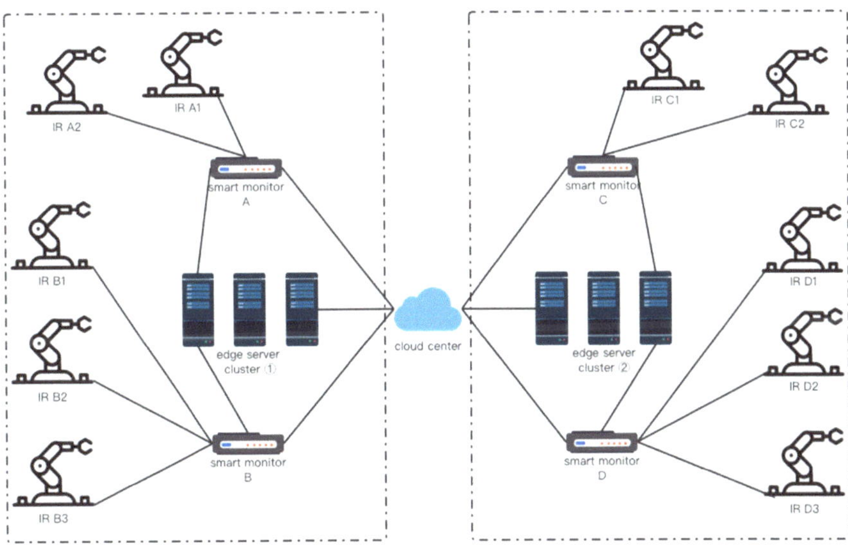

Figure 1. IRMS Edge–Cloud Architecture.

In the operation of the monitoring system, when the monitor generates a monitoring workflow that exceeds its own computing capacity, it will allocate the resources for the generated monitoring workflow according to the current computing resources of the local edge–cloud network environment. Subtasks in the workflow will then run in the allocated computing resources, and each subtask will transfer the completed processed data to the next subtask, and when the last subtask is completed, it will return the computation results to the monitor. Real-time tasks in industrial environments are executing actions or sensing data [24]. Here all monitoring workflow tasks are assumed to be non-real-time or soft real-time monitoring tasks which performed based on standard communication protocols (TCP or UDP/IP) such as switching data, preprocessing data, compression, extracting features. The engineers program and compile the subtask logic code in advance according to the monitoring task characteristics, and program the work order and data dependencies between the subtasks, and they run in the memory of the computing resources as docker processes. The monitoring system does not involve and interfere with the industrial robot's own real-time control system.

3.2. Application Motivating Example: Comprehensive Assessment Workflow for Industrial Robot Monitoring

The Figure 2 shows a comprehensive assessment workflow of industrial robot monitoring. The first step is to acquire data, where the industrial robot transmits internal and external sensor data to the monitor through cables or optical fibers; the next step is to extract characteristics such as time domain, frequency domain, and time-frequency domain from the sensor data respectively; the next step is to normalize the characteristic data; the next step is to perform condition monitoring and life estimation of the industrial robot respectively; and the last step is to make a comprehensive assessment for industrial robots. The subtasks in the workflow have different data and computational characteristics, for example, the extraction of features in the time domain, frequency domain, and time-frequency domain requires a large amount of data input and simple computation to complete, so such tasks are more suitable for deployment in local or edge servers; while tasks such as industrial robot condition monitoring and life estimation require complex computational models, such as dynamic prediction Neural Networks [11] or Deep Learning Algorithm, to transform the processed characteristics into corresponding metrics, which are more suitable for running in cloud centers or high-performance edge servers. Therefore, industrial robot-monitoring workflows need intelligent algorithms to allocate appropriate computing resources to different types of tasks.

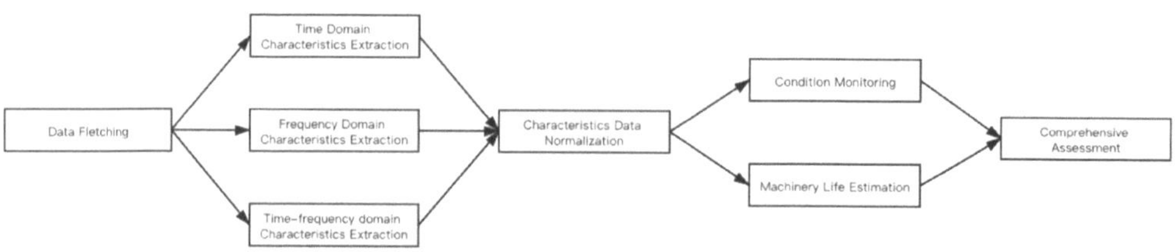

Figure 2. Motivating example: comprehensive assessment workflow.

4. Bi-Objective Optimization Allocation Problem Model and Algorithm

The notations and definitions used in the model for the proposed problem are listed in Table 2.

Table 2. Notations.

Symbol	Definition and Description
t_i	Task i
e_{ij}	The before-and-after relationship between tasks i to j
v_x	Compute node x
f_{xy}	Compute the network link between node x to y
d_{xi}	t_i running on v_x
$instruction_i$	The execution instructions of t_i

Table 2. *Cont.*

Symbol	Definition and Description
$memory_i$	The execution memory space of t_i
$pattern_{ij}$	The workflow pattern of e_{ij}, including Sequence, Parallel Split, Synchronization, Exclusive Choice, Simple Merge
$data_{ij}$	The transfer data of e_{ij}
$comSpeed_x$	The computing speed of v_x, cloud center > edge server > monitor
$latency_x$	The latency between monitor and v_x
$comCost_x$	The computing cost of v_x, edge server > cloud center > monitor = 0
$comCap_x$	The computing capacity of v_x, cloud center > edge server > monitor
$bandwidth_x^{up}$	The upload bandwidth of v_x, cloud center > edge server > monitor
$bandwidth_x^{down}$	The download bandwidth of v_x, cloud center > edge server > monitor
l_{xy}	The communication latency of f_{xy}, monitor to cloud center > monitor to edge server
$commCost_{xy}$	The communication cost of f_{xy}, monitor to cloud center > monitor to edge server
$commCap_{xy}$	The communication capacity of f_{xy}, cloud center to edge server > monitor to cloud center > monitor to edge server

4.1. Industrial Robot-Monitoring Workflow Assignment Problem (IRMWAP) Formulation

Robot-monitoring workflow (RMW), define $G_R = (T, E)$ to denote the graph structure of the RMW, the vertex set T denotes the set of tasks in the workflow, and the edge set E denotes the set of before-and-after relationships between tasks. For vertex $t_i \in T$ can be represented by the triple (id_i, $instruction_i$, $memory_i$), and for edge $e_{ij} \in E$ can be represented by the triple (id_{ij}, $pattern_{ij}$, $data_{ij}$). RMW is the abstract data structure of workflow of Figure 2.

Cloud edge network (CEN), define $G_N = (V, F)$ to represent the graph structure of the CEN, the vertex set V denotes the set of nodes with computing power in the network, and the edge set F denotes the set of network links between nodes. For the vertex $v_x \in V$ can be represented by the seven-tuple (id_x, $latency_x$, $comSpeed_x$, $comCost_x$, $comCap_x$, $bandwidth_x^{up}$, $bandwidth_x^{down}$), and for the edge $f_{xy} \in F$ can be represented by the four-tuple (id_{xy}, $commCap_{xy}$, $commCost_{xy}$, l_{xy}). CEN is the abstract data structure of architecture of Figure 1.

The total optimization objective of latency includes communication latency f^t_{xiyj} and computation latency c^t_{xi}. Communication latency includes network distance delay and bandwidth transmission delay. Computation latency is the time required to execute computation instructions, and computation resources need to communicate with the monitor when the task is started or finished.

$$
\begin{aligned}
T &= \sum f^t_{xiyj} * d_{xi} * d_{yj} + \sum c^t_{xi} * d_{xi} \\
f^t_{xiyj} &= l_{xy} + data_{ij} * \left(\frac{1}{bandwidth_x^{up}} + \frac{1}{bandwidth_y^{updown}} \right) \\
c^t_{xi} &= \begin{cases} latency_x & \text{task i is the start or end,} \\ \frac{instruction_i}{comSpeed_x} & \text{others.} \end{cases}
\end{aligned} \quad (1)
$$

The total optimization objective of cost includes communication cost f^c_{xiyj} and computation cost c^c_{xi}, which depends on the pricing strategy of the service provider and the usage of users.

$$\begin{gathered} C = \sum f^c_{xiyj} * d_{xi} * d_{yj} + \sum c^c_{xi} * d_{xi} \\ f^c_{xiyj} = data_{ij} * commCost_{xy} \\ c^c_{xi} = \frac{instruction_i}{comSpeed_x} * comCost_x \end{gathered} \tag{2}$$

Each task can be allocated only one computing resource and each computing resource can perform multiple tasks in FIFO mode without exceeding its computing capacity. The data dependency between each pair of tasks cannot exceed the limit of communication capacity.

$$d_{xi} = \begin{cases} 1 & t_i \text{ running in } v_x, \\ 0 & \text{others.} \end{cases} \tag{3}$$

$$\sum_{x=1}^{n} d_{xi} = 1 \tag{4}$$

$$\sum d_{xi} * instruction_i \le comCap_x \tag{5}$$

$$\sum d_{xi} * d_{yj} * data_{ij} \le commCap_{xy} \tag{6}$$

The above can be summarized as a Bi-objectives Generalized Quadratic Assignment Problem IRMWAP:

$$\begin{gathered} \min: \ T, C \\ \text{subject to}: (3), (4), (5), (6) \end{gathered} \tag{7}$$

Theorem 1. *The IRMWAP is an NP-hard problem.*

Proof of Theorem 1. From the above equation, it is easy to find that IRMWAP optimizes each objective in accordance with the definition of generalized quadratic assignment problem (GQAP), and GQAP has been proved to be an NP-hard problem [25], then IRMWAP is proved as an NP-hard problem. □

4.2. Improved NSGA2 Based on Transcription Gene and Heuristic Recombination (INSGA2-TGHR)

For how to solve multi-objective optimization problems, there are usually two ideas, one is to use mathematical planning methods to find the exact solution, and the other is to use intelligent computational methods to find the approximate solution. Since this problem is an extension of GQAP and has been proven to be an NP-hard problem, it will become unsolvable when the task and resource size increases, and also commonly used mathematical planning solvers such as gurobi [26] only support mixed integer linear programming and cannot solve quadratic planning problems. On the other hand, intelligent computing is commonly used in non-dominated sorting genetic algorithms (NSGA-II) [12], strength Pareto evolutionary algorithm II (SPEA2) [27], Pareto archival evolutionary strategy (PAES) [28] and multi-objective particle swarm optimization (MOPSO) [29], among which NSGA-II and its improvements performs better in finding a diverse set of solutions and in converging to near the true Pareto-optimal set compared with others [16]. For example, Ghasemi-Falavarjani et al. [17] used the benchmark NSGA-II algorithm to solve the time and energy bi-objective optimization problem in mobile cloud computing, M.A.B. Al-Tarawneh [18] took the application criticality and security as optimization objectives and modeled them as a bi-backpack problem using benchmark NSGA-II to solve, and Mahbuba Afrin [16] proposed Augmented NSGA-II to achieve good results in a multi-objective optimization problem for smart factory workflow resource allocation.

Therefore, for the IRMWAP problem, this work proposes the INSGA2-TGHR algorithm. This algorithm is a heuristic genetic algorithm that is redesigned based on NSGA-II, combining the previous ideas for solving GQAP and the workflow assignment problem. Figure 3 shows that when the monitor generates a RMW that needs to be offloaded, it will obtain the CEN resources information to run the INSGA2-TGHR and allocate computing resources to the tasks. As shown in Figure 4, Transcription Gene is used to create the initial population and Heuristic Recombination is used as a crossover operator to generate new offspring. The idea and detail of proposed algorithm will be discussed as follows:

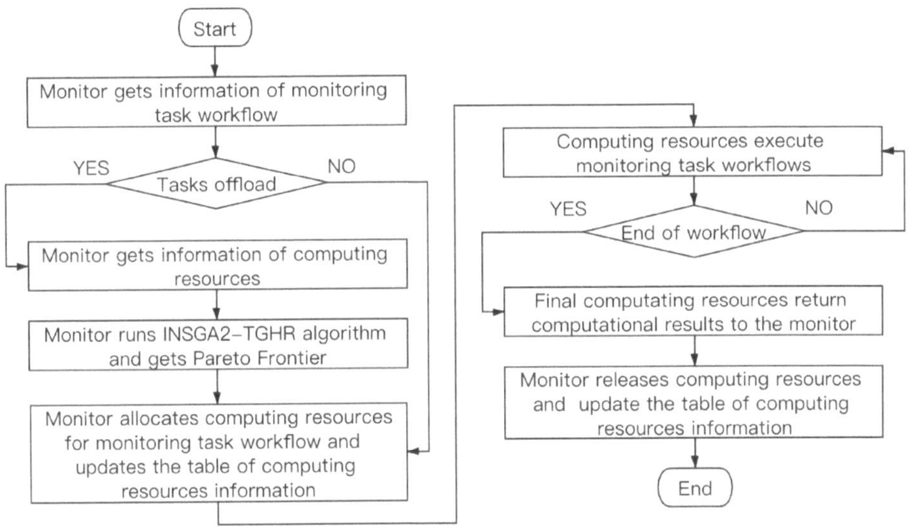

Figure 3. Trigger mechanism flow chart for INSGA2-TGHR.

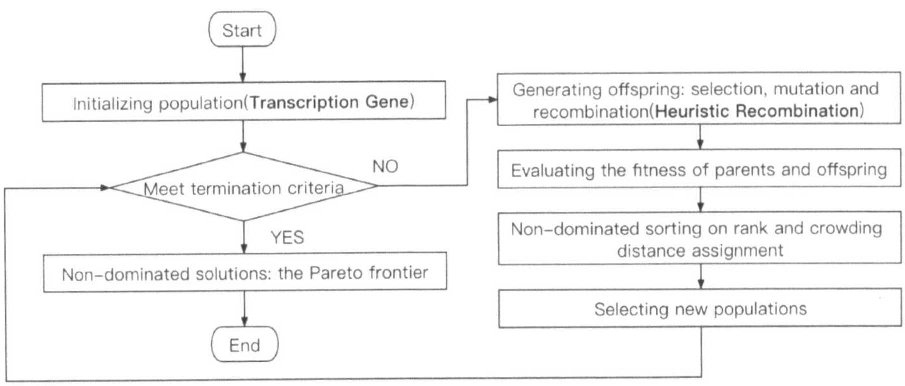

Figure 4. Flow chart of INSGA2-TGHR.

4.2.1. Genetic Factors: Transcription Gene

In traditional genetic algorithms solving workflow assignment problems, all genetic operators are designed as integer arrays of task number length, where each gene made index symbolizes a specific task and gene refers to a specific resource, and matching tasks and resources by random assignment. This has the advantage of simplifying the relationship between tasks, resources, genes and fitness, but the problem is that the number of heterogeneous resources (monitors, edges, clouds) in CEN may not be equal, and then the probability of assigning tasks to different kinds of resources is also not equal. Therefore, the genetic factors are redesigned, and the concept of DNA transcription to generate RNA in biology is used here. "DNA", a set of random real numbers of task length from 0 to N, where N is the type of heterogeneous resources, will be generated first. Then, the "DNA" produces "RNA" by "transcription". For example, a certain edge–cloud network (CEN) is composed of (6, 4, 2) heterogeneous resources, which has 6 local monitors, 4 edge servers and 2 cloud servers and numbered sequentially from #1 to #12. A random real number "0.375402761769494", its integer part "0", corresponds to the 1st group (monitor), its fractional part "0.375402761769494" multiplied by the number of resources of the 1st group "6" and then rounded to "2", corresponds to the third one, then its determined computing resources correspond to the third local monitors, #3. Similarly, a random real number "2.501432671965211" determined computing resource corresponds to the second cloud servers, #12, a random real number "1.5705781532276744" determined computing resource corresponds to the third edge servers, #9. the "RNA" is similar to traditional chromosome structure and can participate in evolution to produce offspring, as well as translate the allocation of resources and calculate fitness based on the encoding of genetic information. The "transcription" process divides the real numbers from 0 to N into an integer part and a decimal part, where the integer part determines the type of heterogeneous resources, and the decimal part determines the number of heterogeneous resources. The specific process is represented by Algorithm 1 and Figure 5, where "DNA" is a pre-obtained array, "Resource" is a dictionary nested array, the "Key" of the dictionary is the resource type and "Value" of the dictionary is an array of resource numbers of that type. In addition, the design of the genetic factors implicitly satisfies (3) (4), each subtask is assigned to only one computing resource.

Algorithm 1: Transcription

input : DNA, Resource
output: RNA
init RNA;
$i \leftarrow 0$;
while $i \leq len(\text{DNA})$ **do**
  $ResourceType \leftarrow int(\text{DNA}[i])$;
  $ResourceIndex \leftarrow int((\text{DNA}[i] - ResourceType) \times len(\text{Resource}[ResourceType]))$;
  $\text{RNA} \leftarrow \text{Resource}[ResourceType][ResourceIndex]$;
  $i \leftarrow i + 1$;
end

4.2.2. Mutation Operator

The mutation operator selects the swap mutation operator, and during mutation the mutation operator selects genes from inside the chromosome for mutation. The resource number with more occurrences inside the chromosome may correspond to the most suitable resource in the current environment, so the swap mutation has a higher probability of mutating a gene into these resources, thus increasing the concentration of resource usage and reducing the migration between subtasks. In addition, if the mutated new chromosome exceeds a certain resource limit, it will be eliminated later.

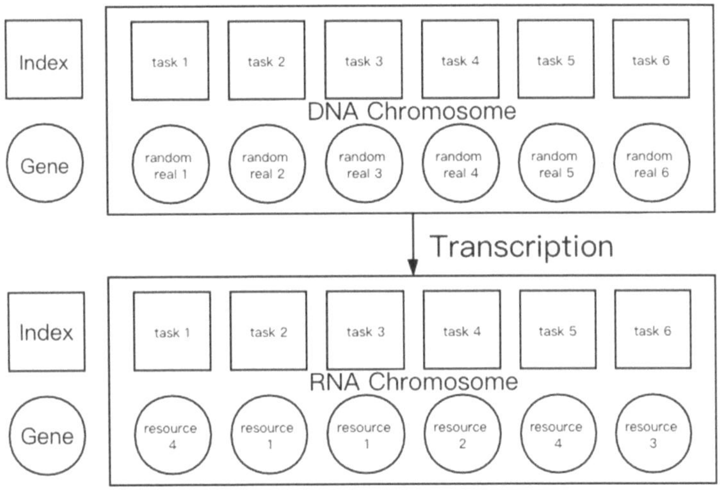

Figure 5. DNA Transcription RNA.

4.2.3. Crossover Operator: Heuristic Recombination

Two-point crossover refers to the random setting of two crossover points in the chromosomes of an individual followed by partial gene exchange. The specific procedure of the two-point crossover is: setting two random crossover points in the coding strings of two individuals paired with each other; exchanging the part of chromosomes of two individuals between the set two crossover points.

The recombination operator is a two-point crossover recombination operator based on the graph partitioning strategy, using the graph partitioning algorithm METIS [30] to group the workflows based on the amount of computation of subtasks in the workflow and the amount of data passed between subtasks. Each group consists of several subtasks, and the gene fragments corresponding to subtasks within the grouping are involved in the crossover as a whole when the recombination variation is performed. When subtasks are assigned to the same resource within a group, the data transfer between several subtasks takes place only in memory and does not involve network communication between computing resources, which effectively reduces the latency and cost of data transfer and thus exhibits a high degree of fitness, and this situation can be called "advantageous gene fragment". For example, task A of data preprocessing is input-heavy and output-heavy, while task B of data feature extracting is input-heavy and output-light type, if these two tasks are executed sequentially on the same computing unit. The process of A's output-heavy and B's input-heavy only needs to be executed in the memory of that computing unit, without network transfer. This significantly reduces the transfer latency and bandwidth costs. The process of Heuristic Recombination is to use the graph partitioning algorithm to partition A and B tasks into the same group, forming an "advantageous gene fragment" as a whole for crossover recombination.

Then the "advantaged gene fragment" may be recombined to the new chromosome during the crossover process. If the entire new chromosome has a higher fitness, it will be reserved during the selection process. The "ordinary gene fragments" that are not assigned to the same resource may be replaced by "dominant gene fragments" in the process of crossover recombination, or may mutate in the process of mutation, and finally the algorithm will decide whether to reserve them for the next round of evolution according to the fitness.

The process is represented in Algorithm 2 and Figure 6, where the METIS [30] algorithm is used for graph partitioning, first, two crossover sites "x1" and "x2" are obtained, and the indexes of other genes consistent with the grouping within the sites are obtained according to the graph partitioning results. The above results were represented by the 0–1 array "col". The recombined newRNA1 uses the gene of RNA1 on the index with "col" of 1 and the gene of RNA2 on the index with "col" of 0. newRNA2 is exactly the opposite.

Algorithm 2: HeuristicDoublePointRecombination

```
input  : RNA1, RNA2, G_R
output: newRNA1, newRNA2
init newRNA1, newRNA2, col;
partitionResult ← METISgraphpartition(G_R);
x1 ← random(0, len(RNA1)), x2 ← random(0, len(RNA1));
if x1 ≥ x2 then
    swap(x1, x2);
end
s ← set(partitionResult[x1:x2]);
i ← 0;
while i ≤ len(partitionResult) do
    if partitionResult[i] in s then
        col[i] ← 1;
    end
    else
        col[i] ← 0;
    end
end
j ← 0;
while j ≤ len(col) do
    if col[j] ≡ 1 then
        newRNA1[j] ← RNA1[j];
        newRNA2[j] ← RNA2[j];
    end
    else
        newRNA1[j] ← RNA2[j];
        newRNA2[j] ← RNA1[j];
    end
end
```

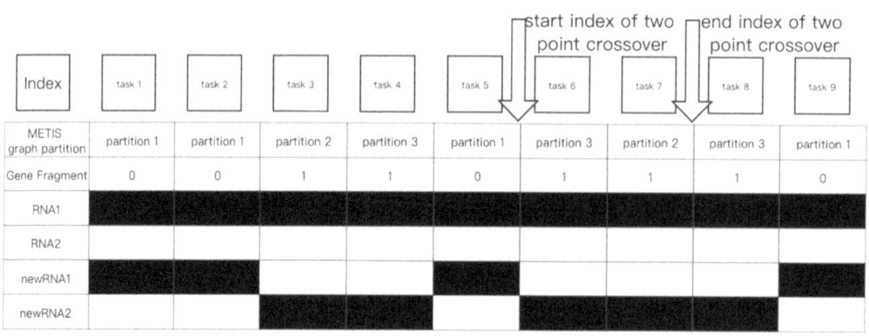

Figure 6. Process of Heuristic Recombination.

4.2.4. Fitness Calculation

For the bi-objective optimization problem, the fitness functions for both optimization objectives latency and cost are defined and obey the conditions of (1) and (2), respectively. In addition, the genetic factors (5) and (6) that do not satisfy the constraints will be directly eliminated.

5. Experiments

In this section, experiments will analyze and compare the performance of the proposed bi-objective resource allocation algorithm by evolving generations, different types and numbers of tasks, different types and numbers of computing resources and the numbers of task instructions. The workflow in Table 3 is similar in structure to the motivational example in Figure 2, the quantification of its parameter instruction, memory, and data dependency are driven by real-world parameters (e.g., the runtime memory of the docker instance) and related references [15,18,31].

Table 3. Experimental Parameters.

Params Type		Params Value
workflow	Number of subtasks	30–80
	instruction	10–800 MI
	memory	100–500 MB
	data dependency	10–100 kB
Local Monitor	CPU	0.8 Ghz*1
	memory	1 GB
	bandwidth	400 Mbps
Edge Server [1]	CPU	3.0 Ghz*4
	memory	8 GB
	bandwidth	2 Gbps
	cost	2 RMB/h
Cloud Server [1]	CPU	3.0 Ghz*8
	memory	16 GB
	bandwidth	4 Gbps
	cost	2 RMB/h
Local to Edge	cost	0.25 RMB/GB
	Latency	10 ms
Edge to Cloud	cost	0.75 RMB/GB
	Latency	30 ms
Local to Cloud	cost	1 RMB/GB
	Latency	40 ms

[1] Price reference is Huawei Cloud ECS and Huawei Cloud IEC [32].

5.1. Simulation Environment

The experimental environment is based on python and geatpy [33], and the reference objectives are selected benchmark NSGA-II, INSGA2-TG(Transcription Gene) algorithm, INSGA2-HR(Heuristic Recombination) algorithm and extremums. Table 4 shows the Ablation of INSGA2-TGHR, where benchmark NSGA-II algorithm has a random initial population, using simulated binary crossover operator and polynomial mutation operator. INSGA2-TG has TG initial population, using simulated binary crossover operator and polynomial mutation operator, INSGA2-HR has random initial population, using swap mutation operator and HR crossover operator, INSGA2-TGHR has TG initial population, using swap mutation operator and HR crossover operator, and extremum of IRMWAP problem in both latency and cost directions are solved respectively by the gurobi [26].

Table 4. Ablation of INSGA2-TGHR.

Algorithm	Initial Population	Mutation and Crossover Operator
Benchmark NSGA-II	random	simulated binary and polynomial mutation
INSGA2-TG	TG	simulated binary and polynomial mutation
INSGA2-HR	random	Swap and HR
INSGA2-TGHR	TG	Swap and HR

Figure 7 compares the Pareto frontier of the four algorithms in the (6, 2, 1) resource setting, which has 6 local monitors, 2 edge servers and 1 cloud server. TGHR algorithm can obtain the solution closest to the extreme value and closer to the true Pareto frontier, and the convergence of its solution set is better than the other three algorithms. The TG algorithm is less exploratory than TGHR and HR, although it has a dominant initial population The HR algorithm contains more monitor genes due to its initial gene population, so that its evolutionary direction will be more likely to favor the low-cost side. Benchmark NSGA-II algorithm does not reach convergence at 200 generations. It shows that the TG and HR algorithms have the advantage of dominant initial populations and evolutionary exploration, respectively, to find non-dominated solutions quickly and close to the Pareto frontier. In addition, the TGHR algorithm has the advantages of both TG and HR, and its solution set is balanced and closer to the real Pareto frontier.

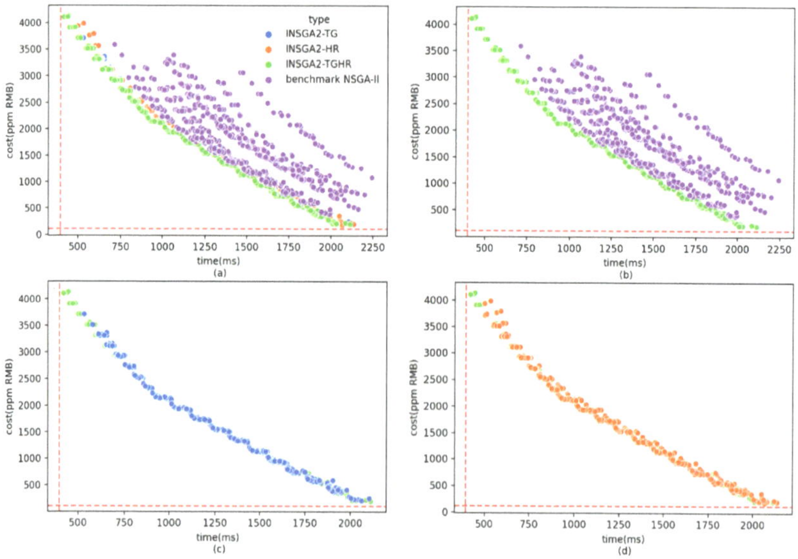

Figure 7. Comparison of Pareto frontier. (**a**) Comparison of four algorithms; (**b**) Comparison of TGHR and benchmark NSGA-II; (**c**) Comparison of TGHR and TG; (**d**) Comparison of a and TGHR and HR.

Impact of Evolution Generation

Figure 8 shows the impact of comparing the number of evolutionary generations of the four algorithms in the (6, 2, 1) resource setting.

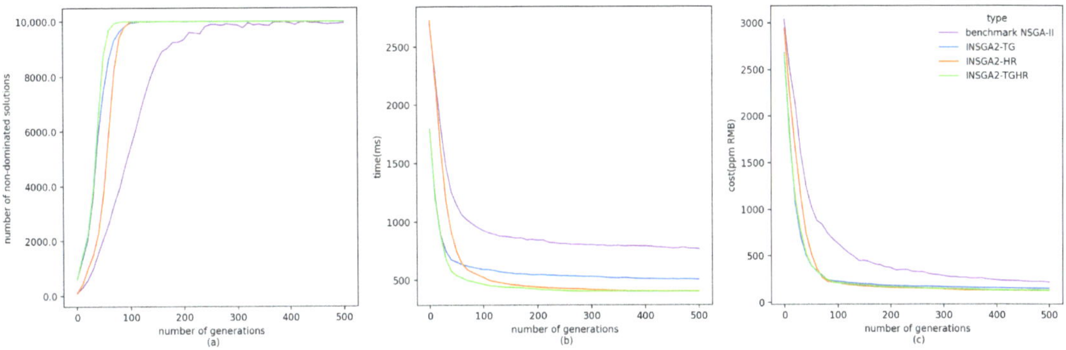

Figure 8. Comparison of evolution generations. (**a**) The relationship between evolutionary generations and non-dominant solutions; (**b**) The relationship between evolutionary generations and low-time Extremum; (**c**) The relationship between evolutionary generations and low-cost Extremum.

The Figure 8a shows the number of non-dominated solutions, as shown in the figure, the TGHR algorithm works best and is the first to obtain close to 100 non-dominated solutions. The TG algorithm is second only to the TGHR algorithm, which is because both transcription gene can obtain more balanced random initial populations. The HR algorithm starts with a lower number of non-dominated solutions, but because the recombination operator of the HR algorithm is based on the variation of the dominant gene fragment, the non-dominated solutions are quickly viewed and retained. Benchmark NSGA-II algorithm is a purely random search, and the number of non-dominated solutions only approaches 100 when the number of iterations reaches 300, and there is some fluctuation after that.

The Figure 8b shows the average of the top 100 low-time solutions, and it can be seen that the TGHR algorithm works best, the HR algorithm is slower than TGHR, but it can still obtain high-quality low-time solutions by 200 generations by relying on the advantage of the HR. TG and Benchmark NSGA-II algorithms do not obtain solutions close to the extremes.

The Figure 8c is comparing the average of the top 100 low-cost solutions. TGHR and HR have relative performance, and HR is slightly better than TGHR because HR has a higher proportion of local genes, so it can find populations containing more local genes and achieve the effect of lower cost. TG algorithm finishes the population through iteration with the advantage of TG in the initial population is not superior. Benchmark NSGA-II algorithm starts to approach the extreme value at around 500 generations.

The TGHR algorithm combines the advantages of both TG and HR: the TG algorithm can obtain a dominant initial population, while the HR algorithm can quickly find and retain a dominant population. It can quickly obtain non-dominance by population advantage in the early evolutionary stage and can approach the extremum in about 100 generations by the recombination operator.

5.2. Impact of the Number of Resources

5.2.1. Impact of Local Resources

Figure 9a–d are comparing different resources in (4, 2, 1), (6, 2, 1), (8, 2, 1), (10, 2, 1) settings, respectively. As shown in the figure, the TGHR algorithm can obtain a more convergent Pareto front with a more balanced distribution under different resource conditions, while the HR algorithm has a solution set that is closer to the extreme value of low cost because its initial population contains more local monitor genes. TG algorithm has weaker convergence than TGHR algorithm and HR algorithm.

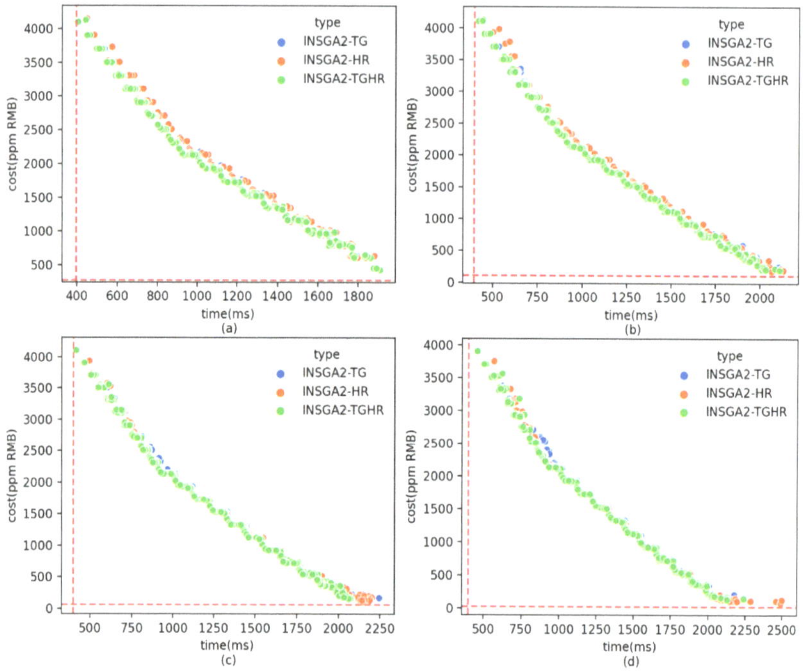

Figure 9. The Pareto frontier in different local resources setting. (**a**), (**b**), (**c**), (**d**) are comparing different resources in (4, 2, 1), (6, 2, 1), (8, 2, 1), (10, 2, 1) settings, respectively.

The box plot Figure 10 shows that as the number of local monitors increases, the solution set provided by the algorithm allocates more resources to the local monitors, thus leading to a slow decrease in cost and a slow increase in time as the number of local monitors increases. At the same time, it can be seen that the TGHR algorithm always finds the solution with the lowest time, and the HR algorithm tends to allocate more resources to the local monitor as the number of local monitors increases due to the characteristics of its initial population, which gives it an advantage in the direction of low cost.

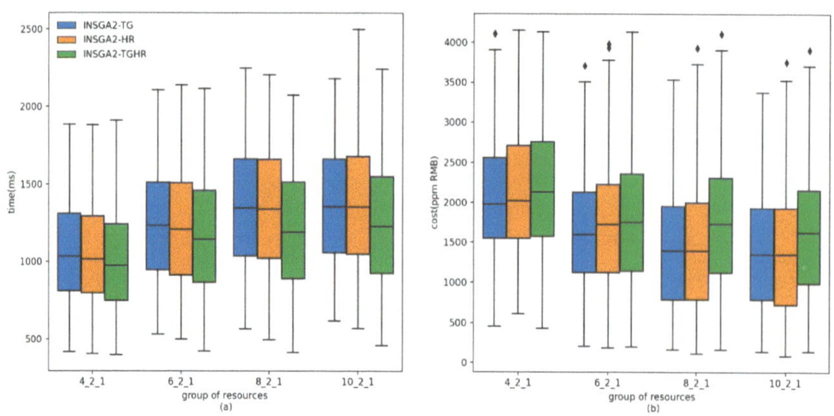

Figure 10. The boxplot in different local resources setting. (**a**) Comparing the effect of local resources setting on time; (**b**) Comparing the effect of local resources setting on cost. The rhombus symbols represent outliers, which are also the same in Figure 12, Figure 14 and Figure 16 in the later text.

5.2.2. Impact of Edge Resources

Figure 11a–d are comparing different resources in (6, 1, 1), (6, 2, 1), (6, 3, 1), (6, 4, 1) settings, respectively. From the figure, it can be seen that when there are fewer edge servers, the TGHR algorithm shows its unique advantage of better convergence of the solution set of the Pareto front to find the closest solution to the extreme value. The TG algorithm and HR algorithm in Figure 11a may fail to complete convergence in 200 generations of evolution due to the inability to find enough non-dominated solutions in the early stage of evolution.

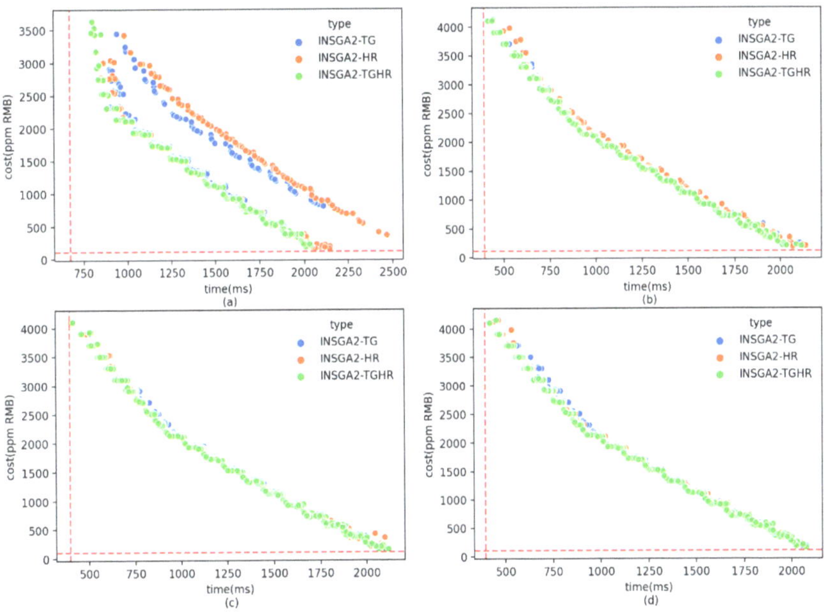

Figure 11. The Pareto frontier in different edge resources setting. (**a**), (**b**), (**c**), (**d**) are comparing different resources in (6, 1, 1), (6, 2, 1), (6, 3, 1), (6, 4, 1) settings, respectively.

The box plot Figure 12 shows that as the TGHR algorithm has stable performance. As the edge server resources increase, the tasks are assigned more on the edge server with higher computational efficiency, so (6, 2, 1) has lower minimum time than (6, 1, 1). However, when the edge resources become abundant and all computational tasks will be offloaded to the edge servers, the minimum time is no longer reduced. As the edge resources increase, the proportion of edge genes in the initial population of the TGHR algorithm increases, and therefore, it no longer has the advantage of low-cost. This can show that the TGHR algorithm is less affected by the change of resource allocation and can perform more stable performance.

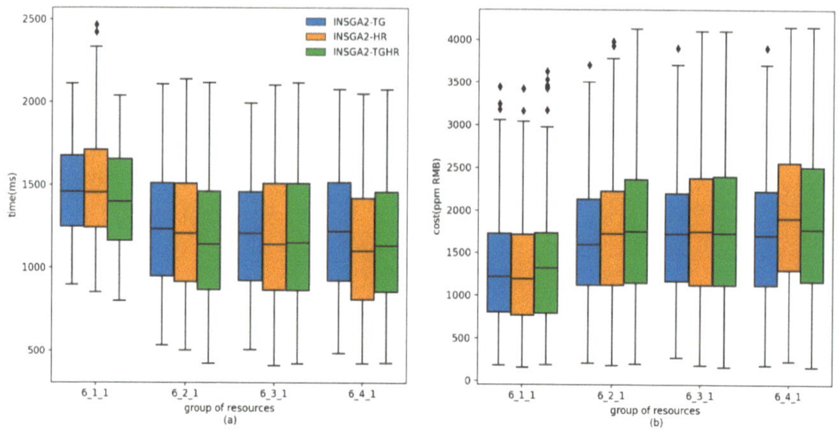

Figure 12. The boxplot in different edge resources setting. (**a**) Comparing the effect of edge resources setting on time; (**b**) Comparing the effect of edge resources setting on cost.

5.3. Impact of the Number of Subtasks

Figure 13 compares the allocation ratio of tasks executed monitor, edge, and cloud for a workflow containing 20 to 80 tasks, and each task occupies 300 M of memory, in (6, 2, 1) resources setting.

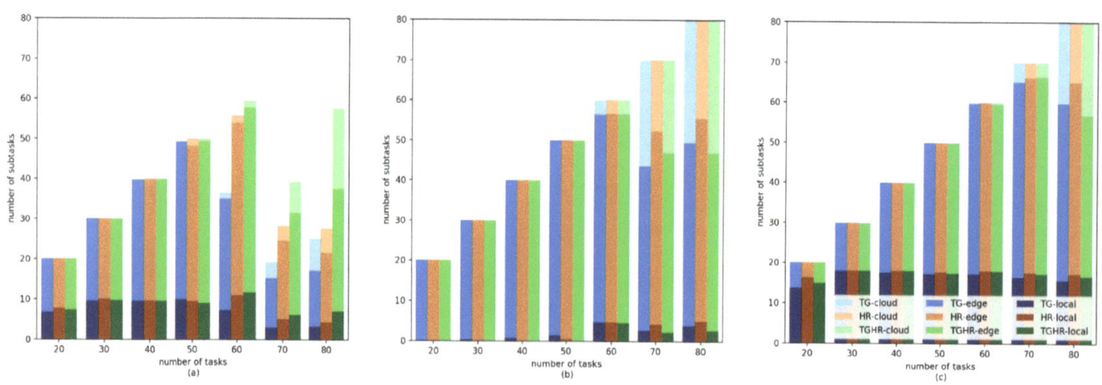

Figure 13. Allocation ratio of resource in different number of tasks. (**a**) Allocation ratio of resource in all Pareto fronts; (**b**) Allocation ratio of resource in near low-time 100 solutions; (**c**) Ratio of resource in near low-cost 100 solutions.

Figure 13a shows the allocation ratio in all Pareto fronts, and the number of local runs in the figure is relatively smooth. As the number of tasks continues to increase, the algorithm gradually assigns tasks to the cloud as the system approaches the edge server operating load. As the number of subtasks increases, when the number of subtasks exceeds 60, the 200 generations evolution starts to unable to meet the demand, and the performance of the TG algorithm starts to degrade and cannot find the non-dominated solution corresponding to the number of populations, and when the number of tasks exceeds 70, all algorithms cannot find the non-dominated solution corresponding to the number of populations. However, it can be seen that the performance of the TGHR algorithm is still the best, and the number of non-dominated solutions found is much more than that of the TG algorithm and the HR algorithm. The figure also shows that the performance of the HR algorithm is slightly better than that of the TG algorithm, which is due to the fact that the HR algorithm speeds up the speed of finding non-dominated solutions.

Figure 13b,c shows the allocation ratio in 100 solutions near low-time and near low-cost in the solution set, respectively. There will always be fewer local machines using low performance in low-time, and as the number of tasks increases, more and more tasks will be allocated in the cloud, gradually reaching a one-to-one ratio of resources on the edge–cloud. As many tasks as possible in low-cost will be assigned first in local monitor, and the maximum capacity of running tasks that local can provide is 18, and the figure has basically approached the maximum load of the local monitor.

From Figure 14, it can be seen that the TGHR algorithm always finds the solution with the lowest time, and the HR algorithm can find the solution with the lowest cost in some cases, which is due to the fact that the initial population of the HR algorithm has more genes representing the assignment to the local, and it can find the non-dominated solution with low cost faster, but its comprehensive performance is not as good as that of the TGHR algorithm.

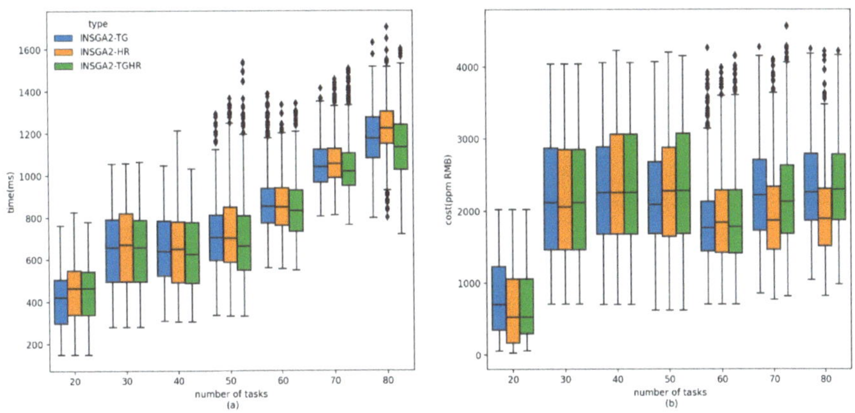

Figure 14. The boxplot in different number of tasks. (**a**) Comparing the effect of number of tasks on time; (**b**) Comparing the effect of number of tasks on cost.

5.4. Impact of the Number of Task Instructions

Figure 15 compares the impact of the number of instructions for different tasks. Where each workflow contains 30 tasks (subtasks), each task occupies 300 M of memory, and the number of instructions to be executed for each task is 50 to 900, in (6, 2, 1) resources setting. It shows that the TGHR algorithm is more sensitive to the number of instructions of subtasks. As the number of instructions of tasks increases, the advantage of cloud side of high-performance computing is emphasized, and the TGHR algorithm reduces the assignment of tasks available at the edge side and assigns more tasks to the cloud side.

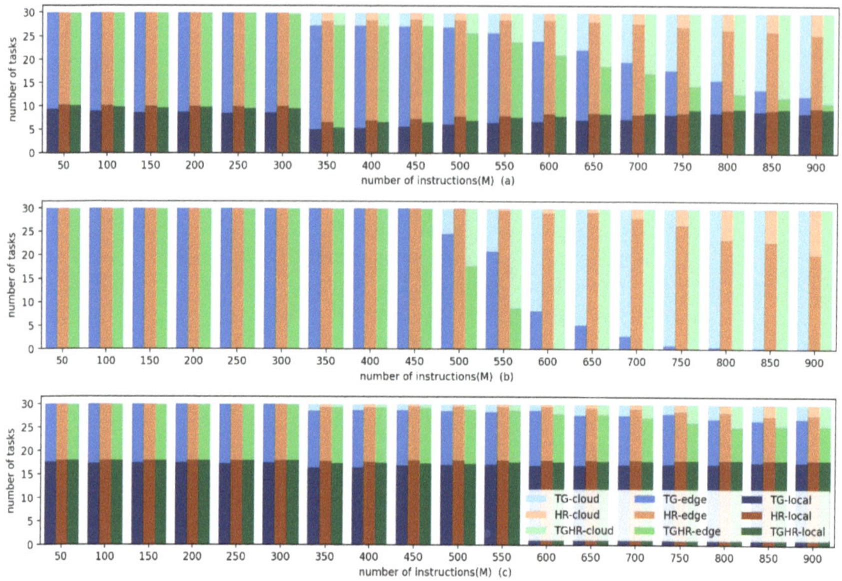

Figure 15. Allocation ratio of resource in different number of task instructions. (**a**) Allocation ratio of resource in all Pareto fronts; (**b**) Ratio of resource in near low-time 100 solutions; (**c**) Allocation ratio of resource in near low-cost 100 solutions.

Figure 15a shows the allocation ratio in all Pareto fronts. It indicates that the share of cloud server increases as the number of instructions increases, due to the fast speed of cloud computing, whose less computation time offsets the impact of bandwidth latency. Figure 15b,c shows the allocation ratio in 100 solutions near low-time and near low-cost in the solution set, respectively. It can be seen that the low-time strategy is more sensitive to time, and low-performance local monitors are not used, while the share of cloud servers with high computational speed increases rapidly. On the contrary, the low-cost strategy is more sensitive to price and basically keeps the local monitors running at full load. The share of cloud servers also increases slowly with the number of task instructions.

As shown in Figure 16, because of the lower number of subtasks, almost all algorithms find low-time and low-cost extremes. HR algorithm has a slightly larger low-time minimum solution than others, which is because its initial population has more local monitor genes and thus fewer edge and cloud genes obtained by evolution. The performance of TGHR algorithm is also more robust as seen in the figure. Starting from instruction number 350, the advantages of cloud computing start to reveal, so the average time of the Pareto frontier for the three algorithms decreases and the average cost increases.

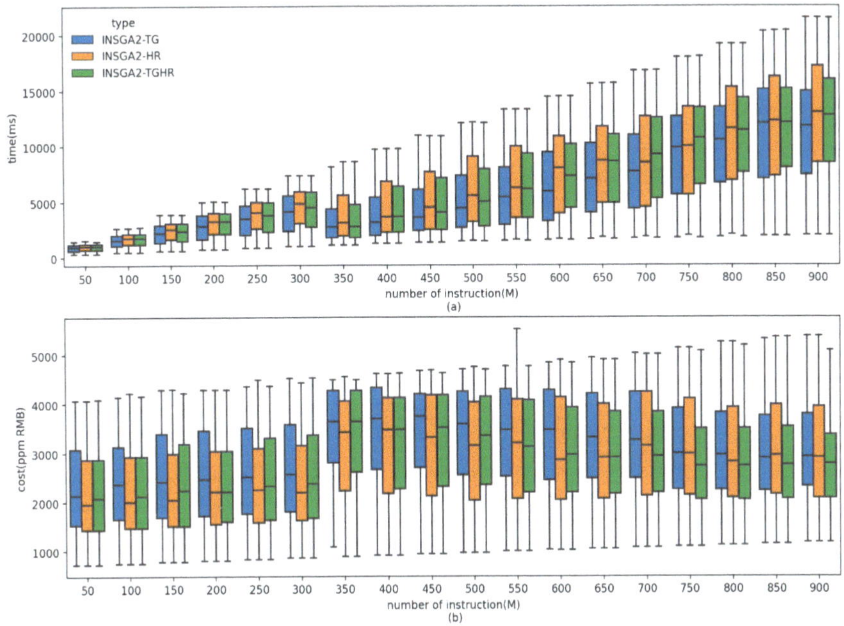

Figure 16. The boxplot in different number of task instructions. (**a**) Comparing the effect of number of task instructions on time; (**b**) Comparing the effect of number of task instructions on cost.

5.5. Results and Discussions

The ablation experiments show that TG can produce a more balanced initial population, HR can accelerate the iteration efficiency. The INSGA2-TGHR algorithm has the advantages of both, and it can find the non-dominated solution set quickly and keep approaching the true Pareto front through evolution. Its performance is stable, and it can find suitable allocation schemes among different types and numbers of computing resources.

6. Conclusions and Future Directions

In this paper, we design a local edge–cloud industrial robot-monitoring system (IRMS) architecture, define the Industrial Robot-Monitoring Workflow Resource Allocation Problem (IRMWAP), and propose an improved NSGA-II algorithm (INSGA2-TGHR) based on the characteristics of IRMSs, using workflows to accomplish industrial robot-monitoring tasks. The experimental results show that the INSGA2-TGHR algorithm has a more balanced initial population and can retain the "dominant gene fragment" in the evolutionary iterations to quickly obtain a non-dominated solution set and a more convergent Pareto frontier through evolutionary iterations, and its Pareto frontier obtains time minima and cost minima that are 4.66% and 15.52% more accurate than benchmark NSGA-II's respectively in 200 generations. The performance of INSGA2-TGHR algorithm is stable on different types and number of computing resources sets, sensitive to the number of instructions of the tasks, and able to offload computationally intensive tasks to more clouds.

In future work, we plan to further validate and extend the applicability of this algorithm in IRMSs. In this paper, we only consider the monitoring of fixed-position industrial robots; but in real industrial factory scenarios, there are also mobile robots involved in activities, and their network environments may include various environments such as Bluetooth, Wi-Fi, and 5G. Therefore, such working conditions as mobility and complex network environment are still the direction of future research. In addition, how the monitor itself can make autonomous decisions and choose the appropriate solution in the Pareto frontier through artificial intelligence algorithms is also a problem to be considered in the future.

Author Contributions: Conceptualization, X.W., Q.H. and X.X.; methodology, Q.H. and X.X.; software, X.W. and X.X.; validation, X.W., Q.H. and X.X.; formal analysis, X.W., Q.H. and X.X.; investigation, X.W.; resources, Q.H.; data curation, X.X.; writing—original draft preparation, X.X. writing—review and editing, X.W. and Q.H.; visualization, X.X.; supervision, Q.H.; project administration, Q.H.; funding acquisition, Q.H. All authors have read and agreed to the published version of the manuscript.

Funding: This work was funded by the National Key R&D Program of China under the grant of 2018YFB1306101.

Institutional Review Board Statement: Not applicable.

Informed Consent Statement: Not applicable.

Acknowledgments: The authors gratefully acknowledge the financial support offered by the National Key R&D Program of China under the grant of 2018YFB1306101.

Conflicts of Interest: The authors declare no conflict of interest.

References

1. Chen, Y.; Dong, F. Robot Machining: Recent Development and Future Research Issues. *Int. J. Adv. Manuf. Technol.* **2013**, *66*, 1489–1497. [CrossRef]
2. Hou, Z.; Chen, J. Research Review of Remote Monitoring and Fault Diagnosis of Industrial Robots. *Mach. Tool Hydraul.* **2018**, *46*, 172–176, 168.
3. Yang, H.; Sun, Z.; Jiang, G.; Zhao, F.; Lu, X.; Mei, X. Cloud-Manufacturing-Based Condition Monitoring Platform With 5G and Standard Information Model. *IEEE Internet Things J.* **2021**, *8*, 6940–6948. [CrossRef]
4. Atmoko, R.A.; Yang, D. Online Monitoring & Controlling Industrial Arm Robot Using MQTT Protocol. In Proceedings of the 2018 IEEE International Conference on Robotics, Biomimetics, and Intelligent Computational Systems (Robionetics), Bandung, Indonesia, 8–10 August 2018; pp. 12–16. [CrossRef]
5. Chen, W.; Yaguchi, Y.; Naruse, K.; Watanobe, Y.; Nakamura, K.; Ogawa, J. A Study of Robotic Cooperation in Cloud Robotics: Architecture and Challenges. *IEEE Access* **2018**, *6*, 36662–36682. [CrossRef]
6. Hu, G.; Tay, W.; Wen, Y. Cloud Robotics: Architecture, Challenges and Applications. *IEEE Network* **2012**, *26*, 21–28. [CrossRef]
7. Kehoe, B.; Patil, S.; Abbeel, P.; Goldberg, K. A Survey of Research on Cloud Robotics and Automation. *IEEE Trans. Automat. Sci. Eng.* **2015**, *12*, 398–409. [CrossRef]
8. Wan, J.; Tang, S.; Yan, H.; Li, D.; Wang, S.; Vasilakos, A.V. Cloud Robotics: Current Status and Open Issues. *IEEE Access* **2016**, *4*, 2797–2807. [CrossRef]
9. O'Donovan, P.; Leahy, K.; Bruton, K.; O'Sullivan, D.T.J. An Industrial Big Data Pipeline for Data-Driven Analytics Maintenance Applications in Large-Scale Smart Manufacturing Facilities. *J. Big Data* **2015**, *2*, 25. [CrossRef]
10. Cloud Edge Computing: Beyond the Data Center. OpenStack Is Open Source Software for Creating Private and Public Clouds. Available online: https://www.openstack.org/use-cases/edge-computing/cloud-edge-computing-beyond-the-data-center/?lang=en_US (accessed on 24 September 2021).
11. Pan, Y.; Er, M.J.; Li, X.; Yu, H.; Gouriveau, R. Machine Health Condition Prediction via Online Dynamic Fuzzy Neural Networks. *Eng. Appl. Artif. Intell.* **2014**, *35*, 105–113. [CrossRef]
12. Deb, K.; Pratap, A.; Agarwal, S.; Meyarivan, T. A Fast and Elitist Multiobjective Genetic Algorithm: NSGA-II. *IEEE Trans. Evol. Computat.* **2002**, *6*, 182–197. [CrossRef]
13. Xuhong, M.; Zhizeng, L. The Design and Implement of Embedded Remote Control System in Industrial Robot. In Proceedings of the 2012 International Conference on Computer Science and Electronics Engineering, Hangzhou, China, 23–25 March 2012; pp. 332–335. [CrossRef]
14. Yin, H.L.; Wang, Y.M.; Xiao, N.F.; Jiang, Y.R. Real-Time Remote Manipulation and Monitoring Architecture of an Industry Robot. In Proceedings of the 2008 IEEE International Joint Conference on Neural Networks (IEEE World Congress on Computational Intelligence), Hong Kong, China, 1–8 June 2008; pp. 1228–1234. [CrossRef]
15. Chen, W.; Yaguchi, Y.; Naruse, K.; Watanobe, Y.; Nakamura, K. QoS-Aware Robotic Streaming Workflow Allocation in Cloud Robotics Systems. *IEEE Trans. Serv. Comput.* **2021**, *14*, 544–558. [CrossRef]
16. Afrin, M.; Jin, J.; Rahman, A.; Tian, Y.-C.; Kulkarni, A. Multi-Objective Resource Allocation for Edge Cloud Based Robotic Workflow in Smart Factory. *Future Gener. Comput. Syst.* **2019**, *97*, 119–130. [CrossRef]
17. Ghasemi-Falavarjani, S.; Nematbakhsh, M.; Shahgholi Ghahfarokhi, B. Context-Aware Multi-Objective Resource Allocation in Mobile Cloud. *Comput. Electr. Eng.* **2015**, *44*, 218–240. [CrossRef]

18. Al-Tarawneh, M.A.B. Bi-Objective Optimization of Application Placement in Fog Computing Environments. *J. Ambient. Intell. Hum. Comput.* **2021**. [CrossRef]
19. Rahman, A.; Jin, J.; Cricenti, A.; Rahman, A.; Panda, M. Motion and Connectivity Aware Offloading in Cloud Robotics via Genetic Algorithm. In Proceedings of the GLOBECOM 2017-2017 IEEE Global Communications Conference, Singapore, 4–8 December 2017; pp. 1–6. [CrossRef]
20. Xie, Y.; Guo, Y.; Mi, Z.; Yang, Y.; Obaidat, M.S. Loosely Coupled Cloud Robotic Framework for QoS-Driven Resource Allocation-Based Web Service Composition. *IEEE Syst. J.* **2020**, *14*, 1245–1256. [CrossRef]
21. Li, S.; Zheng, Z.; Chen, W.; Zheng, Z.; Wang, J. Latency-Aware Task Assignment and Scheduling in Collaborative Cloud Robotic Systems. In Proceedings of the 2018 IEEE 11th International Conference on Cloud Computing (CLOUD), San Francisco, CA, USA, 2–7 July 2018; pp. 65–72. [CrossRef]
22. Hirsch, M.; Mateos, C.; Zunino, A.; Majchrzak, T.A.; Grønli, T.-M.; Kaindl, H. A Task Execution Scheme for Dew Computing with State-of-the-Art Smartphones. *Electronics* **2021**, *10*, 2006. [CrossRef]
23. Sanabria, P.; Tapia, T.F.; Neyem, A.; Benedetto, J.I.; Hirsch, M.; Mateos, C.; Zunino, A. New Heuristics for Scheduling and Distributing Jobs under Hybrid Dew Computing Environments. *Wirel. Commun. Mobile Comput.* **2021**, *2021*, 8899660. [CrossRef]
24. Pallasch, C.; Wein, S.; Hoffmann, N.; Obdenbusch, M.; Buchner, T.; Waltl, J.; Brecher, C. Edge Powered Industrial Control: Concept for Combining Cloud and Automation Technologies. In Proceedings of the 2018 IEEE International Conference on Edge Computing (EDGE), San Francisco, CA, USA, 2–7 July 2018; pp. 130–134. [CrossRef]
25. Hahn, P.M.; Kim, B.-J.; Guignard, M.; Smith, J.M.; Zhu, Y.-R. An Algorithm for the Generalized Quadratic Assignment Problem. *Comput. Optim. Appl.* **2008**, *40*, 351–372. [CrossRef]
26. Gurobi. The Fastest Solver—Gurobi. Available online: https://www.gurobi.com/ (accessed on 24 September 2021).
27. Zitzler, E.; Laumanns, M.; Thiele, L. SPEA2: Improving the Strength Pareto Evolutionary Algorithm. *TIK-Rep.* **2001**. [CrossRef]
28. Knowles, J.; Corne, D. The Pareto Archived Evolution Strategy: A New Baseline Algorithm for Pareto Multiobjective Optimisation. In Proceedings of the Proceedings of the 1999 Congress on Evolutionary Computation-CEC99 (Cat. No. 99TH8406), Washington, DC, USA, 6–9 July 1999; pp. 98–105. [CrossRef]
29. Coello Coello, C.A.; Lechuga, M.S. MOPSO: A Proposal for Multiple Objective Particle Swarm Optimization. In Proceedings of the Proceedings of the 2002 Congress on Evolutionary Computation, CEC'02 (Cat. No.02TH8600), Honolulu, HI, USA, 12–17 May 2002; Volume 2, pp. 1051–1056. [CrossRef]
30. Karypis, G.; Kumar, V. A Fast and High Quality Multilevel Scheme for Partitioning Irregular Graphs. *SIAM J. Sci. Comput.* **1998**, *20*, 359–392. [CrossRef]
31. He, S.; Lyu, X.; Ni, W.; Tian, H.; Liu, R.P.; Hossain, E. Virtual Service Placement for Edge Computing Under Finite Memory and Bandwidth. *IEEE Trans. Commun.* **2020**, *68*, 7702–7718. [CrossRef]
32. Price Calculator. HUAWEI CLOUD. Available online: https://www.huaweicloud.com/pricing.html (accessed on 24 September 2021).
33. Geatpy: The Genetic and Evolutionary Algorithm Toolbox with High Performance in Python. Available online: http://www.geatpy.com/ (accessed on 24 September 2021).

applied sciences

MDPI

Article

An Improved Similarity Trajectory Method Based on Monitoring Data under Multiple Operating Conditions

Jiancheng Yin, Yuqing Li, Rixin Wang and Minqiang Xu *

Deep Space Exploration Research Center, Harbin Institute of Technology, Harbin 150001, China; wdyydy@163.com (J.Y.); bradley@hit.edu.cn (Y.L.); wangrx@hit.edu.cn (R.W.)
* Correspondence: xumq@hit.edu.cn

Abstract: With the complexity of the task requirement, multiple operating conditions have gradually become the common scenario for equipment. However, the degradation trend of monitoring data cannot be accurately extracted in life prediction under multiple operating conditions, which is because some monitoring data is affected by the operating conditions. Aiming at this problem, this paper proposes an improved similarity trajectory method that can directly use the monitoring data under multiple operating conditions for life prediction. The morphological pattern and symbolic aggregate approximation-based similarity measurement method (MP-SAX) is first used to measure the similarity between the monitoring data under multiple operating conditions. Then, the similar life candidate set, and corresponding weight are obtained according to the MP-SAX. Finally, the life prediction results of equipment under multiple operating conditions can be calculated by aggregating the similar life candidate set. The proposed method is validated by the public datasets from NASA Ames Prognostics Data Repository. The results show that the proposed method can directly and effectively use the original monitoring data for life prediction without extracting the degradation trend of the monitoring data.

Keywords: similarity trajectory; multiple operating conditions; life prediction; monitoring data; morphological pattern; symbolic aggregate approximation

Citation: Yin, J.; Li, Y.; Wang, R.; Xu, M. An Improved Similarity Trajectory Method Based on Monitoring Data under Multiple Operating Conditions. *Appl. Sci.* **2021**, *11*, 10968. https://doi.org/10.3390/app112210968

Academic Editor: Jasiulewicz-Kaczmarek Małgorzata

Received: 17 October 2021
Accepted: 17 November 2021
Published: 19 November 2021

1. Introduction

As an important task of condition-based maintenance (CBM), health prognostics has become more and more popular which can predict the remaining useful life (RUL) of equipment according to historical data or observed degradation trend, which can make effective maintenance plan for equipment to improve the reliability of equipment and reduce the loss and safety problems caused by equipment failure [1,2].

With the rapid development of machine learning and deep learning technology, artificial intelligence (AI) based on machine learning and deep learning were gradually applied to fault diagnosis and life prediction [3–8]. Such as She et al. [9] proposed a bidirectional gated recurrent unit prediction method based on bootstrap to solve the problem that the uncertainty of the prediction. Wang et al. [10] proposed a new framework named recurrent convolutional neural network to address the limitations of the convolutional neural network, which different degradation states did not consider the temporal dependencies and the prediction results were uncertain. A novel neural network called quantum recurrent encoder-decoder neural network was proposed by Chen et al. [11] to improve the prediction accuracy in the degradation trend prediction of rotating machinery. Figueroa et al. [12] proposed a framework for feature selection embedded in deep neural networks (DNN) for PHM to addresses the accuracy interpretability tradeoff. For turbofan engines, Muneer et al. [13] proposed four data-driven prognostic models based on deep neural networks, and analyzed the influence of its network structure on generalization abilities. Furthermore, Muneer et al. [14] proposed a new attention-based deep convolutional neural network incorporating the time window to predict the RUL of turbofan engines.

However, although the excellent prediction results of equipment can be obtained by the life prediction method based on machine learning and deep learning, the accuracy of prediction results mostly depends on the design of network structure and the selection of parameters. At this time, the similarity trajectory method (STM) based on the case-based reasoning (CBR) which can be regarded as a form of intradomain analogy, and a branch of machine learning shown obvious advantages. CBR can solve the problem of a new instance only by measuring the similarity between a new instance and the historical instance and reusing the information and knowledge of the historical instance [15]. Therefore, STM can realize life prediction without constructing a specific prediction model and selecting parameters, which can reduce the influence of network structure design and parameter selection on the accuracy of prediction results. Therefore, STM has been widely discussed and applied to life prediction since it was proposed by Wang et al. [16]. For example, You et al. [17] pointed out that the traditional Euclidean distance cannot highlight the importance of the recent samples, and on the basis of the Euclidean distance, the decay coefficient was introduced to make the recent degradation samples have a larger weight. Liang et al. [18,19] improved the prediction accuracy of STM by improving the deficiency of degradation indicators construction. Cannarile et al. [20] proposed evidential similarity-based regression for life prediction and related uncertainty based on both complete and incomplete degradation trajectories. Yang et al. [21] proposed an integrated prediction model based on the similarity trajectory method and the differential evolution support vector regression for predicting the tool wear and life. A general data-driven based similarity-based approach was proposed by Li et al. [22] to predict the RUL of the electromagnetic relay. In addition, the similarity-based approach which was based on the framework of CBR was used for the fight trajectory prediction [23], pan-Arctic and regional sea ice area and volume anomalies prediction [24], and so on.

Although STM can effectively use the trajectory of historical data to predict the life of the equipment, the original monitoring data were usually smoothed to reduce the impact of random fluctuations on the similarity measurement [25]. However, for multiple operating conditions, the monitoring data cannot be smoothed effectively to extract the degradation trend due to the influence of operating conditions (the detailed description was employed in Section 4.3).

Aiming at the abovementioned problem that the degradation trend cannot be extracted effectively under multiple operating conditions, this paper proposes a novel prediction scheme for the life prediction of equipment under multiple operating conditions based on morphological pattern and symbolic aggregate approximation-based similarity measurement method (MP-SAX) and STM. According to the characteristics that the equipment performance degradation is reflected in the trend change of monitoring data, while the changes of operating conditions and environment are reflected in the detailed change of monitoring data, the MP-SAX is first used to measure the similarity between the monitoring data under multiple operating conditions. Then, the similar life candidate set, and corresponding weight are obtained according to the MP-SAX. Finally, the life prediction results of equipment under multiple operating conditions can be calculated by aggregating the similar life candidate set.

The rest of this paper is organized as follows. The background knowledge of the similarity measurement method based on morphological pattern and symbolic aggregate approximation (MP-SAX) is described in Section 2. In Section 3, the proposed method is described in detail. The dataset and problem in life prediction under multiple operating conditions are illustrated in Section 4. In Section 5, the results and discussion of the proposed method are explained to verify the effectiveness of the proposed method. Finally, the conclusion of this paper is drawn in Section 6. The meaning of all acronyms are listed in Table A1 in Appendix A.

2. Background of MP-SAX

For the whole life cycle of equipment, the performance degradation of equipment often leads to the change of trend component of monitoring signal, while the change of operating condition or environment often causes the change of detail component of monitoring signal. The similarity measurement method based on the morphological pattern and symbolic aggregate approximation (MP-SAX), which measured the similarity of time series by measuring the similarity of trend component and detail component respectively can effectively measure the similarity of the time series with the changes both in trend and detail [26]. Therefore, the MP-SAX can be used to measure the similarity of monitoring signal during equipment degradation under multiple operating conditions. For two time series $X = (x_1, x_2, \cdots, x_n)$ and $Y = (y_1, y_2, \cdots, y_m)$, the MP-SAX of the two-time series can be obtained as follows:

Step 1: According to empirical mode decomposition (EMD) [27,28], the time series $X = (x_1, x_2, \cdots, x_n)$ and $Y = (y_1, y_2, \cdots, y_m)$ are decomposed to IMFs $\{IMF_i^X\}_{i=1}^{N_1}$ and $\left\{IMF_j^Y\right\}_{j=1}^{N_2}$.

Step 2: According to multi-scale permutation entropy (MPE) [29], the MPEs $\{MPE_i^X\}_{i=1}^{N_1}$ and $\left\{MPE_j^Y\right\}_{j=1}^{N_2}$ of each IMFs can be calculated. Then, the IMFs $\{IMF_i^X\}_{i=1}^{N_1}$ and $\left\{IMF_j^Y\right\}_{j=1}^{N_2}$ are reconstructed into trend component (TR^X and TR^Y) and detail component (DE^X and DE^Y) according to the MPEs as follows:

$$TR^X = \sum_{i \in \{i | MPE_i^X < 0.4\}} IMF_i^X \tag{1}$$

$$TR^Y = \sum_{j \in \{j | MPE_j^Y < 0.4\}} IMF_j^Y \tag{2}$$

$$DE^X = \sum_{i \in \{i | MPE_i^X \geq 0.4\}} IMF_i^X \tag{3}$$

$$DE^Y = \sum_{j \in \{j | MPE_j^Y \geq 0.4\}} IMF_j^Y \tag{4}$$

where $TR^X = (tx_1, tx_2, \cdots, tx_n)$, $TR^Y = (ty_1, ty_2, \cdots, ty_m)$, $DE^X = (dx_1, dx_2, \cdots, dx_n)$, $DE^Y = (dy_1, dy_2, \cdots, dy_m)$.

Step 3: The trend component (TR^X and TR^Y) are converted into morphological pattern (MP) symbol sequences MC^X and MC^Y as follows (take MC^X as an example):

$$MC_i^X = \begin{cases} 3, & (tx_i - tx_{i-1})/t > 1 \\ 2, & (tx_i - tx_{i-1})/t = 1 \\ 1, & (tx_i - tx_{i-1})/t < 1 \\ 0, & tx_i = tx_{i-1} \\ -1, & (tx_i - tx_{i-1})/t > -1 \\ -2, & (tx_i - tx_{i-1})/t = -1 \\ -3, & (tx_i - tx_{i-1})/t < -1 \end{cases} \tag{5}$$

where t is the time interval between two consecutive sample points.

Step 4: The detail component (DE^X and DE^Y) are converted into symbolic aggregate approximation (SAX) symbol sequences SC^X and SC^Y as follows (take SC^X as an example):

Step 4.1: The detail component $DE^X = (dx_1, dx_2, \cdots, dx_n)$ is first normalized as follows:

$$NDE^X = \frac{DE^X - \mu}{\sigma} \tag{6}$$

where NDE^X is the normalized series of DE^X, μ is the mean value of DE^X, and σ is its standard deviation.

Step 4.2: The normalized series NDE^X is divided into w equal-sized segments, then the normalized series can be represented by the average of each segment as follows:

$$\overline{Ndx_k} = \frac{w}{n} \sum_{i=\frac{n}{w}(k-1)+1}^{\frac{n}{w}k} Ndx_i \tag{7}$$

where $\overline{Ndx_k}$ is the average of the kth segment of normalized series NDE^X.

Step 4.3: The distribution space of $\overline{Ndx_k}$ on the amplitude is divided into α equiprobable regions, the breakpoints β refers to the lookup table in [30,31].

Step 4.4: The SAX symbol sequences SC^X can be obtained by assigning symbols to each region which is determined by breakpoints.

Step 5: The symbol sequences similarity of trend component STR and detail component SDE are measured by the longest common subsequence (LCS) [32] respectively.

Step 6: The similarity Sim_{tol}^{MP-SAX} between the two-time series X and Y can be obtained as follows:

$$Sim_{tol}^{MP-SAX} = W_{TR} \cdot STR + W_{DE} \cdot SDE \tag{8}$$

where W_{TR} and W_{DE} are the weight of trend component and detail component of time series respectively, the determination of weight refers to [26].

3. The Background and Proposed Method

3.1. The Description of Background

The similarity trajectory method (STM), as a life prediction method without fitting historical curves or constructing a specific prediction model, was widely used in the life prediction of equipment [16,20,21,25]. However, the traditional STM usually used Euclidean distance to measure the similarity between the test sample and historical samples. Although Euclidean distance was the simplest similarity measurement method, there are many limitations in the measurement process [33]. In addition, taking a set of turbofan engine simulation data from NASA Ames Prognostics Data Repository (Details of the simulation data are described in Section 4) as an example, the influence of operating conditions or equipment operating environment on monitoring data is illustrated as shown in Figure 1.

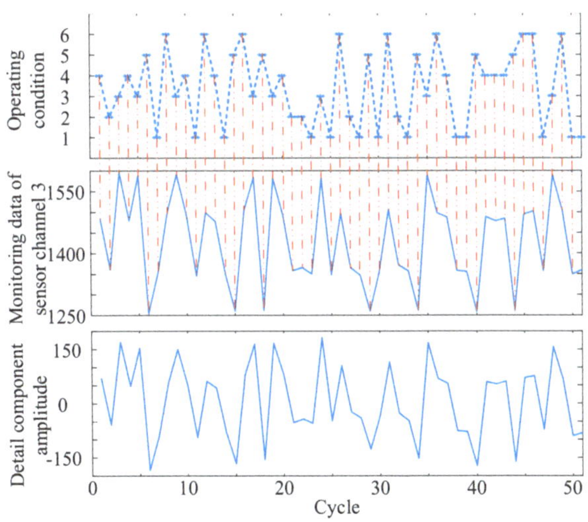

Figure 1. The influence of multiple operating conditions on the detail component of monitoring data.

According to Section 2, the detail component of monitoring data of Sensor Channel 3 is extracted as shown in Figure 1. As shown in Figure 1, the detail component of monitoring data is basically consistent with the original monitoring data. Besides, although there are some differences in the shape between the original monitoring data and the operating condition, the change points of the operating condition are basically the turning points of the original monitoring data amplitude. Thus, it can be seen that the change of operating conditions will affect the shape of the detail components of monitoring data under multiple operating conditions.

3.2. The Proposed Method

Therefore, in order to solve the life prediction problem of equipment under multiple operating conditions, this paper proposes a novel similarity trajectory method based on morphological pattern and symbolic aggregate approximation (MP-SAX-STM) by using morphological pattern and symbolic aggregate approximation similarity measurement method. The diagrammatic sketch of the proposed method is shown in Figure 2. For the test sample X_o, historical samples X_r, where the ith historical sample is X_{ri} and the number of historical samples is N, the process of the proposed method can be described as follows:

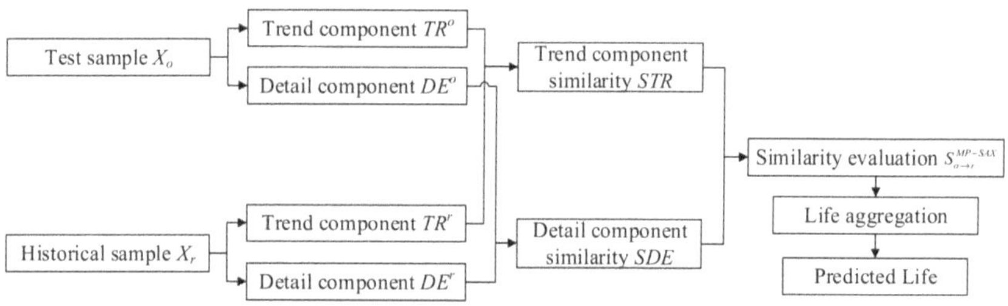

Figure 2. The diagrammatic sketch of the proposed method.

Step 1: According to Step 1 and Step 2 in Section 2, the test sample X_o and historical sample X_{ri} are divided into trend component (TR^o and TR^{ri}) and detail component (DE^o and DE^{ri}).

Step 2: According to Step 3 to Step 5 in Section 2, the similarity of the trend component STR and the detail component SDE can be obtained, then the similarity $S_{o\to ri}^{MP-SAX}$ between X_o and X_{ri} can be calculated by Equation (8).

Step 3: Determine the similar life candidate set and corresponding weight. The life of the historical sample is regarded as a similar life candidate set as follows:

$$SL_{ri} = t_{Ei} \tag{9}$$

where t_{Ei} is the time corresponding to the last set of data of the ith historical sample. And the weight corresponding to the candidate set is as follows:

$$W_i = e_i \bigg/ \sum_{i=1}^{N} e_i \tag{10}$$

where

$$e_i = \left(\sum_{i=1}^{N} S_{o\to ri}^{MP-SAX} \right) / S_{o\to ri}^{MP-SAX} \tag{11}$$

and $\sum_{i=1}^{N} W_i = 1$. Therefore, the RUL of the test sample can be obtained as follows:

Step 4: Aggregate life to obtain predicted life. According to the similar life candidate set and corresponding weight, the predicted life of test sample PL_o can be obtained as follows:

$$PL_o = \sum_{i=1}^{N} W_i \cdot SL_{ri} \quad (12)$$

4. The Description of Dataset and Problem in Life Prediction under Multiple Operating Conditions

4.1. The Description and Analysis of the Dataset

The dataset used in this part to verify the effectiveness of the proposed method was the simulation data of turbofan engine from NASA Ames Prognostics Data Repository which was built by Saxena et al. based on the Commercial Modular Aero-Propulsion System Simulation (CMAPSS) [34]. As the simulation model accurately reflected the degradation law of the turbofan engine, the simulation data can effectively reflect the degradation of the turbofan engine under different failure modes. There were four sets of life cycle data in the simulation data, the information of operating condition and failure mode simulated by each set of data were shown in Table 1. Further, each set of data contained two subsets, namely the train set, and test set, the train set contained the complete life cycle of turbofan engines. In this part, only the train set of Set #2 and #4 was used to verify the proposed method. According to [25], the meaning of each column of data of each life cycle data in Set #2 and #4 was shown in Table 2. In addition, only the monitoring data of Sensor Channels 2, 3, 4, 7, 11, 12, 15, 20, and 21 were sensitive to the fault of the turbofan engine and had an obvious degradation trend [16,25]. However, there may be information redundancy among the monitoring data of each sensor channel. Therefore, cross calculate the correlation between the monitoring data of Sensor Channels 2, 3, 4, 7, 11, 12, 15, 20, and 21, and the results were shown in Figure 3.

Table 1. The information of operating condition and failure mode simulated by each set of data.

Set	Operating Condition	Failure Mode
#1	1	1
#2	6	1
#3	1	2
#4	6	2

Table 2. Brief description of one life cycle data of simulation data.

Cycle	OP 1	OP 2	OP 3	S 1	S 2	...	S 21
1	42.0049	0.8400	100	445.00	549.68	...	6.3670
2	20.0020	0.7002	100	491.19	606.07	...	14.6550
⋮	⋮	⋮	⋮	⋮	⋮		⋮
5	25.0063	0.6207	60	462.54	536.10	...	8.6754
6	34.9996	0.8400	100	449.44	554.77	...	8.9057
7	0.0019	10^{-4}	100	518.67	641.83	...	23.4578
⋮	⋮	⋮	⋮	⋮	⋮		⋮
17	9.9989	0.2506	100	489.05	603.80	...	17.1975
⋮	⋮	⋮	⋮	⋮	⋮		⋮
321	42.0058	0.8400	100	445.00	549.71	...	6.4590

OP—operating parameters; S—sensor.

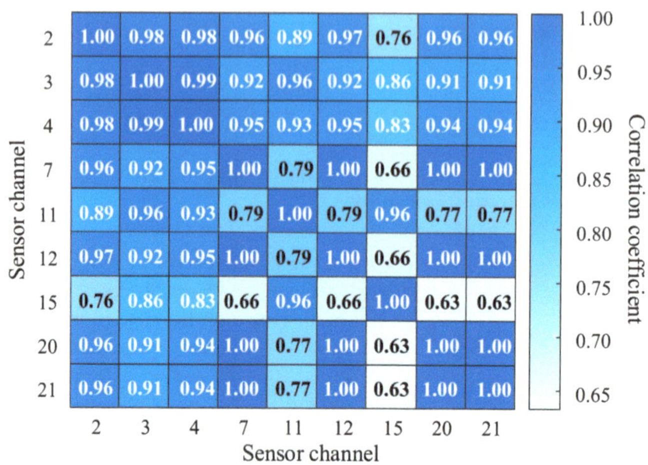

Figure 3. Correlation coefficient between the monitoring data of each sensor.

As shown in Figure 3, there was a high correlation between the monitoring data of each sensor channel. And the correlation between the monitoring data of Sensor Channels 3, 4 and other sensor channel was the highest. Therefore, only the monitoring data of Sensor Channel 3 in Set #2 and #4 was taken as an example to verify the effectiveness of the proposed method.

4.2. Evaluation Indicators of Prediction Results

In order to more intuitively evaluate the performance of the proposed method from the quantitative perspective, we used mean absolute error (MAE), root mean square error (RMSE), mean absolute percentage error (MAPE), and mean absolute scaled error (MASE) to evaluate the prediction accuracy under each prediction experiment. MAE, RMSE, MAPE, and MASE have been proved to still have good performance in measuring prediction accuracy in some scenarios [35]. The MAE, RMSE, MAPE, and MASE can be obtained as following [35]:

$$\text{MAE} = \frac{1}{l}\sum_{i=1}^{l}|s_i - yc_i| \tag{13}$$

$$\text{RMSE} = \sqrt{\frac{1}{l}\sum_{i=1}^{l}(s_i - yc_i)^2} \tag{14}$$

$$\text{MAPE} = \frac{1}{l}\sum_{i=1}^{l}\left|\frac{s_i - yc_i}{s_i}\right| \times 100 \tag{15}$$

$$\text{MASE} = \frac{l-1}{l}\frac{\sum_{i=1}^{l}|s_i - yc_i|}{\sum_{j=2}^{l}|s_j - s_{j-1}|} \tag{16}$$

where l is the number of sample points, s_i is the actual life, yc_i is the predicted life.

4.3. The Description of Problems in Life Prediction under Multiple Operating Conditions

The traditional STM usually extracted the degradation model (degradation trend) of the monitoring signal to eliminate the influence of noise and other random fluctuations on the overall degradation trend of equipment. However, as shown in Figure 4a, for some monitoring signals, when the equipment was operating under multiple operating conditions, the monitoring signals were affected by different operating conditions. Further, the obvious degradation trend cannot be seen from the monitoring signal. Therefore, according to [25], the influence of operating conditions on the degradation trend of monitoring signal was eliminated by normalization of operating conditions as follows:

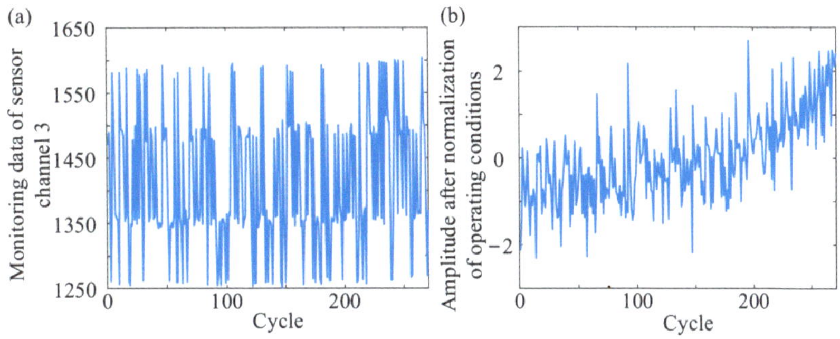

Figure 4. The simulation data of turbofan engine. (**a**) Monitoring data of Sensor Channel 3. (**b**) Data after normalization of operating conditions.

It was assumed that each life cycle point x can be divided into one of the P operating regions O_P, that was $\{x\}^{(P)} = \{x_i | u_i \in O_P\}$, where u_i was the operating condition corresponding to each point x_i. After dividing the operating conditions, the mean and variance of each operating condition were as follows:

$$\widetilde{x}^{(P)} = Mean\left(\{x\}^{(P)}\right) \tag{17}$$

$$s^{(P)} = Std\left(\{x\}^{(P)}\right) \tag{18}$$

Then the normalization of operating conditions $\{y\}^{(P)}$ can be obtained as follows:

$$\{y\}^{(P)} = \frac{\{x\}^{(P)} - \widetilde{x}^{(P)}}{s^{(P)}} \tag{19}$$

Finally, according to the time position of each point in the original life cycle, the normalized monitoring signal of the operating condition can be obtained, as shown in Figure 4b.

Through the operating condition normalization, although the normalized monitoring signal of the operating condition can show an obvious degradation trend, the mean and variance of each operating condition needed to be known in the process of operating condition normalization. Therefore, this brought a problem, for the equipment in service, how to determine the mean and variance of each operating condition. Generally, there were two solutions: (1) the mean and variance were from each operating condition of service equipment in-service stage; (2) the mean and variance were from each operating condition of historical life cycle data. However, the mean and variance obtained by the two solutions were different from those of each operating condition after the complete failure of service equipment. As shown in Figure 5, the degradation model of the first half of Life Cycle 1 was obtained by the mean and variance obtained from the abovementioned two solutions respectively. As shown in Figure 5b,c, the degradation model of the first half of Life Cycle 1 obtained by the mean and variance which came from the abovementioned two solutions was obviously different from that of Life Cycle 1 obtained by the mean and variance of each operating condition of the whole life. This difference existed not only in shape but also in amplitude. Therefore, for the service equipment, the degradation model of monitoring data in-service stage cannot be accurately obtained under multiple operating conditions, which was an important problem encountered in using STM for life prediction under multiple operating conditions.

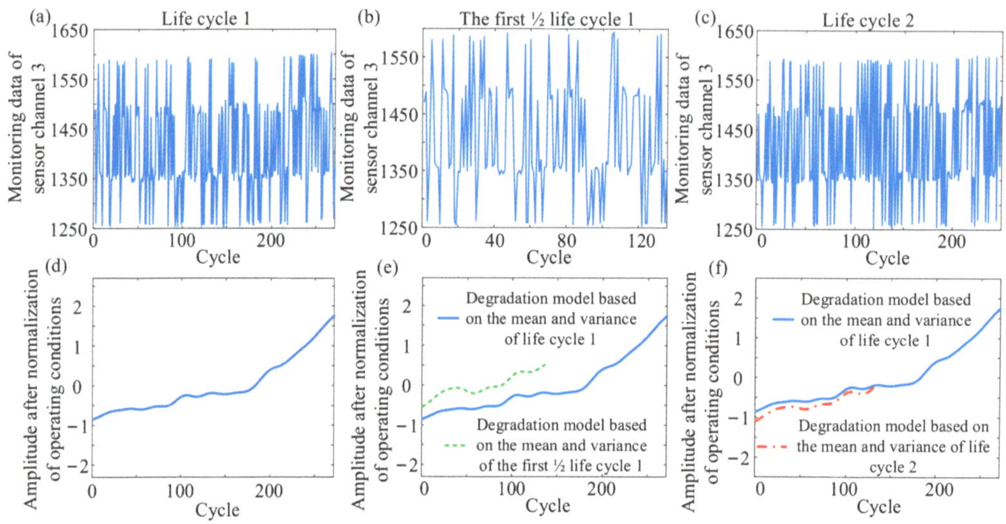

Figure 5. Degradation model comparison based on different mean and variance. (**a**) Life cycle monitoring Data 1 of Sensor Channel 3. (**b**) The first half monitoring Data 1 of Sensor Channel 3. (**c**) Life cycle monitoring Data 2 of Sensor Channel 3. (**d**) Degradation model based on the mean and variance of life cycle monitoring data. (**e**) Degradation model comparison based on the mean and variance of the first half of monitoring Data 1. (**f**) degradation model comparison based on the mean and variance of life cycle monitoring data 2.

5. Results and Discussions

In order to further explain the influence of the degradation model which cannot be accurately obtained in-service stage on the prediction results, we randomly selected 30 sets of life cycle data from Set #2 and Set #4 respectively, then the first 1/2, 2/3, 3/4, and 4/5 of the selected life cycle data were used as the test samples. Since there was not one similar life candidate set selected in the process of life prediction by STM, the degradation model was calculated only according to the mean and variance which were from each operating condition of service equipment in-service stage. Finally, STM was used to predict the life based on the degradation model and original monitoring data respectively, and the MAE, RMSE, MAPE, and MASE were calculated as shown in Figures 6 and 7.

As shown in Figures 6 and 7, in the same scenario that the life was predicted by STM, the evaluation indicators of prediction results based on the degradation model were significantly larger than those based on original monitoring data. Moreover, the evaluation indicators of prediction results based on the degradation model were not regular under the different lengths of test samples, while those based on original monitoring data shown a downward trend with the increase of test sample length. Therefore, in the scenario of multiple operating conditions, the degradation model obtained only based on part of the monitoring data of the service equipment cannot be completely equivalent to that obtained by whole life, and the accurate life prediction results cannot be obtained in the process of life prediction.

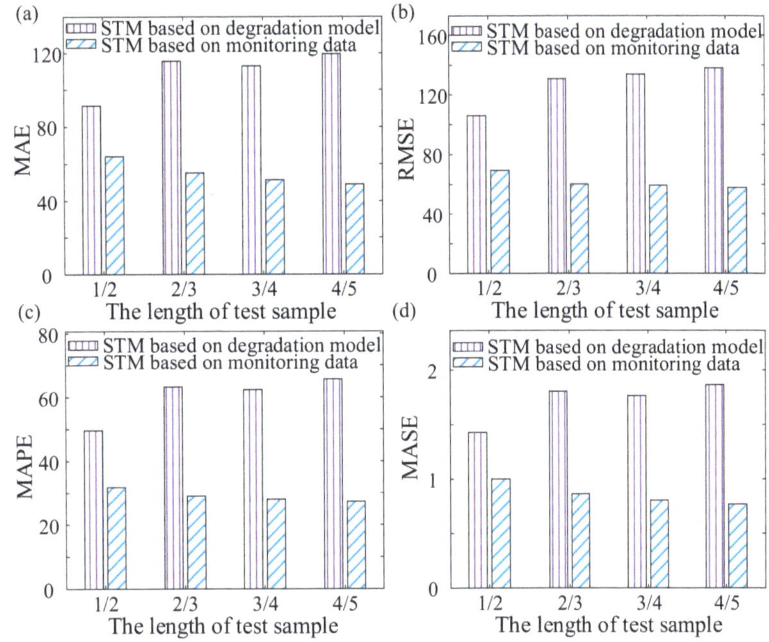

Figure 6. MAE, RMSE, MAPE, and MASE results of Set #2 based on the different test dataset. (**a**) MAE. (**b**) RMSE. (**c**) MAPE. (**d**) MASE.

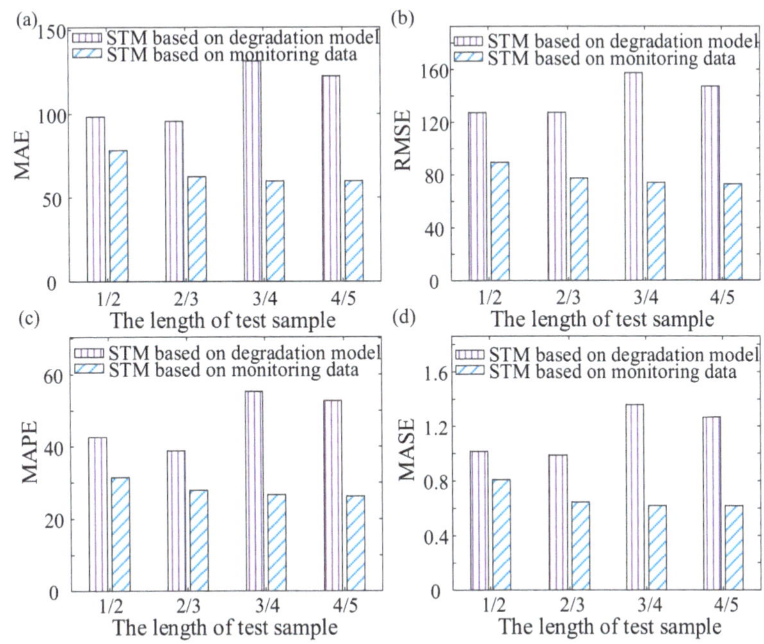

Figure 7. MAE, RMSE, MAPE, and MASE results of Set #4 based on the different test dataset. (**a**) MAE. (**b**) RMSE. (**c**) MAPE. (**d**) MASE.

Consequently, the proposed method improved the life prediction results under multiple operating conditions by improving the accuracy of direct measurement of monitoring data similarity. Likewise, 30 sets of life cycle data from Set #2 and Set #4 were selected randomly respectively, then the first 1/2, 2/3, 3/4, and 4/5 of the selected life cycle data were used as the test samples. Finally, MP-SAX-STM was used for life prediction and compared with the prediction results of STM which was based on original monitoring data, and the MAE, RMSE, MAPE, and MASE were calculated as shown in Figures 8 and 9 Where the subjective weight determination method was selected when calculating the similarity by MP-SAX, in which the weight of trend component is 0.75 and the weight of detail component is 0.25.

As shown in Figures 8 and 9, when only monitoring data were used for life prediction under multiple operating conditions, the evaluation indicators of MP-SAX-STM were lower than those of STM. Furthermore, with the increase of the length of the test sample, the evaluation indicators of MP-SAX-STM and STM showed a downward trend. Furthermore, for Set #2 and Set #4, the relative reduction of each indicator of MP-SAX-STM compared with STM was shown in Tables 3 and 4, respectively.

Table 3. Relative reduction of each indicator of MP-SAX-STM compared with STM of Set #2.

Indicators	1/2	2/3	3/4	4/5
MAE	23.71%	24.48%	22.28%	21.74%
RMSE	22.35%	22.32%	23.20%	21.78%
MAPE	25.21%	25.13%	22.50%	22.63%
MASE	23.71%	24.48%	22.28%	21.74%

Table 4. Relative reduction of each indicator of MP-SAX-STM compared with STM of Set #4.

Indicators	1/2	2/3	3/4	4/5
MAE	9.30%	13.69%	13.51%	16.70%
RMSE	8.65%	11.38%	11.49%	12.36%
MAPE	8.67%	13.45%	12.82%	13.30%
MASE	9.30%	13.69%	13.51%	16.70%

As shown in Tables 3 and 4, there was little difference among the relative change of each indicator under the same length. For Set #2, there was little difference in the change of the same indicator under different lengths, while, for Set #4, the variation under same indicator and different lengths increased approximately with the increase of length. Therefore, the prediction results of STM can be effectively improved by improving the accuracy of the similarity measurement of traditional STM. Moreover, in the case of only using monitoring data, MP-SAX-STM can also achieve life prediction, and the prediction results were better than the traditional STM. Besides, for Set #4, the improvement effect of MP-SAX-STM was also approximately enhanced with the increase of the know data length. To further illustrate the life prediction results of the MP-SAX-STM and STM, one set was randomly selected from the 30 sets of life cycle data from Set #2 and Set #4 to predict the remaining useful life (RUL) under different lengths as shown in Figure 10 respectively. And the RUL error was shown in Figure 11.

Compared with STM, as shown in Figure 10, both the MP-SAX-STM RUL prediction results of Set #2 and Set #4 were closer to the actual RUL. Therefore, under multiple operating conditions, the life prediction can be effectively realized by MP-SAX-STM even if the degradation model was not extracted and only the monitoring data which was sensitive to the operating condition was used. In addition, as shown in Figure 11, for Set #2, the RUL error of MP-SAX-STM fluctuated around 0, while the error of STM was much less than 0 and gradually approached 0. For Set #4, although the overall trend of the error of MP-SAX-STM passed through 0, the error of STM was always greater than 0. Therefore, for the life prediction problem under multiple operating conditions, MP-SAX-STM can improve the prediction accuracy compared with STM.

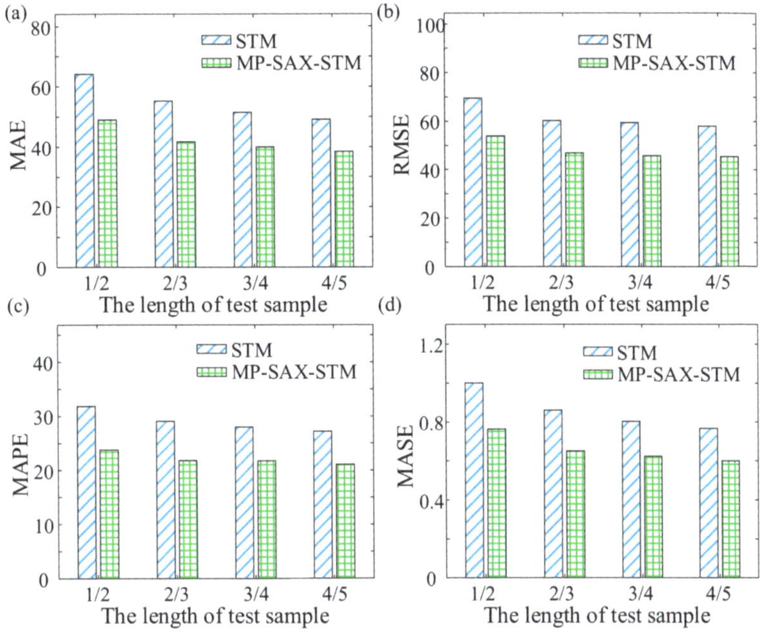

Figure 8. MAE, RMSE, MAPE, and MASE results of Set #2. (**a**) MAE. (**b**) RMSE. (**c**) MAPE. (**d**) MASE.

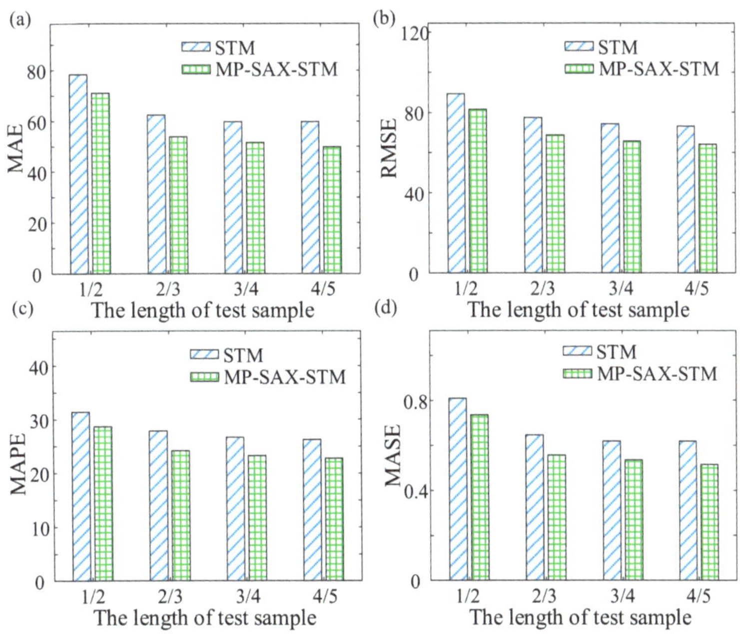

Figure 9. MAE, RMSE, MAPE, and MASE results of Set #4. (**a**) MAE. (**b**) RMSE. (**c**) MAPE. (**d**) MASE.

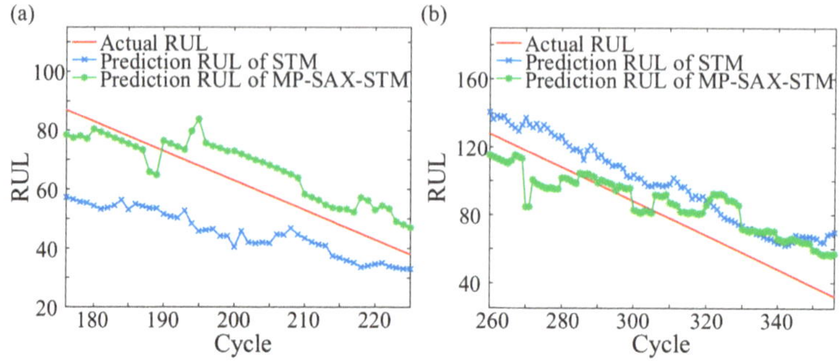

Figure 10. RUL result of the single tested sample. (**a**) Set #2. (**b**) Set #4.

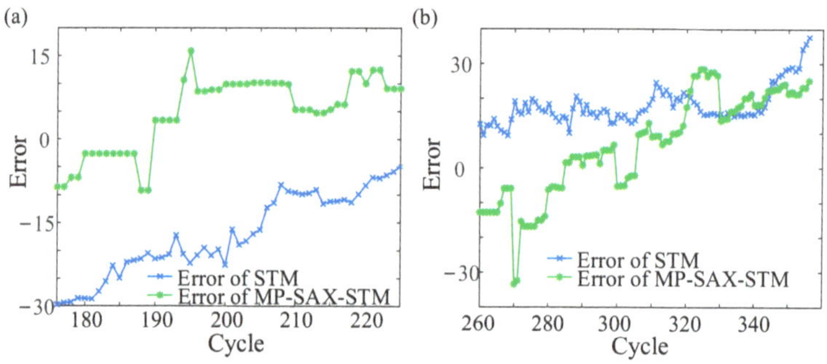

Figure 11. RUL error of the single tested sample. (**a**) Set #2. (**b**) Set #4.

Altogether, the prediction result of STM can be effectively improved by improving the similarity measurement accuracy of STM. In addition, in the scenario of multiple operating conditions, MP-SAX-STM can effectively solve the life prediction problem that the degradation model of monitoring data cannot be accurately obtained, and effectively realized the life prediction.

6. Conclusions

This paper proposes an improved STM based on the morphological pattern and symbolic aggregate approximation-based similarity measurement method, which can directly and effectively use the original monitoring data for life prediction without extracting the degradation trend of the monitoring data. According to the characteristics that the equipment performance degradation is reflected in the trend change of monitoring data, while the changes of operating conditions and environment are reflected in the detail change of monitoring data, the MP-SAX is first used to measure the similarity between the monitoring data under multiple operating conditions. Then, the similar life candidate set and corresponding weight are obtained according to the MP-SAX. Finally, the life prediction results of equipment under multiple operating conditions can be calculated by aggregating the similar life candidate set. Through the analysis and verification of public datasets of the turbofan engine from the NASA Ames Prognostics Data Repository, it is proved that the proposed method can achieve life prediction only using original monitoring data without extracting degradation trend of monitoring data. In addition, the prediction result of STM can be effectively improved by improving the similarity measurement accuracy of STM.

Although some important techniques associated with the proposed method have been investigated in this paper, the STM can only obtain the prediction point of life and do not consider the uncertainty in the prediction.

Author Contributions: Conceptualization, J.Y.; formal analysis, J.Y.; investigation, J.Y.; methodology, J.Y.; project administration, M.X.; software, J.Y. and R.W.; supervision, M.X.; validation, Y.L.; writing—original draft, J.Y.; writing—review and editing, J.Y., Y.L., R.W. and M.X. All authors have read and agreed to the published version of the manuscript.

Funding: This research received no external funding.

Institutional Review Board Statement: Not applicable.

Informed Consent Statement: Not applicable.

Data Availability Statement: Publicly available datasets were analyzed in this study. This data can be found here: [https://ti.arc.nasa.gov/m/project/prognostic-repository/CMAPSSData.zip].

Conflicts of Interest: The authors declare no conflict of interest.

Appendix A

Table A1. Meaning of acronyms.

Acronyms	Meaning	Acronyms	Meaning
AI	Artificial intelligence	MP-SAX	Morphological pattern and symbolic aggregate approximation-based similarity measurement method
CBM	Condition based maintenance	MP-SAX-STM	Similarity trajectory method based on morphological pattern and symbolic aggregate approximation
CBR	Case-based reasoning	OP	Operating parameters
CMAPSS	Commercial modular aero-propulsion system simulation	RMSE	Root mean square error
DE	Detail component	RUL	Remaining useful life
EMD	Empirical mode decomposition	S	Sensor
IMF	Intrinsic mode function	SAX	Symbolic aggregate approximation
LCS	Longest common subsequence	SC	Symbolic aggregate approximation symbol sequences
MAE	Mean absolute error	SDE	Similarity of detail component
MAPE	Mean absolute percentage error	STM	Similarity trajectory method
MASE	Mean absolute scaled error	STR	Similarity of trend component
MC	Morphological pattern symbol sequences	TR	Trend component
MP	Morphological pattern	-	-

References

1. Lei, Y.; Li, N.; Guo, L.; Li, N.; Yan, T.; Lin, J. Machinery health prognostics: A systematic review from data acquisition to RUL prediction. *Mech. Syst. Signal Process.* **2018**, *104*, 799–834. [CrossRef]
2. Jardine, A.K.; Lin, D.; Banjevic, D. A review on machinery diagnostics and prognostics implementing condition-based maintenance. *Mech. Syst. Signal Process.* **2006**, *20*, 1483–1510. [CrossRef]
3. Rezamand, M.; Kordestani, M.; Carriveau, R.; Ting, D.S.-K.; Orchard, M.E.; Saif, M. Critical Wind Turbine Components Prognostics: A Comprehensive Review. *IEEE Trans. Instrum. Meas.* **2020**, *69*, 9306–9328. [CrossRef]
4. Bhargava, C.; Sharma, P.K.; Senthilkumar, M.; Padmanaban, S.; Ramachandaramurthy, V.K.; Leonowicz, Z.; Blaabjerg, F.; Mitolo, M. Review of Health Prognostics and Condition Monitoring of Electronic Components. *IEEE Access* **2020**, *8*, 75163–75183. [CrossRef]

5. Nath, A.G.; Udmale, S.S.; Singh, S.K. Role of artificial intelligence in rotor fault diagnosis: A comprehensive review. *Artif. Intell Rev.* **2020**, *54*, 2609–2668. [CrossRef]
6. Lin, S.-L. Intelligent Fault Diagnosis and Forecast of Time-Varying Bearing Based on Deep Learning VMD-DenseNet. *Sensors* **2021**, *21*, 7467. [CrossRef]
7. Tsai, C.-M.; Wang, C.-S.; Chung, Y.-J.; Sun, Y.-D.; Perng, J.-W. Multi-Sensor Fault Diagnosis of Underwater Thruster Propeller Based on Deep Learning. *Sensors* **2021**, *21*, 7187. [CrossRef]
8. Cui, L.; Tian, X.; Shi, X.; Wang, X.; Cui, Y. A Semi-Supervised Fault Diagnosis Method Based on Improved Bidirectional Generative Adversarial Network. *Appl. Sci.* **2021**, *11*, 9401. [CrossRef]
9. She, D.; Jia, M. A BiGRU method for remaining useful life prediction of machinery. *Measurement* **2021**, *167*, 108277. [CrossRef]
10. Wang, B.; Lei, Y.; Yan, T.; Li, N.; Guo, L. Recurrent convolutional neural network: A new framework for remaining useful life prediction of machinery. *Neurocomputing* **2020**, *379*, 117–129. [CrossRef]
11. Chen, Y.; Li, F.; Wang, J.; Tang, B.; Zhou, X. Quantum recurrent encoder–decoder neural network for performance trend prediction of rotating machinery. *Knowl. Based Syst.* **2020**, *197*, 105863. [CrossRef]
12. Barraza, J.F.; Droguett, E.L.; Martins, M.R. Towards Interpretable Deep Learning: A Feature Selection Framework for Prognostics and Health Management Using Deep Neural Networks. *Sensors* **2021**, *21*, 5888. [CrossRef] [PubMed]
13. Muneer, A.; Taib, S.M.; Naseer, S.; Ali, R.F.; Aziz, I.A. Data-Driven Deep Learning-Based Attention Mechanism for Remaining Useful Life Prediction: Case Study Application to Turbofan Engine Analysis. *Electronics* **2021**, *10*, 2453. [CrossRef]
14. Muneer, A.; Taib, S.M.; Fati, S.M.; Alhussian, H. Deep-Learning Based Prognosis Approach for Remaining Useful Life Prediction of Turbofan Engine. *Symmetry* **2021**, *13*, 1861. [CrossRef]
15. Aamodt, A.; Plaza, E. Case-Based Reasoning: Foundational Issues, Methodological Variations, and System Approaches. *AI Commun.* **1994**, *7*, 39–59. [CrossRef]
16. Wang, T.; Yu, J.; Siegel, D.; Lee, J. A similarity-based prognostics approach for Remaining Useful Life estimation of engineered systems. Proceedings of 2008 International Conference on Prognostics and Health Management, Denver, CO, USA, 6–9 October 2008; pp. 1–6. [CrossRef]
17. You, M.-Y.; Meng, G. A generalized similarity measure for similarity-based residual life prediction. *J. Process. Mech. Eng.* **2011**, *225*, 151–160. [CrossRef]
18. Zeming, L.; Jianmin, G.; Hongquan, J.; Xu, G.; Zhiyong, G.; Rongxi, W. A similarity-based method for remaining useful life prediction based on operational reliability. *Appl. Intell.* **2018**, *48*, 2983–2995. [CrossRef]
19. Liang, Z.; Gao, J.; Jiang, H.; Gao, X.; Gao, Z.; Wang, R. A Degradation Degree Considered Method for Remaining Useful Life Prediction Based on Similarity. *Comput. Sci. Eng.* **2018**, *21*, 50–64. [CrossRef]
20. Cannarile, F.; Baraldi, P.; Zio, E. An evidential similarity-based regression method for the prediction of equipment remaining useful life in presence of incomplete degradation trajectories. *Fuzzy Sets Syst.* **2019**, *367*, 36–50. [CrossRef]
21. Yang, Y.; Guo, Y.; Huang, Z.; Chen, N.; Li, L.; Jiang, Y.; He, N. Research on the milling tool wear and life prediction by establishing an integrated predictive model. *Measurement* **2019**, *145*, 178–189. [CrossRef]
22. Li, L.L.; Ma, D.J.; Li, Z.G. Residual Useful Life Estimation by a Data-Driven Similarity-Based Approach. *Qual. Reliab. Eng. Int.* **2017**, *33*, 231–239. [CrossRef]
23. Lin, Y.; Yang, B.; Zhang, J.; Liu, H. Approach for 4-D Trajectory Management Based on HMM and Trajectory Similarity. *J. Mar. Sci. Technol.* **2019**, *27*, 246–256. [CrossRef]
24. Comeau, D.; Giannakis, D.; Zhao, Z.; Majda, A.J. Predicting regional and pan-Arctic sea ice anomalies with kernel analog forecasting. *Clim. Dyn.* **2019**, *52*, 5507–5525. [CrossRef]
25. Wang, T. *Trajectory Similarity Based Prediction for Remaining Useful Life Estimation*; ProQuest Dissertations Publishing: Cincinnati, OH, USA, 2010; pp. 41–70.
26. Yin, J.; Wang, R.; Zheng, H.; Yang, Y.; Li, Y.; Xu, M. A New Time Series Similarity Measurement Method Based on the Morphological Pattern and Symbolic Aggregate Approximation. *IEEE Access* **2019**, *7*, 109751–109762. [CrossRef]
27. Yu, D.; Cheng, J.; Yang, Y. Application of EMD method and Hilbert spectrum to the fault diagnosis of roller bearings. *Mech. Syst. Signal Process.* **2005**, *19*, 259–270. [CrossRef]
28. Li, Y.; Xu, M.; Wei, Y.; Huang, W. An improvement EMD method based on the optimized rational Hermite interpolation approach and its application to gear fault diagnosis. *Measurement* **2015**, *63*, 330–345. [CrossRef]
29. Li, Y.; Xu, M.; Wei, Y.; Huang, W. A new rolling bearing fault diagnosis method based on multiscale permutation entropy and improved support vector machine based binary tree. *Measurement* **2016**, *77*, 80–94. [CrossRef]
30. He, X.; Shao, C.; Xiong, Y. A non-parametric symbolic approximate representation for long time series. *Pattern Anal. Appl.* **2016**, *19*, 111–127. [CrossRef]
31. Sun, Y.; Li, J.; Liu, J.; Sun, B.; Chow, C. An improvement of symbolic aggregate approximation distance measure for time series. *Neurocomputing* **2014**, *138*, 189–198. [CrossRef]
32. Hsu, W.J.; Du, M.W. Computing a longest common subsequence for a set of strings. *BIT Numer. Math.* **1984**, *24*, 45–59. [CrossRef]
33. Batista, G.E.; Keogh, E.J.; Tataw, O.M.; De Souza, V.M. CID: An efficient complexity-invariant distance for time series. *Data Min. Knowl. Discov.* **2014**, *28*, 634–669. [CrossRef]

34. Saxena, A.; Kai, G.; Simon, D.; Eklund, N. Damage propagation modeling for aircraft engine run-to-failure simulation. In Proceedings of the International Conference on Prognostics and Health Management, Denver, CO, USA, 6–9 October 2008; pp. 1–9. [CrossRef]
35. Hyndman, R.; Koehler, A.B. Another look at measures of forecast accuracy. *Int. J. Forecast.* **2006**, *22*, 679–688. [CrossRef]

applied sciences

MDPI

Article

Method of Selecting the Means of Transport of the Winning, Taking into Account Environmental Aspects

Arkadiusz Kowalski [1,*] and Robert Waszkowski [2,*]

1 Faculty of Mechanical Engineering, Wroclaw University of Science and Technology, 50-371 Wrocław, Poland
2 Cybernetics Faculty, Military University of Technology, 00-908 Warszawa, Poland
* Correspondence: arkadiusz.kowalski@pwr.edu.pl (A.K.); robert.waszkowski@wat.edu.pl (R.W.)

Abstract: The transport of the winning in deep mines, using the room and pillar mining system, is most often performed with bucket loaders and haul trucks. In the era of attempts to stop rapid climate change, it is crucial to choose the transport means for the winning both in terms of efficiency and cost-effectiveness and to consider its environmental aspect. Permissible levels of pollutant emissions in exhaust gases are defined for this type of means of transport by the EU Stage Standards. There is a discernible need to develop a multi-criteria method supporting the decision-making process, which should reward loaders and haul trucks that meet more stringent emission standards. The article proposes an innovative idea of taking environmental aspects into account when selecting loaders and haul trucks for excavated material transport tasks, so that the amount of pollutants emitted by them in exhaust gases, e.g., the sum of hydrocarbons and nitrogen oxides ($HC+NO_x$), is also taken into consideration when assigning means of transport to particular tasks. Based on simulation studies for a specific case, it was found that a 20% reduction of $HC+NO_x$ emission is possible with only a 2% increase in the transport costs of the winning. For this purpose, an objective function was used formulated on the basis of two criteria: minimization of the transport cost of the winning and the level of pollutant emissions in the exhaust gases. Since dozens of mining machines are operated continuously in deep mines of non-ferrous metal ores, the application of the proposed method would significantly reduce the emission of pollutants in the used air coming out of ventilation shafts.

Keywords: method of selecting means of transport; stage emissions standards; simulation modeling; reduction of pollutant emissions; multi-criteria optimization

Citation: Kowalski, A.; Waszkowski, R. Method of Selecting the Means of Transport of the Winning, Taking into Account Environmental Aspects. *Appl. Sci.* **2021**, *11*, 5512. https://doi.org/10.3390/app11125512

Academic Editor: Jasiulewicz-Kaczmarek Małgorzata

Received: 12 May 2021
Accepted: 10 June 2021
Published: 14 June 2021

1. Introduction—Significance of the Problem

The transport of the winning in non-ferrous metal ore mines, using a room and pillar mining system, is most often conducted using bucket loaders and haul trucks. The permissible levels of pollutant emissions in the exhaust gases for this type of transport modes are specified in the Stage standards, for the engines used in new non-road mobile machinery (NRMM). These standards have been structured as progressively more stringent levels from Stage I to V. From Stage V, Regulation (EU) 2016/1628 of the European Parliament and of the Council [1] lays down emission requirements for all categories of diesel and spark ignition non-road mobile engines, the applicability of $HC+NO_x$ and particulate matter (PM) emission levels. Details are presented in Table 1.

Although the standard currently in force in Europe is Stage V, the loaders and haul trucks used to transport the winning in non-ferrous ore mines often scarcely meet Stage II or IIIA exhaust emission standards, introduced ten or more years earlier. Why is this happening and what is the reason for that? The situation may be explained by the fact that the economically justified life cycle of a loader or a haulage truck in deep mines is usually six to eight years. It might be surprising that it is still possible to buy from manufacturers loaders and haul trucks meeting barely Stage II and IIIA standards, which can be easily determined when analyzing the available catalogues of currently offered mining machines [2]. At the same time, some manufacturers provide mining machines that meet the newer exhaust emission standards [3]. Long transition periods and the principles of operation on the market of previously approved structures come to the assistance of mining machinery manufacturers. There are also manufacturers of mining machinery who do not provide information on what emission standards their machines meet [4], apparently considering this parameter to be irrelevant for their potential customers.

Table 1. Stage Standards in the EU, changes to emission limit values in 2009–2020.

<table>
<tr><th rowspan="2">Power Range (kW)</th><th>2009</th><th>2010</th><th>2011</th><th>2012</th><th>2013</th><th>2014</th><th>2015</th><th>2016</th><th>2017</th><th>2018</th><th>2019</th><th>2020</th></tr>
<tr><th colspan="12">Emission Limits [g/kWh]</th></tr>
<tr><td>19–37</td><td colspan="10">$HC+NO_x$ = 7.5 PM = 0.6</td><td colspan="2" rowspan="5">Stage V [2] $HC+NO_x$ = 0.4 PM = 0.015</td></tr>
<tr><td>37–56</td><td colspan="3" rowspan="2">Stage IIIA $HC+NO_x$ = 4.7 PM = 0.4</td><td colspan="7">$HC+NO_x$ = 4.7 PM = 0.025</td></tr>
<tr><td>56–75</td><td colspan="2" rowspan="2">Stage IIIB $HC+NO_x$ = 3.3 PM = 0.025</td><td colspan="5" rowspan="3">Stage IV [1] $HC+NO_x$ = 0.4 PM = 0.025</td></tr>
<tr><td>75–130</td><td colspan="3">$HC+NO_x$ = 4.0 PM = 0.3</td></tr>
<tr><td>130–560</td><td colspan="2">$HC+NO_x$ = 4.0 PM = 0.2</td><td colspan="3">$HC+NO_x$ = 2.0 PM = 0.025</td></tr>
</table>

[1] For the power range 56–130 kW, Stage IV was in force in the EU from 1 August 2014. [2] For the power range 56–130 kW, Stage V provides for an annual moratorium.

In the era of attempts to stop rapid climate change, it is crucial to choose the transport modes for the winning in deep mines both in terms of efficiency and cost-effectiveness as well as considering their environmental aspect. As an environmental aspect, it is proposed to use the permissible level of pollutant content in the exhaust gases from the engines of loaders and haul trucks, determined under the Stage standards as the sum of hydrocarbons and nitrogen oxides ($HC+NO_x$) masses or the mass of particulate matter (PM).

There is a need to develop a multi-criteria method supporting the decision-making process on the type of transport mode to choose for transporting the winning in the room and pillar mining systems. Taking into account the environmental aspect, the method should reward loaders and haul trucks that comply with more stringent emission standards.

2. Modeling of Logistic Systems in the Field of Mining

The problem of providing the necessary resources for the flow of cargo in logistics systems is widely described in the available literature. For the selection of variants of different solutions in logistics systems, multi-criteria assessment methods are applied; they use unit costs for this purpose but can also include various technical and operational or ergonomic criteria. Experiments and computer simulations concerning supply chains are carried out [5], and also in the context of gradually implementing potential economic, social and environmental factors for development of sustainable products [6]. However, classical optimization methods focus on one criterion for evaluating solution variants. The selection of transport modes with a limited capacity based on one optimization criterion [7] is possible for the specified transport tasks. The very potential of the logistics system can also be optimized [8]. In order to assess the potential, it is necessary to take into account the dynamics of logistics processes and their stochastic nature [9,10]. One of the important steps in the optimization process is the sufficiently detailed identification of cost drivers in the supply chain area [11] when costs are used as an evaluation criterion [12].

The transport of the winning during the exploitation of a mining field is one example of a logistics problem. Methods of multi-criteria evaluation of logistics systems and their optimization could also be used in the mining industry, after necessary modifications.

On the other hand, the first computational and simulation models encountered in mining are most commonly the examples of queuing theory, such as models of haul trucks transporting the winning movement. E. Koenigsberg was one of the first to solve the problem of determining the production for a fixed number of crews working on several walls in an underground coal mine. The mathematical solution was compared with the actual results obtained in coal mines in Illinois [13,14].

The next step in development was the use of software languages for building computer simulation models. P. Harvey developed a model of rail transport and used the GPSS programing language for simulation [14–16]. The case was taken from an underground molybdenum ore mine. The simulation model was used to determine the optimal number of trainsets to transport a given amount of the winning to the crusher. The loaded trains had to queued for unloading and to wait for an empty single track and for the crusher area to be free. The construction of simulation models of belt conveyors in underground coal mines was a particular challenge. R. Sanford seems to be the first to tackle this problem, using Fortran language to simulate the operation of a belt conveyor system [17,18].

S. Suboleski and J. Lucas developed a program in Fortran language that simulated mining operations for room and pillar systems [14,19]. T. O'Neil and C. Manula applied the simulation model to handle the transport of the winning in the opencast mine [20]. Another example of the first simulation models was the model of transport processes in an opencast copper ore mine. It described the operations of loading the mining product onto the haul trucks with the use of five loaders. The winning was first dumped into an ore crusher and then onto a waste pile or a leaching area. Based on the simulation model, it was determined how the dispatcher should direct the haul trucks to different loaders to minimize the queuing time. The results of the model written in Fortran language showed that the decisions made on its basis would significantly improve the loading process [21].

In the following years, the number of works based on simulation models, built using various simulation languages and commercial software in the field of mining, increased rapidly.

3. Description of Mining Processes in Copper Ore Mines According to Business Process Model and Notation (BPMN)

The room and pillar mining system, used in underground mines of non-ferrous ores, is adapted to the geological and mining characteristics of the deposit, in particular its thickness. The exploitation of the non-ferrous metal ore deposits with a thickness of up to 5 m is carried out with the use of explosives. For this purpose, the exploitation field is cut with rooms and drifts with the separation of protective pillars—this is the cross-cutting stage. The cross-cutting of the field leaves protective pillars, the task of which is to protect the ceiling. The shorter dimension of the protective pillar is most often located perpendicularly to the works site. As the front progresses, the pillars from the last rows in front of the extraction line are cut into smaller ones, minimizing the losses of the winning from the exploited field—this is the retreating stage. The size of the residual pillars depends on the local geological and mining conditions as they should provide appropriate stable support for the ceiling in the area of the liquidated field. Over time, the ceiling gradually settles on the remaining pillars, closing the exploited space.

After blasting the mine face, the mining product is extracted with the use of production loaders (LHD), which load it onto haul trucks (HT) or transport it directly to the dumping point built on a belt conveyor. Then, the excavated material is transported by a network of belt conveyors to the main retention reservoirs at the mining shafts. In the mining system using explosives, technological tasks are carried out in a closed cycle of technological operations (Figure 1):

- Mine face control and measurements;
- Drilling blast holes;
- Loading explosives;
- Blasting;
- Airing;
- Loading and haulage of the winning (Figure 2);
- Roof dressing;
- Roof anchoring.

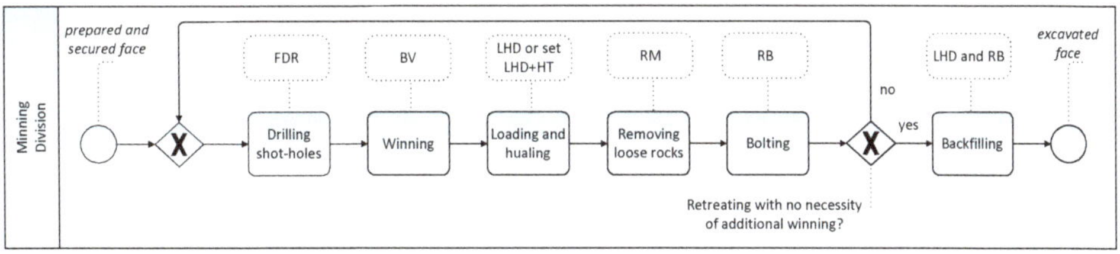

Figure 1. Copper ore mining process recorded in business process model and notation (BPMN) [22].

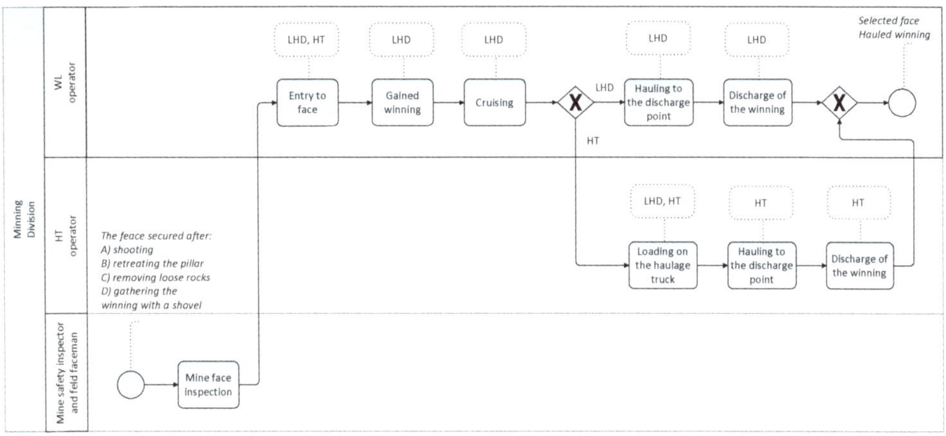

Figure 2. The process of loading and unloading the winning recorded in business process model and notation (BPMN) [22].

After the last step of the cycle, the whole process in the mine face starts over. It would not be possible without carrying out several auxiliary operations on an ongoing basis.

In deep mines, using explosives for the extraction, mining technologies are based on the use of mobile mining machines on wheeled chassis. In the implementation of mining processes, self-propelled machines of various construction are used for the following groups of works:

- Mining of the deposit, i.e., drilling trucks (FDR) and blasting vehicles (BV);
- Securing and liquidating mining excavations, i.e., bolting rig (RB) or rock ripping machine (RM);
- Loading and transport of excavated material, i.e., bucket loaders (LHD) and haul trucks of various sizes (HT);
- Transport of personnel, materials, devices and machines as well as tools (including transport, fuel and lubrication vehicles);
- Preparation of transport roads and auxiliary works (including bulldozers, drainage trucks).

A preliminary analysis of the costs of the winning haulage process from the site indicates that a significant cost of the process is associated with the mining product transport, called tire haulage. The selection of a machine set (direct haulage with a loader or a set of a loader and haul trucks), in relation to the amount of the product hauled per shift and the distance to dumping points is crucial for the cost optimization of this process [23].

4. Selection of Transport Modes according to the Economic Efficiency Criteria

The method of selecting the transport modes proposed in the book [24] is based on the observation that the distance to be covered by the means of transport when transporting the excavated material for the **cross-cutting stage** can be described as the product of the distance between the loading and unloading points for individual mine faces multiplied by the number of journeys to be made to haul the material out of the blasted mine faces.

Importantly, the duration of this task in the non-ferrous metal ore mine is limited to the duration of one shift since the primary goal is to ensure an appropriate output level. The distance between the dumping point and the loading point of the material d_{ij} located at the earlier intersection of the crosscut and the corridor (which results from the technology of loading the winning between the loader and haul trucks) can be expressed by the following Formula [24]:

$$d_{ij}= i\left(S_f+S_{ch}\right) + q + j(D_f + S_{ch}) \tag{1}$$

where S_f—width of the technological pillar [m], D_f—length of the technological pillar [m], S_{ch}—width of the drift [m], i—number of corridors, and j—number of crosscuts. By q the distance [m] between the discharge point and the beginning of the exploitation field is marked, resulting from the fact that the field of exploitation is contoured with a bundle of drifts in which the belt conveyor is placed.

The places of loading points, conventionally marked as PZ, are shown in Figure 3, which includes a fragment of the exploitation field.

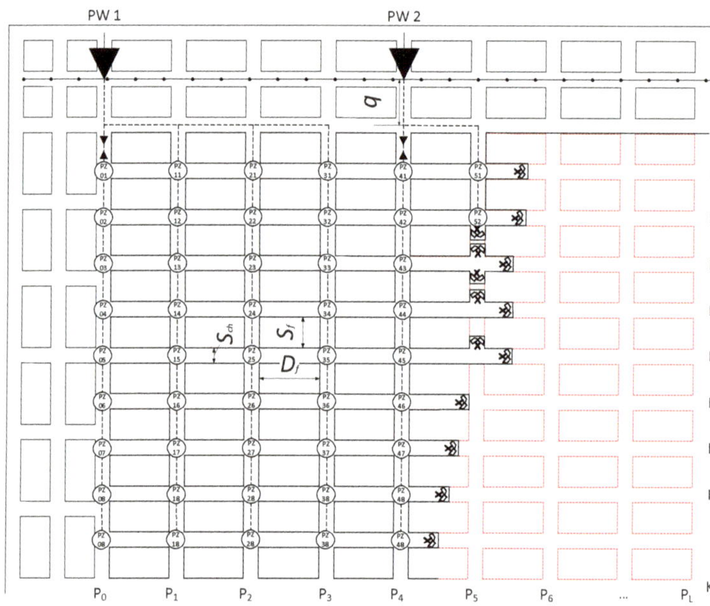

Figure 3. Characteristic dimensions of the exploitation field and a diagram of the location of the winning loading points PZ at the cross-cutting stage [24].

It was taken into account that the point of loading the winning is always at the closest intersection of the corridor and the crosscut as space is needed for the loader to manoeuvre while collecting the product into the bucket (the product is scattered after the wall blasting) and for loading onto the haul truck. The loader cannot turn around in a tight excavation; to load HT the LHD loader must reverse to the nearest intersection. In the case of the winning haulage with the use of loaders only, minor movements of the loader while scooping the material onto a bucket have been omitted. Hence, both cases—the haulage of the output with the use of loaders and the sets of a loader plus haul trucks—have been described in a standardized manner. Then, using n, n_g, n_d, n_l, n_p the number of journeys for the winning and with it, between the discharge point PW and the loading point PZ, from individual types of chambers have been marked. For the calculation of the volume of the excavation sites made during the cross-cutting stage, for the entire field of exploitation, chambers in the shape of a "cross" were proposed. Additionally, it turned out to be necessary to describe the space remaining at the boundaries of the field with the following nomenclature: "bottom" and "top", "left" and "right". In addition, these spaces are characterized by the common name "the remainder". The proposed shapes of chambers and additional spaces, when summed up, make it possible to describe the total volume of workings at the cross-cutting stage, which was the main idea of such a division, details are shown in Figure 4.

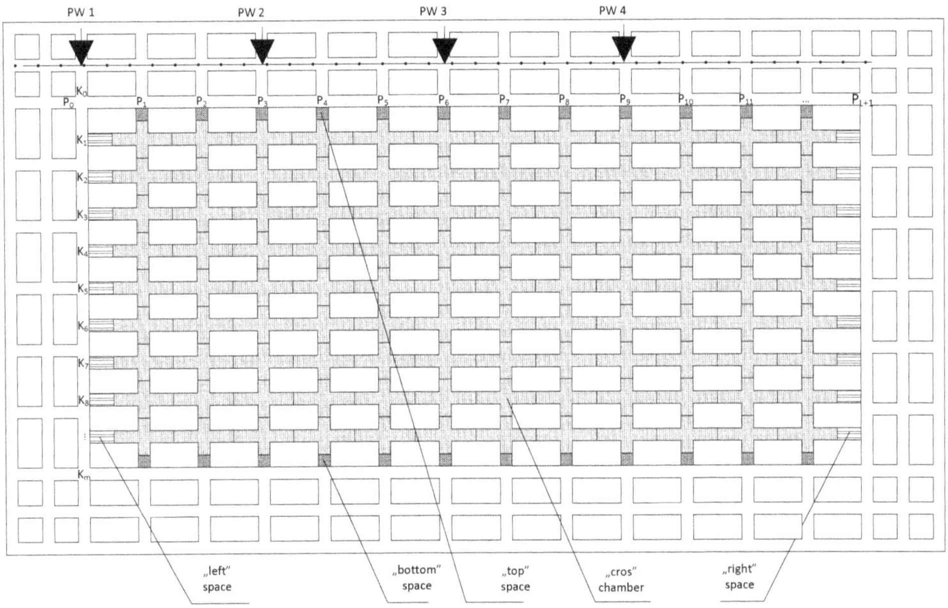

Figure 4. The method of dividing the field of exploitation into the "cross" chambers and additional "bottom" and "top", "left" and "right" space [24].

The number of journeys of the transport means depends on the volume of the product that can be obtained from a single area of a given type and the volume of the winning that can be transported at one time. The discharge points are opened in synchronisation with the movement of the front line of the crosscut.

The construction of discharge points located above the belt conveyors is associated with incurring certain costs by deep mines. Most often, deep mines try to build and activate the mentioned discharge points at the last possible moment: when the workings line reaches the planned location of the discharge point. Earlier construction of the discharge points for the entire field of exploitation means a capital freeze for up to two or three years, depending on the pace of exploitation and the size of the field. Currently, the most commonly used strategy is to launch successive discharge points on a par with the cross-cutting line.

Based on the dimensions and shapes of the rooms, it is possible to establish their volume, and thus the mass of the excavated material, and then determine the number of transport journeys necessary to haul the excavated material with the selected means of transport. Details are shown in Figure 5. In the cross-section, the drifts and crosscuts have the shape of an inverted trapezoid; at the stage of designing the exploitation method of the deposit, its characteristic dimensions are known: h—thickness of the exploitation gangway and angle α—the slope of the sidewall (the sidewall of the mining excavation in the useful mineral or possibly in the gangue), most often $\alpha = 10°$.

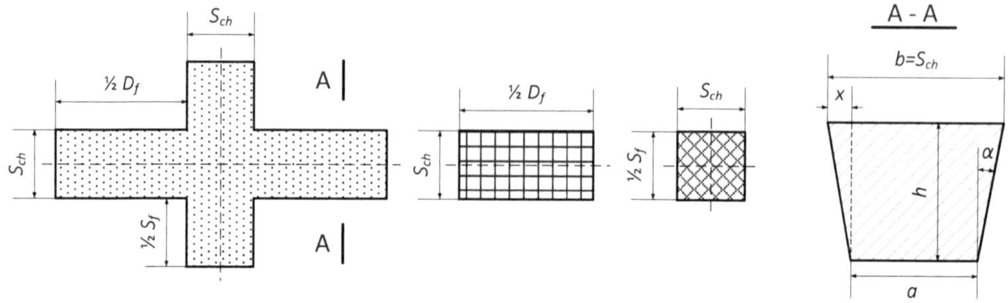

Figure 5. Geometric figures describing the volume of the excavation for the entire field of exploitation during the cross-cutting stage, along with a cross-section: (**a**) "cross"; (**b**) "right" and "left"; (**c**) "top" and "bottom"; (**d**) trapezoidal excavation [24].

The number of journeys n of the transport mode, taking into account the fact that all the winning obtained from the area must be transported (the number of journeys is, therefore, a whole number, obtained after rounding up) and the specific volume $V_{ŚrT}$ that can be taken by the transport mode at one time, for the "cross" chamber is:

$$n = \left\lceil \frac{V}{V_{SrT}} \right\rceil = \left\lceil \frac{h(S_{ch} - h\tan\alpha)\left(D_f + S_{ch} + S_f\right)}{V_{SrT}} \right\rceil, \tag{2}$$

and for other spaces [24]:

$$n_g = n_d = \left\lceil \frac{V_g}{V_{SrT}} \right\rceil = \left\lceil \frac{V_d}{V_{SrT}} \right\rceil = \left\lceil \frac{\frac{1}{2}h(S_{ch} - h\tan\alpha)S_f}{V_{SrT}} \right\rceil, \tag{3}$$

$$n_l = n_p = \left\lceil \frac{V_l}{V_{SrT}} \right\rceil = \left\lceil \frac{V_p}{V_{SrT}} \right\rceil = \left\lceil \frac{\frac{1}{2}h(S_{ch} - h\tan\alpha)D_f}{V_{SrT}} \right\rceil. \tag{4}$$

At this point, all the components that allow the calculation of the distance to be covered by the means of transport during the winning haulage for the cross-cutting stage are already determined; based on the aforementioned product of the distance d_{ij} to be covered between the loading and unloading points for individual mine faces according to Formula (1) and the number of journeys n according to Formula (2) (or the number of journeys n_g *and* n_d according to Formula (3) and the number of journeys n_l *and* n_p according to Formula (4)).

The **retreating** stage in the room and pillar systems consists in cutting the technological pillars into smaller ones. The retreating line remains synchronised with the progress of the cross-cut. The last rows of pillars in front of the goaf are cut with undercuts, usually about 7 m wide. Figure 6 shows the described situation.

Generally speaking, the minimum area of the remaining pillars measured under the excavation roof is r = 12 m^2, which most often ensures the stability of the roof. Using the expression for the volume of excavated material to be transported from the technological pillar $V_{filara\ tech.}$, the number of journeys of the transport mode n_{Likw} for the retreating stage can be described as [24] as:

$$n_{Likw} = \left\lceil \frac{V_{filara\ tech.}}{V_{SrT}} \right\rceil = \left\lceil \frac{\frac{1}{2}\left(2D_f - 4\frac{r}{S_f} - 6h\tan\alpha\right)hS_f}{V_{SrT}} \right\rceil. \tag{5}$$

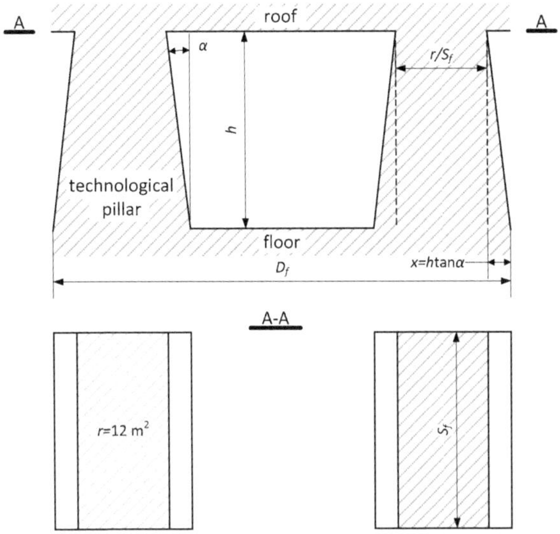

Figure 6. The dimensions of the technological pillar with the undercut marked; there will be two remaining pillars with the cross-section area r = 12 m^2 under the roof [24].

Using Formula (5), it is possible to calculate the distance to be covered by the means of transport for the cross-cutting stage, again based on the product d_{ij} of the distance to be covered between the loading points for liquidated technological pillars and the unloading point according to Formula (1) and the number of journeys n_{Likw} according to Formula (5).

To meet the condition of a limited time of transport activities to the duration of one shift—for the removal of the entire excavated material—it is crucial to determine the number of transport modes necessary to complete the task. Knowing the dependence determining the distance d_{ij} Formula (1), it is possible to describe the time needed to carry out the haulage cycle t_{ij} taking into account the basic relationship between the distance, velocity and time for uniform straight-line traffic (loading time t_{zal} and unloading time t_{roz}) were also taken into account [24]:

$$t_{ij} = t_{zal} + \frac{d_{ij}}{V_u} + \frac{d_{ij}}{V} + t'_{roz}, \tag{6}$$

where: t_{ij}—haulage cycle time for loading points PZ (h), t_{zal}—loading time (h), t'_{roz}—unloading time (h) ($t'_{roz} = t_{roz}$ for loaders and a set of a loader and one haul truck, $t'_{roz} = 0$ for sets consisting of a loader and two or more haul trucks, due to their parallel nature of work), V_u—the speed of travel with the winning from the loading point to the discharge point (km/h), V—the speed from the discharge point to the loading point, without the winning (km/h).

The acceleration of the haul truck or the loader and cornering at reduced speed was considered negligible. They will be taken into account indirectly by the average speed of the loaders and haul trucks—the maximum catalogue speeds of loaders and haul trucks, possible to achieve only under ideal working conditions, will not be used in the calculations. Knowing the value of the effective time t_{ef} (h) available during one shift, it is possible to determine the number of journeys between the discharge point and the loading point in the mine face, which can be completed by the loader or haul truck operator [24]:

$$Kurs_{ij} = \frac{t_{ef}}{t_{ij}} = \frac{t_{ef}}{t_{zal} + \frac{d_{ij}}{V_u} + \frac{d_{ij}}{V} + t'_{roz}}, \tag{7}$$

where the $Kurs_{ij}$ denotes the number of journeys possible for a single transport mode during one shift for a specific effective time t_{ef} and distance d_{ij}.

With Formulas (6) and (7), we can determine, based on the load capacity of M loaders or haul trucks working in sets, the load that can be transported in t_{ef} time, for a single transport mode [24]:

$$M_{ij} = \frac{t_{ef}}{t_{ij}} M, \tag{8}$$

where: M_{ij}—maximum amount (mass) of the winning that can be transported from the loading point PZ (Mg), M—load capacity of the loader bucket (Mg) or a haul truck.

The calculation of the number of loaders or sets consisting of a loader and a different number of haul trucks n_{ij} necessary for the implementation of the planned tasks does not present any major difficulties [24]:

$$n_{ij} = \left\lceil \frac{M_{plan}}{M_{ij}} \right\rceil, \tag{9}$$

where: n_{ij}—number of necessary means of transport, M_{plan}—planned mass of the winning to be transported (Mg) during one working shift.

5. Building a Mining Field Exploitation Model—Schedule, Parameterization

The schedule of mining works for the cross-cutting and retreating stages should make it possible to obtain the sum of the costs of road haulage and the emission of pollutants from internal combustion engines related to the transport of the excavated material during the exploitation of the mining field. It is assumed that the model will count the upper level of HC+NO_X emissions resulting from individual Stage standards. The basic assumption when building the simulation model is to include in the exploitation model of the mining field the geometric dimensions of drifts and pillars. The collected data will be used to calculate the distance covered by the transport means. Additionally, it is necessary to define the location of the planned discharge points, their location in the bundle of contouring drifts, in a specific intersection and the drift.

The size of the tested field of exploitation is determined by specifying the number of drifts and intersections and numbering them accordingly. The distances for which the excavated material is transported for the entire field of exploitation are calculated. Similar activities are carried out for the retreating stage, additionally taking into account the location of the protective pillars.

The next step in building the model is to describe the tire haulage measures that are to be included in the model. The loaders are described by the following parameters: effective working time, loading and unloading time, load capacity, driving speed with and without the material. For the sets consisting of a loader and a variable number of haul trucks, additional parameters include the number of buckets needed to load the winning onto the haul truck and the number of HT. The entered data characterize machines with the symbols LHD2, LHD3, LHD4 (the main difference between them is the load capacity and operating cost) and the LHD2 + 1×HT, LHD2 + 2×HT, LHD2 + 3×HT and LHD2 + 4×HT sets. Other important parameters are the amount of the winning planned for transport during one shift, rock density of the excavated material and the operating cost of loaders and haul trucks per shift. A table of haulage costs for a given amount of the winning from the haulage distance will be calculated. The condition is checked whether a given loader/set can complete the task within the effective working time; if not, the task can only be performed with a larger number of machines/sets. On this basis, the most cost-effective centre of the tire haulage for a given distance and amount of excavated material is found (what will be explained in detail in Section 6).

The designated work schedule for the analyzed mining field covers separately cross-cutting and retreating stages. The commencement date of works for the cross-cutting and retreating stages should be specified, necessarily taking into account the weekdays and so-called "Black Saturdays". Only the shifts during which mining is conducted are taken into account. To plan the progress of works for the mining field, it is necessary to remember about the necessity to plan the level of extraction for the entire period in which the analyzed field is intended to be exploited. For further calculation steps, the daily output from the so-called volumetric changes is evenly divided between the cross-cutting and retreating. Determining the planned amount of output between shifts makes it possible to divide the daily production, e.g., on Saturday, 30% of the daily extraction is performed on the first shift, 70% on the second, shifts 3 and 4 do not carry out mining works. To take into account the sequence of works carried out for the cross-cutting stage, and thus to reproduce the behaviour of the wings of the extraction front, the order of operations is defined manually by assigning consecutive numbers until all intersections of the drifts in the exploitation field are described. Then, the planned sequence of retreating of individual pillars, remaining after the cross-cutting stage, is also introduced. The result of the calculations performed in the schedule will be not only the designation of the end of operation date for the cross-cutting and retreating stage (Figure 7) but also the assignment of the means of transport with the highest value of the objective function for each shift, until the completion of mining works.

Figure 7. Planned schedule of mining works in the room and pillar mining system.

To calculate the schedule of the progress of mining works, it is also necessary to propose the dates of construction of reloading points on conveyor belts that collect the excavated material transported using wheeled haulage. Thanks to the synchronization of the dates of availability of reloading points with the extraction front, the greatest benefits from the shortening of the tire haulage routes might be achieved.

The set of calculation parameters is presented in Tables 2 and 3. The parametric nature of the model should be emphasized; the calculations are carried out based on the geometric dimensions of the mining field, the quantities characterizing various types of means of transport and the method of field exploitation (the level of extraction, mining works on specific days of the week).

Table 2. Summary of computational geometric parameters of the mining field in the room and pillar system for the needs of simulation experiments.

Geometric Parameters of the Mining Field in the Room and Pillar System [1]					
parameter name	S_{ch}—width of the drift [m]	D_f—length of the pillar [m]	S_f—width of the pillar [m]	i—number of corridors	j—number of crosscuts
value	7.0	7.5	15.0	32	48
parameter name	q—initial distance to the discharge point [m]	h—thickness of the exploitation gangway [m]	α—sidewall slope angle [°]	r—minimum area of residual pillars [m^2]	
value	36	3.0	10	12	

[1] The data was collected during the R&D project implementation "Adaptation and implementation of Lean in the copper mines", financed by NCBiR (No. 09-0011-10/2011).

Table 3. List of computational parameters of loaders and haul trucks for the needs of simulation experiments.

	Means of Transport [1]						
Parameter name	LHD4	LHD3	LHD2	1×LHD2 + 1×HT	1×LHD2 + 2×HT	1×LHD2 + 3×HT	1×LHD2 + 4×HT
M—mass of the transported the winning (Mg)	10.67	8.12	4.64	13.92	27.84	41.76	55.68
t_{zal}—loading time (min)	1.0	0.5	0.5	1.5	1.5	1.5	1.5
t_{roz}—unloading time (min)	1.5	1.5	1.5	1.5	1.5	1.5	1.5
V_u—speed of travel with the winning (km/h)	7.0	6.0	5.0	8.0	8.0	8.0	8.0
V—speed of travel without the winning (km/h)	9.0	8.0	7.0	10.0	10.0	10.0	10.0
Operating cost per 1 shift (EUR)	840	680	525	1050	1575	2100	2625

[1] The data was collected during the R&D project implementation "Adaptation and implementation of Lean in the copper mines", financed by NCBiR (No. 09-0011-10/2011).

In the simulation model, the starting dates of the cross-cutting and retreating stage were adopted, bearing in mind that the retreating stage takes place immediately after the cross-cutting. Residual pillars must not lose their load-bearing capacity, decreasing over time, which ensures the stability of the roof. The weekly schedule of mining works includes 4 shifts, five workdays a week and selected working Saturdays. The retreating is not carried out during the night shift, the effective time for mining works was assumed to be t_{ef} = 3.5 h. The value of the winning density, based on the example of copper ore, was assumed to be 2.32 Mg/m^3, it was necessary to recalculate the relation between weight and volume.

The production plan M_{plan} took into account the average planned mass of the winning to be transported—obtained from the mine faces—by the stage of cross-cutting and retreating, was 3 615 Mg. A certain extraction fluctuation between the individual months of mining was also introduced (to increase the model's adequacy), the average deviation was approximately 70 Mg.

When calculating the schedule, data on the costs of tire haulage for the cross-cutting and retreating stages as well as HC+NO_x emissions are collected, as shown in Figure 8.

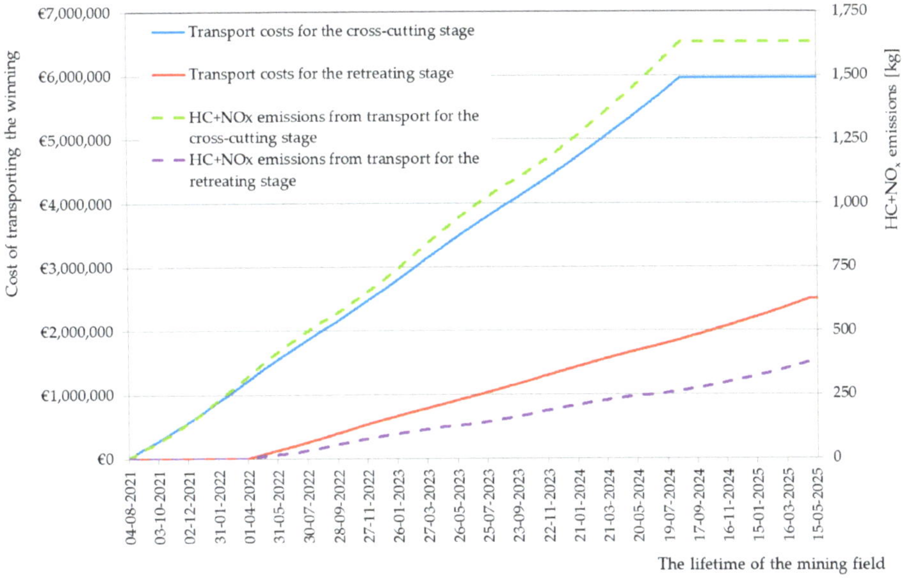

Figure 8. The cost of transporting the excavated material with wheeled haulage during the exploitation of the mining field in a room and pillar mining system (means of transport, complying with the Stage IV standard, selected according to the minimum cost).

A slight fluctuation in the rate of increase in the total costs of transporting the excavated material and the amount of HC+NO_x emissions during the exploitation of the mining field, shown in Figure 8, results from the increase in the distance between the mine faces and discharge points as the extraction front line moves. At the moment of launching the next discharge point, the distance for which the winning is transported gets reduced, thus the rate of increase in costs and the amount of HC+NO_x emissions decreases. Then, the distance begins to increase again, together with moving away from the discharge point. The costs of transporting the excavated material are limited by the mechanism of selecting the least expensive means of transport; the obtained results are shown in Figure 9.

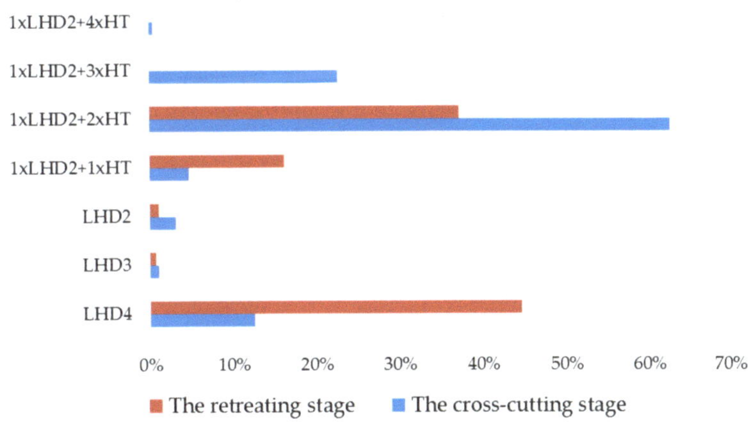

Figure 9. Selection of the means of transport for the winning, depending on the stage of exploitation of the mining field.

For the cross-cutting stage, the most frequently chosen means of transport is the set of 1×LHD2 + 2×HT (62%) and LHD4 (12%), while during the retreating stage these proportions change: LHD4 (45%) and 1×LHD2 + 2×HT (37%). Other types of means of transport are chosen much less frequently. The crucial difference between these two stages while selecting the means of transport is the smaller amount of the winning to be obtained from the pillars at the retreating stage.

6. Method of Selecting Means of Transport, Taking into Account Environmental Aspects

For the simulation studies, the idea of a means of transport selection mechanism was proposed to select the type of a loader or a set not only according to the lowest cost criterion, but also taking into account the differences in HC+NO_x emissions between them suggested to be used as the second selection criterion.

The method proposed in the book [24] allowed us to determine the most cost-effective type of transport for a given distance and amount of the winning to be transported between the loading and the discharge points:

$$K_{ij} = \frac{n_{ij} \cdot koszt}{M_{plan}}, \quad (10)$$

where: K_{ij}—the total cost of transport between the loading point and the discharge point, converted into (EUR/Mg), *koszt*—the sum of costs incurred for the operation of the machine during one shift (EUR). The cost variable should be understood as the sum of the costs of.

- Operation of the loader: necessary fuel and oils, consumables;
- Purchase costs, taken into account as depreciation;
- Maintenance services;
- Operator's salary with margins.

The most cost-effective type of transport is certainly the cheapest resource, marked as the Z_{opt} variable [24]:

$$Z_{optkoszt} = \text{argmin}_{z \in Z} K_{ij}(z), \quad (11)$$

where Z marks the set of types of transport means.

The emission of HC+NO_x pollutants, which occurred during the transport of the winning during one shift for a specific means of transport, was calculated in the field exploitation model during the scheduling of mining works from Section 5, based on the formula:

$$E_{ij} = n_{ij} t_{ef} P(z) \eta E_{stage}, \quad (12)$$

where E_{ij}—the calculated amount of HC+NO_x pollutant emissions for one shift [g/working shift] for the selected number of means of transport, E_{stage}—HC+NO_x emission levels according to individual Stage standards [g/kWh], η—engine power utilization during a working shift (%). Similarly to the methods of estimates [2,3] determining fuel consumption used by manufacturers of means of transport, it was assumed that the loader uses 40% of the engine power and the haul truck in 50%. The formula also takes into account the number of means of transport n_{ij} necessary to ensure transport capacity and the effective time t_{ef} of their work.

For the loader plus haulage vehicles, Formula (12) was used, for each of the machines that make up the set, the emissions from different types of means of transport were summed up. It was assumed that the means of transport the least harmful for the environment is a loader or a set of loader and haulage vehicles with the lowest HC+NO_x emission, according to the formula:

$$Z_{optemisja} = \text{argmin}_{z \in Z} E_{ij}(z). \quad (13)$$

The objective function of the method of selecting the means of transport for the winning in the room and pillar mining systems was based on the expression:

$$Y_{opt} = w_1 Z_{opt\ koszt} + w_2 Z_{opt\ emisja}, \quad (14)$$

where w_1 and w_2 are the weights determining the shares of criteria in the objective function. Additionally, the function of the weights is to balance the impact of criteria in the case of a large difference in their numerical values.

Using the field scheduling model, it was then decided to determine HC+NO_x emission levels for loaders and haul trucks meeting Stage II, Stage IIIA, Stage IIIB and Stage IV emission standards.

7. Results of Simulation Studies

Based on the expressions presented in the above sections, a simulation model was prepared to schedule the progress of mining works in the room and pillar systems. In this model, the selection of means of transport was based on the method proposed in Section 6, taking into account environmental aspects. The calculated HC+NO_x pollutant emission level for the transport of the winning with tire haulage was used as one of the criteria for evaluating the solution (Formula (13)) as a function of the objective (Formula (14)).

For the first simulation tests, the implementation of transport with vehicles meeting the Stage II emission standard was assumed. For the calculations, the pairs of weights in the range $w_1 = 0$ and $w_2 = 1$ (the situation in which the cost of transport is negligible, the result in 100% is determined by the minimum level HC+NO_x) up to $w_1 = 1$ and $w_2 = 0$ (the situation in which the result in 100% depends on the cost of transport, while HC+NO_x emissions are not taken into account) were adopted. The calculation results are presented in Figure 10.

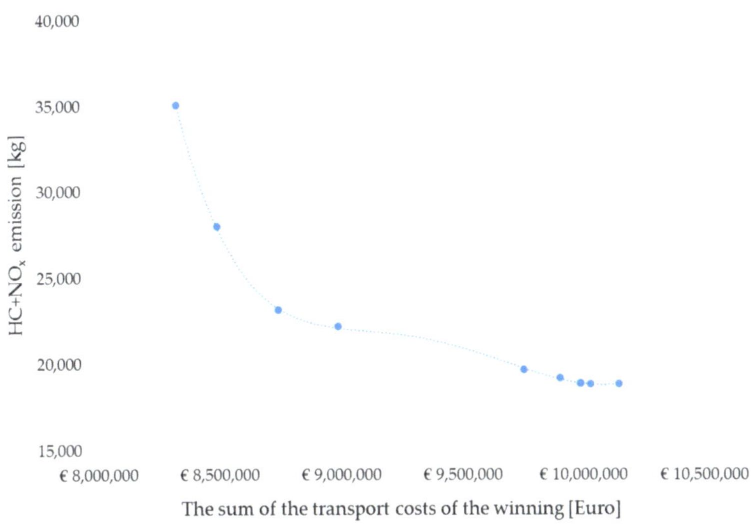

Figure 10. Dependence of the HC+NO_x emission level on the transport costs of the winning, depending on the adopted weights of optimization criteria as a function of the objective.

It is not surprising that the highest level of HC+NO_x emission was recorded with the minimum cost of the winning transport during the exploitation of the mining field. It is interesting, however, that it is possible to significantly reduce the level of HC+NO_x emissions (from 35,110 to 18,860 kg over more than 3 years of operation) by the skilful selection of the type of means of transport, even if all loaders and haul trucks meet the same Stage II standard. The potential of a possible reduction of HC+NO_x emissions is better illustrated by the presentation of percentage values—Figure 11.

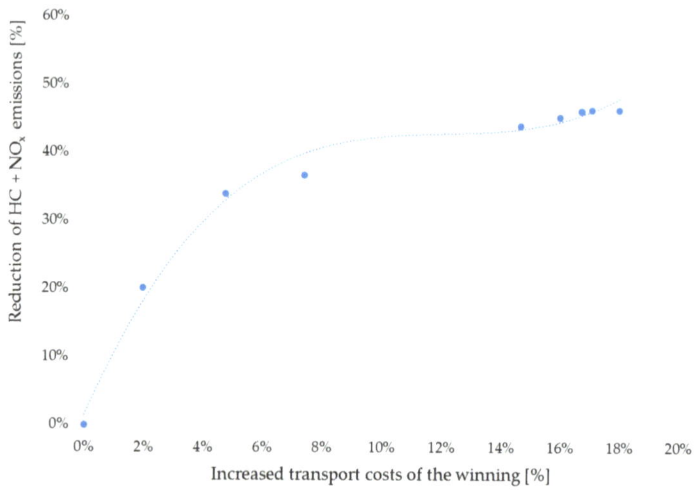

Figure 11. The potential to reduce HC+NO$_x$ emissions from the transport costs of the winning depending on the adopted weights of optimization criteria as a function of the objective.

The obtained dependence shows that a 20% reduction of HC+NO$_x$ emission is possible using the proposed method of selecting the means of transport, with a 2% increase in the cost of the winning transport. The nature of the dependence is similar to the well-known Pareto principle, although an 80% reduction of emissions turned out to be impossible to achieve, regardless of the costs of transporting the output.

Further simulation tests were carried out with the assumption that the deep mine replaces the means of transport for those complying with Stage IIIA, Stage IIIB and Stage IV emission standards. Analogous simulation tests on the method of selecting the types of transport means according to the objective function (expression 14) were carried out, as presented for Stage II standards. The results of the calculations are presented in Figure 12 to facilitate comparison.

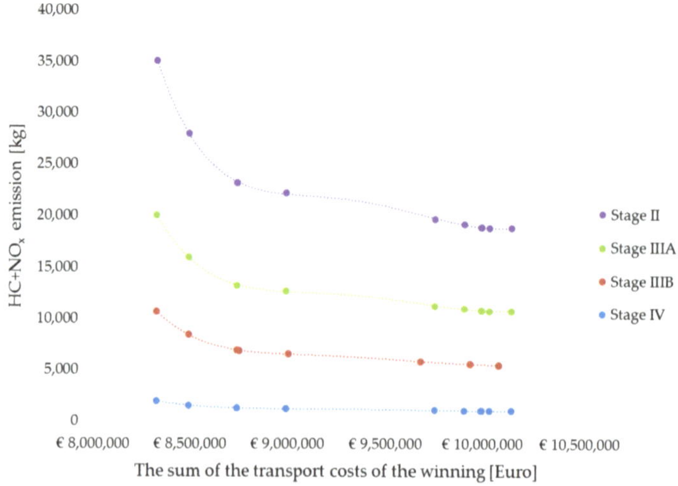

Figure 12. The level of HC+NO$_x$ emissions during transport of the winning depending on the standards met by loaders and haulage vehicles; In each case, the proposed method of selecting means of transport was used.

The proposed method has proven successful in the case of using loaders and haulage trucks meeting Stage IIIA, Stage IIIB and Stage IV emission standards for the transport of the winning. The calculations show a reduction in HC+NO_X emissions during the transport of the winning from 35,110 kg for the Stage II standard to 1077 kg for the Stage IV standard. This is a very large reduction that confirms how restrictive the current standards are, while the permissible level of HC+NO_X emission for the combustion engines in the power range from 56 to 560 kW is the same for Stage IV and Stage V.

The driving style of the loader and hauler operators also has an impact on exhaust emissions. Unfortunately, we did not have portable emissions measurement systems (PEMS), nor did we have permission from the mine management to make such measurements. The solution to this problem was to use for calculations the permissible emission values resulting from Stage standards, being aware that the actual emission of pollutants during ore transportation is even higher than it results from the presented simulation calculations.

8. Discussion and Conclusions

The proposed method makes it possible to optimize the selection of the type of the winning transport in deep mines using the room and pillar exploration system, on the basis of the objective function based on two criteria: minimizing the cost of transporting the winning and the level of pollutant emissions in the exhaust gas (on the example of HC+NO_X). The method is based on scheduling the progress of mining works carried out according to the room and pillar exploration system, taking into account the stages of cross-cutting and retreating. Its basic assumption is to take into account the geometric dimensions of drifts and pillars as well as various means of transporting the excavated material in order to be able to perform calculations for different cases of mining fields. Calculated 20% reduction in HC+NO_X emissions with a 2% increase in transport costs shows the great potential of the developed method.

The advantages of the presented method and algorithms include the fact that PM emissions can be counted in the same way using the Stage or Tier emission standards set by the United States Environmental Protection Agency (EPA), as well as the mass of exhaust gases from internal combustion engines of mining machines. For this purpose, theoretical assumption that the combustion engine of the mining machine is powered by a mixture with twice the excess air and burning 1 kg of diesel oil means taking about 30 kg of air from the environment, might be applied. With an average fuel consumption of about 30 kg per hour of work, this means taking about 900 kg of air during this time and emission of a similar mass of exhaust gases to the excavation site [25], which can be easily recalculated into the mass of exhaust gases emitted during one working shift.

The idea of selecting the means of transporting the output (loaders and haul trucks) not only for efficiency and economic reasons, but also using the ecological criterion, may be useful not only during the operational management of transport processes. There is a potential possibility of using it when deciding which means of transport to supplement the existing machinery park in deep mines with; at purchase of new means of transport, preceded by an analysis of costs and levels of pollutant emissions from internal combustion engines. Another potential direction for the expansion of the mining field exploitation model is the possibility of taking into account the emission of pollutants or the mass of exhaust gases from other types of mining machines used, which should be interesting in the context of issues related to ventilation of workings.

Finally, there is the question of whether the calculated values of pollutant emissions from mining machinery exhaust gases, additionally spread over several years of mining field operation, are significant for the natural environment? When looking for an answer to this question, the scale of the phenomenon should be taken into account: at the same time, in one large mining concern, as many as 150–200 mining fields are operated in a room and pillar system. Deep mines have fiscal obligations related to environmental protection, resulting, inter alia, from the method of their exhaust shafts functioning, and they incur—perhaps too low—environmental charges for the emission of each ton of gas.

Author Contributions: Conceptualization, A.K. and R.W.; methodology, A.K.; validation, R.W.; formal analysis, A.K.; resources, A.K. and R.W.; data curation, A.K.; writing—original draft preparation A.K. and R.W.; writing—review and editing, A.K.; visualization, A.K.; funding acquisition, R.W. Both authors have read and agreed to the published version of the manuscript.

Funding: This research received no external funding.

Institutional Review Board Statement: Not applicable.

Informed Consent Statement: Not applicable.

Data Availability Statement: Not applicable.

Conflicts of Interest: The authors declare no conflict of interest.

References

1. *Regulation (EU) 2016/1628 of the European Parliament and of the Council—of 14 September 2016—on Requirements Relating to Gaseous and Particulate Pollutant Emission Limits and Type-Approval for Internal Combustion Engines for Non-Road Mobile Machinery, Amending Regulations (EU) No 1024/2012 and (EU) No 167/ 2013, and Amending and Repealing Directive 97/ 68/ EC*; Official Journal of the European Union: Brussels, Belgium, 2016.
2. Sandvik Mining and Rock Technology. Technical Specification Sandvik TH430L Underground Truck. Available online: https://www.rocktechnology.sandvik/en/products/underground-loaders-and-trucks/underground-trucks/th430l-underground-truck/ (accessed on 4 January 2021).
3. GHH Fahrzeuge Schmidt Kranz Group. Technical Specification GHH Fahrzeuge SLP-14H. Available online: http://www.ghh-fahrzeuge.de/de/produkte/fahrlader/flache-fahrlader/slp-14h/ (accessed on 4 January 2021).
4. KGHM ZANAM. Technical Specifications LHD LKP-0406. Available online: https://www.kghmzanam.com/en/produkty/mining-machinery/loaders/lhd-lkp-0406/ (accessed on 24 February 2021).
5. Grzybowska, K.; Hoffa-Dabrowska, P. Analysis of the Survival of Complex Systems with an Actions Coordination Mechanism. In Proceedings of the Highlights of Practical Applications of Cyber-Physical Multi-Agent Systems, Porto, Portugal, 21–23 June 2017; Springer International Publishing: Cham, Switzerland, 2017; pp. 197–208.
6. Grzybowska, K. Reconfiguration to Renovation in a Sustainable Supply Chain. In Proceedings of the Intelligent Systems in Production Engineering and Maintenance, Wrocław, Poland, 17–18 September 2018; Springer International Publishing: Cham, Switzerland, 2019; pp. 761–770.
7. Jacyna, M. *Wybrane Zagadnienia Modelowania Systemów Transportowych*; Oficyna Wydawnicza Politechniki Warszawskiej: Warszawa, Poland, 2009; ISBN 978-83-7207-817-9.
8. Bocewicz, G.; Nielsen, P.; Jasiulewicz-Kaczmarek, M.; Banaszak, Z. Dynamic Planning of Mobile Service Teams' Mission Subject to Orders Uncertainty Constraints. *Appl. Sci.* **2020**, *10*, 8872. [CrossRef]
9. Wasiak, M. *Modelowanie Przepływu Ładunków w Zastosowaniu Do Wyznaczania Potencjału Systemów Logistycznych*; Oficyna Wydawnicza Politechniki Warszawskiej: Warszawa, Poland, 2011; ISBN 1230-9265.
10. Kłosowski, G.; Gola, A.; Amila, T. Computational Intelligence in Control of AGV Multimodal Systems. *IFAC Pap.* **2018**, *51*, 1421–1427. [CrossRef]
11. Wasiak, M.; Jacyna-Gołda, I.; Wydawnictwo Naukowe PWN. *Transport Drogowy W Łańcuchach Dostaw. Wyznaczanie Kosztów*; Wydawnictwo Naukowe PWN: Warszawa, Poland, 2016; ISBN 978-83-01-18487-2.
12. Bocewicz, G.; Nielsen, I.; Gola, A.; Banaszak, Z. Reference Model of Milk-Run Traffic Systems Prototyping. *Int. J. Prod. Res.* **2020**, 1–18. [CrossRef]
13. Koenigsberg, E. Cyclic Queues. *J. Oper. Res. Soc.* **1958**, *9*, 22–35. [CrossRef]
14. Sturgul, J. *Mine Design—Examples Using Simulation*; Society for Mining Metallurgy & Exploration (SME): Littleton, CO, USA, 2000; ISBN 978-0-87335-181-2.
15. Harvey, P. Analysis of Production Capabilities. In Proceedings of the 4th APCOM Symposium 1964—Application of Computers and Operations Research in the Mineral Industry, Golden, Colorado, USA, 1964; pp. 713–726.
16. Rist, K. The Solution of a Transportation Problem by Use of a Monte Carlo Technique. In *Proceedings of the 1st International Symposium on Computer Application in Mining (APCOM)*; University of Arizona: Tucson, AZ, USA, 1961; Volume L2, pp. 1–15.
17. Sanford, R. Stochastic Simulation of a Belt Conveyor System. In *Proceedings of the International Symposium on Computer Application in Mining (APCOM)*; University of Arizona: Tucson, AZ, USA, 1965; Volume D, pp. 1–18.
18. Sturgul, J.R. *Discrete Simulation and Animation for Mining Engineers*; CRC Press: Boca Raton, FL, USA, 2015; ISBN 978-1-4822-5442-6.
19. Suboleski, S.; Lucas, J. Simulation of Room and Pillar Face Mining System. In *Proceedings of the International Symposium on Computer Application in Mining (APCOM)*; Society of Mining Engineers of the American Institute of Mining: Salt Lake City, UT, USA, 1969; pp. 373–384.

20. O'Neil, T.J.; Manula, C.B. *Computer Simulation of Materials Handling in Open Pit Mining*; Pennsylvania State Univ., University Park, Dept. of Mining: University Park, PA, USA, 1966.
21. Saayman, P. Optimization of an Autonomous Vehicle Dispatch System in an Underground Mine. Master's Thesis, University of Pretoria, Pretoria, South Africa, 2005.
22. Kowalski, A.; Chlebus, T. Multiagent Managerial Model of Technical Infrastructure Used at Ore Mining. In *Highlights of Practical Applications of Cyber-Physical Multi-Agent Systems*; Bajo, J., Vale, Z., Hallenborg, K., Rocha, A.P., Mathieu, P., Pawlewski, P., Del Val, E., Novais, P., Lopes, F., Duque Méndez, N.D., et al., Eds.; Springer International Publishing: Cham, Switzerland, 2017; Volume 722, pp. 187–196, ISBN 978-3-319-60284-4.
23. Kowalski, A.; Sobol-Wojciechowska, J.; Szwancyber, Ł.; Śliwiński, P. Komputerowe Techniki Symulacyjne do Optymalizacji Procesu Odstawy Urobku w Kopalniach KGHM "Polska Miedź" SA. *Wiadomości Górnicze*. 2013. Available online: http://yadda.icm.edu.pl/yadda/element/bwmeta1.element.baztech-830ea633-31da-4502-a541-4de3ee36c838 (accessed on 12 May 2021).
24. Kowalski, A. *Metodyka Planowania Efektywności Wydobycia Urobku w Komorowo-Filarowych Systemach Eksploatacji*; Oficyna Wydawnicza Politechniki Wrocławskiej: Wrocław, Poland, 2019; ISBN 978-83-7493-072-7.
25. Szlązak, N.; Borowski, M.; Sporysz, G. Analiza emisji składników spalin z maszyn z silnikami spalinowymi w kopalniach podziemnych. *Górnictwo Geol.* **2013**, *8*, 101–114.

applied sciences

MDPI

Article

Improve the Energy Efficiency of the Cooling System by Slide Regulating the Capacity of Refrigerator Compressors

Javier Cárcel-Carrasco *, Manuel Pascual-Guillamón and Fidel Salas-Vicente

ITM, Institute of Materials Technology, Universitat Politècnica de València, 46022 Valencia, Spain; mpascual@mcm.upv.es (M.P.-G.); fisavi@doctor.upv.es (F.S.-V.)
* Correspondence: fracarc1@csa.upv.es; Tel.: +34-963-87-7000; Fax: +34-963-87-9459

Abstract: A fundamental part of the electric consumption of the main industries of the food sector comes from the refrigeration production, needed in all production phases. Therefore, every measure that aims to optimize the electric consumption and increase the efficiency of centralized industrial refrigeration systems will help the energetic waste of the company, improving reliability and maintenance. Acting on the regulation of capacity of power compressors used can be a good way to save energy. This article shows a case studied by the authors in an industrial company in the meat industry in Spain, where the refrigeration systems have a great importance in the productive process. It displays the methodology used, the description of the taken actions and the results obtained. These combined measures brought about an improvement, with an energetic saving value reaching 400 MWh per year, leading to an equivalent in CO_2 emission reduction of 147.9 tons.

Keywords: industrial refrigeration plant; food industry; energy efficiency; refrigeration compressors

Citation: Cárcel-Carrasco, J.; Pascual-Guillamón, M.; Salas-Vicente, F. Improve the Energy Efficiency of the Cooling System by Slide Regulating the Capacity of Refrigerator Compressors. *Appl. Sci.* **2021**, *11*, 2019. https://doi.org/10.3390/app11052019

Academic Editor: Jasiulewicz-Kaczmarek Małgorzata

Received: 28 January 2021
Accepted: 21 February 2021
Published: 25 February 2021

1. Introduction

This work shows the potential opportunities for energy saving in a large industry in its installations of industrial refrigeration, through combining the use of speed regulation of the compressors with the slide control in big refrigerators' compressors. For this purpose, refrigeration production has been studied, considering a large factory in the meat industry situated in the Valencian Community (Spain) [1], analysing electric energy consumption, suggesting the actions to be carried out and executing the necessary measures to achieve maximum efficiency in the production of industrial refrigeration.

The energy consumption is considerable for this type of consumer, and the meat industry has been identified as one of the most suitable segments for the implementation of actions to improve energy demand [2,3]. Heat, ventilation, and refrigeration are among the largest amounts of energy consumed in processes in the meat industry [4]. Electricity consumption is mainly used for cooling and ventilation, while fossil fuels such as natural gas or diesel are generally used for process heating. In the meat industry, refrigeration production and distribution constitute between 45% and 55% of the total final electricity consumption on weekdays [4], making this process more energy intensive for most consumers in this segment. In the meat industries, the energy demand to produce industrial refrigeration can represent a very important percentage of the energy consumption, generally higher than 50%. There are studies that have been carried out [5–7] to evaluate the response of the demand of different sectors (mainly for commercial and industrial sectors), where flexibility has traditionally been related to the capacity of a system to adapt itself to production changes [8,9] or it can absorb these changes depending on any of the system entities or the external environment [10]. Optimizing and improving energy demand in refrigeration can be an important point to achieve in this type of industry.

Improving the energy efficiency of industrial refrigeration systems in the meat industry has been analysed from different aspects [11–15]. In this work, the results of the improvement of a large industrial refrigeration system will be seen, describing a practical case where measures have been implemented for the energy improvement of this system by regulating the speed of electric motors in combination with the variation of the volumetric compression ratio by the physical regulation of the mechanical slide integrated in the compressors, allowing one to evaluate the effectiveness of this technique in such a promising sector as the meat industry.

This article analyses the study of a real case in a first level meat industry located in Spain, where energy consumption in industrial refrigeration is 44.1% of the total. The work is organized as follows: Section 2 describes the main characteristics of the company's industrial refrigeration system. Later (Section 3) the measurements, carried out to establish the improvement process by combining the speed regulation with the mechanical slide regulation of the compressors to adapt the system capacity, are exposed. Once the capacity of the different refrigeration lines has been analysed, Section 4 shows the measures adopted and the results obtained, which show significant energy savings in the refrigeration plant, close to 400 MWh per year.

2. Improve of Regulation Capacity in Refrigeration Compressors. Case Study

The company studied is a large industry located in the Valencian Community in Spain, where it has a complex consisting of a beef processing centre surrounded by the attached facilities necessary for the proper functioning of the main activity. In this factory, two work shifts of 8 hours each are executed daily, plus a third cleaning shift.

This factory was designed to operate with the greatest respect for the environment, as well as to achieve the highest efficiency ratios in operation, having obtained important national and international awards. Electric energy consumption can be seen in the graphs in Figures 1 and 2, where the electricity consumption profile for different times of the year is observed. Within the electrical energy consumption of the factory, the refrigeration systems account for a percentage of 44.1% of the total (Figure 3).

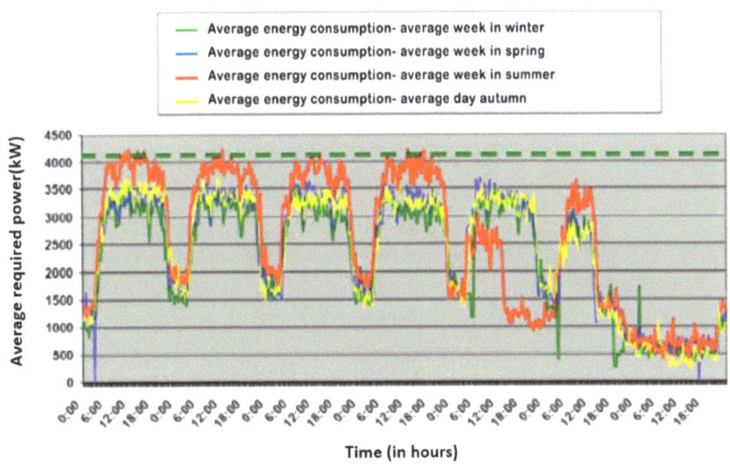

Figure 1. Electricity consumption profile during average week in the factory.

The factory's refrigeration plant (Figures 4 and 5) is made up of nine screw-type refrigeration compressors that use ammonia (NH_3) as a refrigerant, which is distributed through three main lines (Figure 4). Line N°1 at −40 °C of evaporation is associated with the freezing tunnels, line N°2 at −33 °C of evaporation is for the meat treatment processes and freezing chambers, and Line N°3 at −15 °C of evaporation is for chambers and the heated meat processing areas (Figures 4 and 5). The compressors used in the industrial refrigeration system have the technical characteristics of the equipment for the specific operating data of the different operating refrigeration circuits, as described in Table 1.

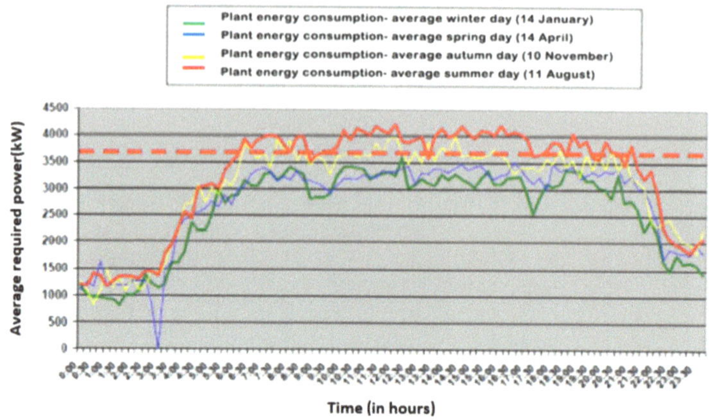

Figure 2. Electricity consumption profile in the factory during four periods.

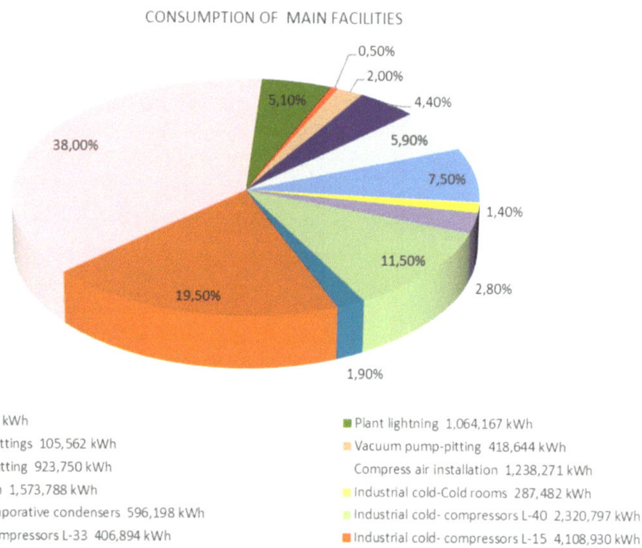

Figure 3. Sectorization of electricity consumption by main facilities.

Table 1. Technical and operational characteristics of refrigeration equipment.

Designation (ID)	Line	Nominal Cold Power (Kw)	Nominal Absorbed Power (Kw)	Cos ϕ	V	Average Annual Capacity (%)		Annual Working Hours (h)
A9/C9		347.2	394.3	0.84	408		74.9	4.318
A1/C1	NH3	227.8	268.5	0.91	408	82	85	8.428
A2/C2	−40 °C	227.8	268.5	0.91	408		84.5	1.931
A3/C3	NH3	110.4	88.6	0.77	408	75.9	77.5	3.630
A4/C4	−33 °C	110.4	88.6	0.77	408		74	3.062
A5/C5		1108.9	336	0.9	408		70.8	5.634
A6/C6	NH3	1108.9	336	0.89	408	62.92	61.3	3.342
A7/C7	−15 °C	1108.9	336	0.89	408		55.8	3.254
A8/C8		1108.9	336	0.89	408		52.2	1.478

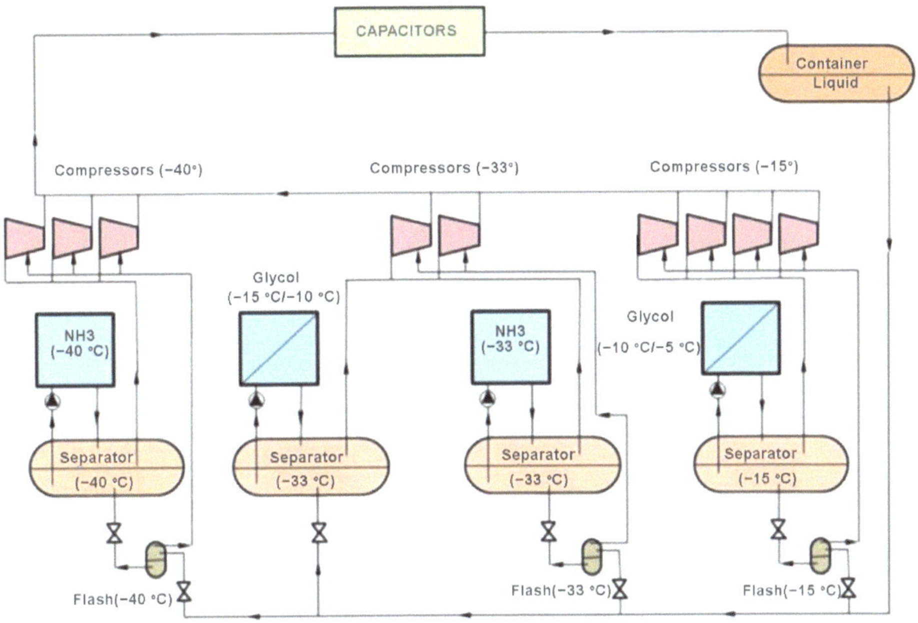

Figure 4. Refrigeration system main diagram.

Figure 5. View of industrial refrigeration compressor room under study.

The automatic regulation of the capacity of the refrigeration compressors is carried out by means of an integrated PID (proportional, integral, derivative) function that modifies the location of the mechanical slide integrated in the compressor to adapt the volumetric compression ratio (Vi) to the existing working conditions (thermal load), that are determined from the variation of the suction pressure (evaporation pressure) of the compressors.

With the variation of the slide, the quantity of refrigerant gas driven is directly affected by the recirculating part of the refrigerant sucked in before being compressed, thereby significantly reducing the associated energy consumption.

With the installation of variable speed drives to the compressor motors, the necessary frequency is delivered based on the suction pressure of the refrigeration system; if it rises the frequency and speed of rotation will rise, increasing the capacity of the compressor until it reaches the pressure corresponding to the required evaporation temperature.

The voltage and power factor values have been extracted from the measurements made during the data collection phase. The cooling capacity values have been obtained from the compressor manufacturer's computer application for the average operating pressure values of the different refrigeration circuits (Table 2):

Table 2. Average operating pressure values of the refrigeration circuits.

	Paspiration	Pcondensation
Circuit NH_3 −40 °C	0.61 bar	12.40 bar
Circuit NH_3 −33 °C	1.20 bar	12.60 bar
Circuit NH_3 −15 °C	2.35 bar	12.20 bar

3. Measurements

In Table 1, it can be seen how the work between the different compressors of the same circuit is quite distributed. This fact, together with the number of annual working hours that is limited since the facility works two shifts and not three, conditions the return on investment for a general speed variation solution.

Figure 6 graphically shows the operating profile obtained from the different groups of compressors in a typical summer week, during a few days of the first week of July. In the same way, more detailed measurements were made for a typical winter day during the month of February (Figure 7).

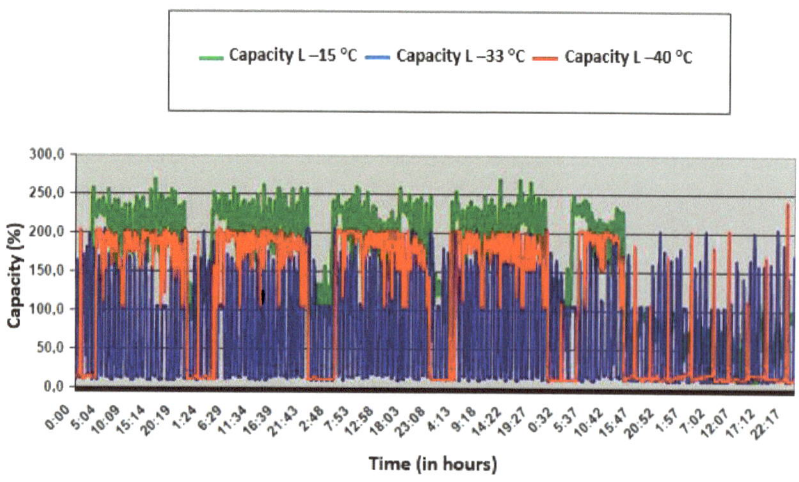

Figure 6. Operation profile of the different groups of compressors in the summer period.

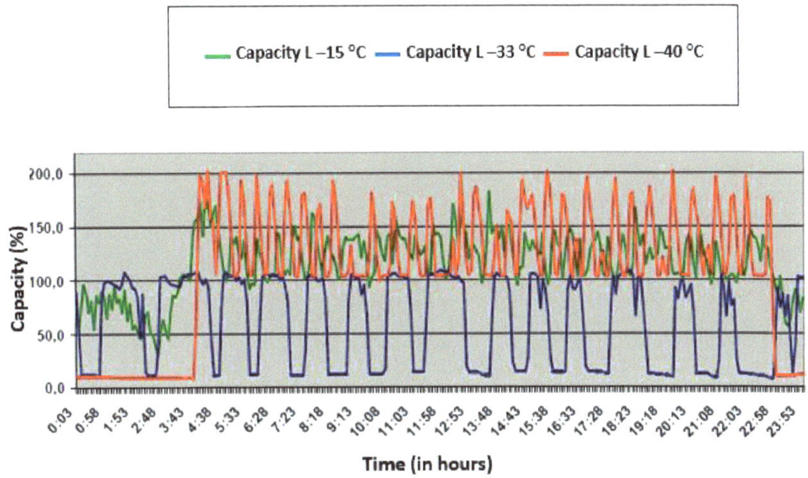

Figure 7. Operation profile for a typical day in February.

The data in Figures 6 and 7 obtained from the operation measurements of the factory facilities control system help to visualize the operation mode and the average capacity level of the facility. With the capacity analysis, it was possible to measure the percentage of time (in a period of 30 days) for a certain level of capacity for each of the three refrigeration circuits (Figure 8).

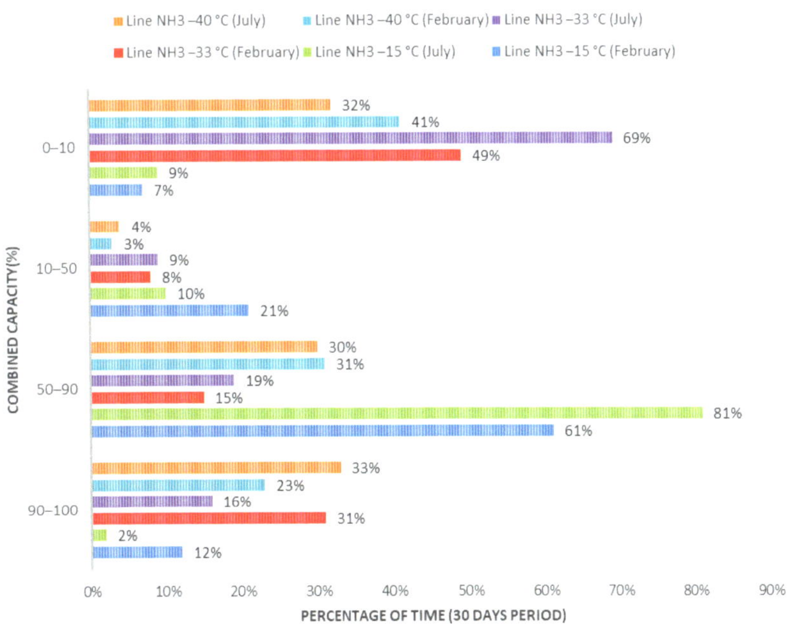

Figure 8. Capacity analysis for refrigeration circuits.

A greater detail of the independent capacity analysis for each of the lines shows the results for each of the lines: the line for the −40 °C circuit formed by three compressors (A1, A2 and A9) (Figure 9); the line for the −33 °C circuit formed by two compressors (A3 and A4) (Figure 10); the line for the −15 °C circuit formed by four compressors (A5, A6, A7 and A8) (Figure 11).

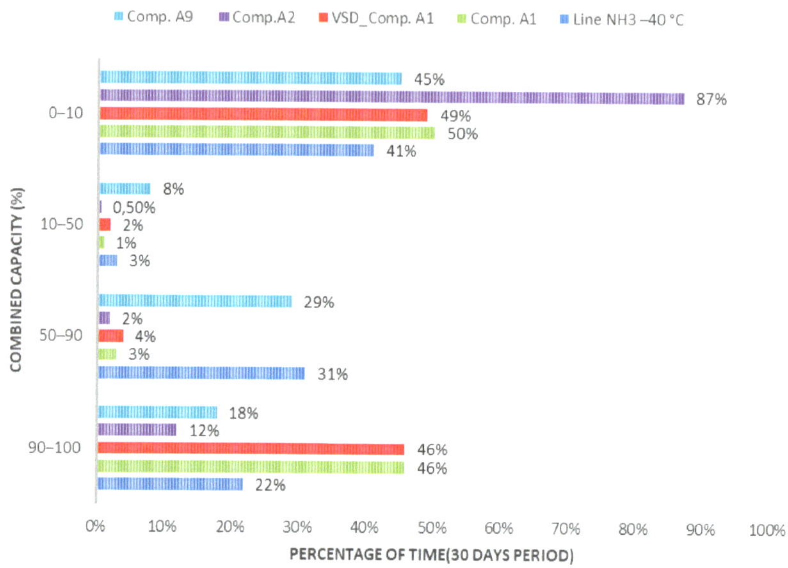

Figure 9. Capacity analysis for the −40 °C circuit.

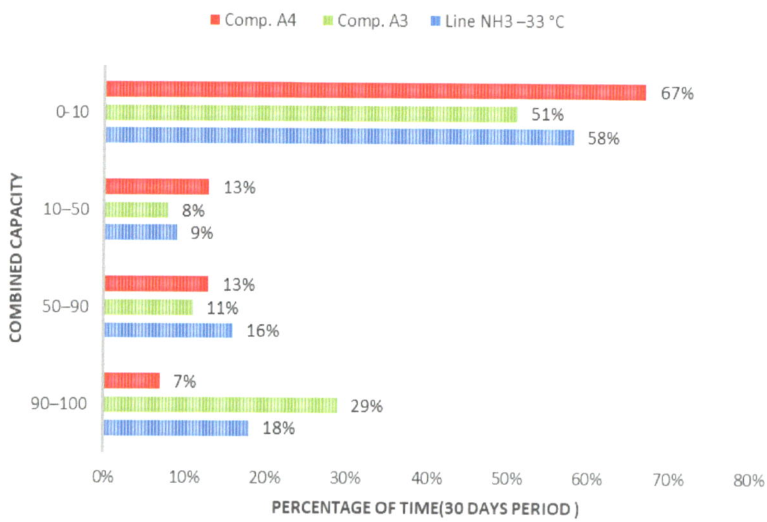

Figure 10. Capacity analysis for the −33 °C circuit.

In Figure 9, the red columns indicate the percentage of time it is at a certain percentage (of capacity) over the nominal revolutions of the compressor, unlike the rest of the columns that the joint capacity axis indicates the average position in which the slide is located and therefore the average load percentage of the compressor.

Based on the results obtained, it is easy to see how only compressor A9 presents an opportunity in terms of capacity regulation through speed variation (VSD), since the rest of compressors, including compressor A1 which already has VSDs, are practically above

90% all the time or stopped, which greatly limits the return on the investment of a solution of this type from the energy point of view.

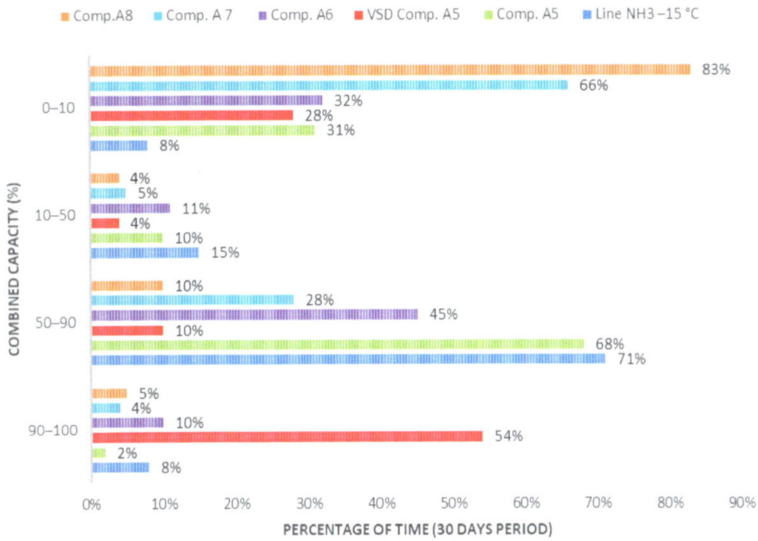

Figure 11. Capacity analysis for the −15 °C circuit.

It should be noted that during the analysis, it is seen that the capacity regulation in compressor A1 is carried out mainly by slide. As can be seen in Figure 9, the capacity regulation by speed variation (red column) occurs in very punctually and in short periods of time, with the result being that 95% of the time, the compressor is stopped or above 90% of the nominal revolutions of the compressor.

In the capacity analysis of the circuit at −33 °C (Figure 10) formed by two compressors (A3 and A4), more than 20% of the time, the compressors in this circuit are at partial load, which means that almost 60% of the time they are unemployed, which is why the capacity regulation solution by speed variation is not economically viable from the point of view of strictly energy saving. Despite this, there may be other technical reasons and reliability that do justify it.

For the circuit at −15 °C, the capacity analysis for the four compressors gives the results reflected in Figure 11.

As in the case of the compressor A1 of the −40 °C circuit, the compressor A5 is stopped for more than 85% of the time or operating above 90% (red columns) of the nominal speed of the compressor (3000 rpm). Despite this, in the same percentage of time, it is between 50% and 90% the of capacity (brown columns), which indicates that the regulation is carried out by the slider practically all the time, with the speed variation operating in a timely manner. This last point indicates that probably, the variable speed drives in both the A1 and A5 cases seem to be programmed mainly for starts, and not to regulate capacity according to energy efficiency criteria.

The data to highlight are that the set of compressors is at an average of 70% of the time between 50% and 90% of capacity, a fact that points to a clear opportunity for energy saving for the proposed energy saving solution.

From this capacity analysis, energy efficiency measures can be established by means of two speed regulation actions in compressors A6 and A9, in combination with capacity regulation by the slide.

4. Improvement of Energy Efficiency by Speed Regulation in Compressors A6 and A9. Results Obtained

To calculate energy savings for the speed variation solution in compressor number 6 the histogram in Figure 11 is considered to determine the number of hours it operates at a certain degree of capacity as well as the COP values (Coefficient of Performance), for the same percentages of capacity, obtained from the manufacturer's data, the result of which can be seen in Table 3. The percentage of time is based on the maximum hours that can be worked per year (8760 h).

Table 3. Energy saving results due to speed variation in compressor No. 6.

COMP.A6 Capacity (%)	Time	Absorbed Power (kW)	Refrigeration Power (kW)	COP (no VSD)	Abs. Power VSD (kW)	COP (VSD)	Specific Saving	Estimated Saving (kWh)
95–100	0.00%	313.20	1230.80	3.93	322.60	3.82	2.00%	0
90–95	9.97%	299.70	1144.40	3.82	299.52	3.79	0.69%	1.800
85–90	7.24%	290.40	1083.70	3.73	283.25	3.77	−0.98%	−1.808
80–85	12.57%	281.30	1023.50	3.64	267.00	3.74	−2.78%	−8.627
75–80	7.08%	272.20	963.20	3.54	250.70	3.71	−4.73%	−7.992
70–75	5.99%	263.30	902.50	3.43	234.40	3.68	−6.94%	−9.591
65–70	3.98%	254.30	840.90	3.31	218.15	3.65	−9.37%	−8.301
60–65	2.92%	245.20	777.90	3.17	201.88	3.61	−12.10%	−7.580
55–60	2.42%	222.90	684.60	3.07	185.60	3.57	−13.85%	−6.554
50–55	3.90%	215.00	638.60	2.97	169.40	3.51	−15.48%	−11.379
45–50	2.22%							
40–45	1.32%							
35–40	1.46%							
30–35	1.40%	149.50	343.90	2.30	118.00	2.51	−8.37%	−12.408
25–30	1.31%							
20–25	1.39%							
15–20	1.09%							
10–15	1.14%							
0–10	32.61%	0.00	0.00	0.00	0.00	0.00	0.00	0.00

The energy saving result for this compressor is 72,440 kWh (sum of all the savings obtained for each percentage of capacity). To estimate the total savings, we have obtained the savings by adding the savings of the rest of the compressors in the circuit, since it is assumed that the compressor with speed variation would remain in the queue, regulating the capacity of the circuit and the rest could be kept at 100% capacity levels, maximizing the COP of all compressors. In the same way, the savings analysis carried out for the rest of the −15 °C line compressors is as follows: compressor A5: 62,865 kWh; compressor A6: 72,440 kWh; compressor A7: 45,804 kWh; compressor A8: 25,663 kWh.

The total annual energy saving with the measurements carried out in the A6 compressor, regulating the capacity of the −15 °C circuit, is 206,772 kWh.

Although, as previously mentioned, the number of operating hours is a limiting factor, it is possible to keep a compressor operating for several hours to maximize energy savings.

Based on the capacity analysis carried out, it was proposed to optimize the capacity management of the A6 compressor of the ammonia circuit at −15 °C, integrating a speed variator and managing the slide regulation in conjunction with the speed variation as a function of the suction pressure (evaporation) so that the system can operate according to the graph in Figure 12 extracted from the characteristics provided by the manufacturer of the refrigeration compressors.

As can be seen in the data in Figure 12, throughout the regulation range (100% to 44%), limits imposed by the working conditions of the motor/compressor (lubrication and ventilation) established by the manufacturer regarding the spin speed (2950 rpm to 1475 rpm), savings of up to 31% can be achieved. It is also worth noting that for the upper regulation range (100% to 80%), the capacity control by means of the slide acts in a similar way to the speed variation conditions, with the consequent reduction in energy saving opportunities.

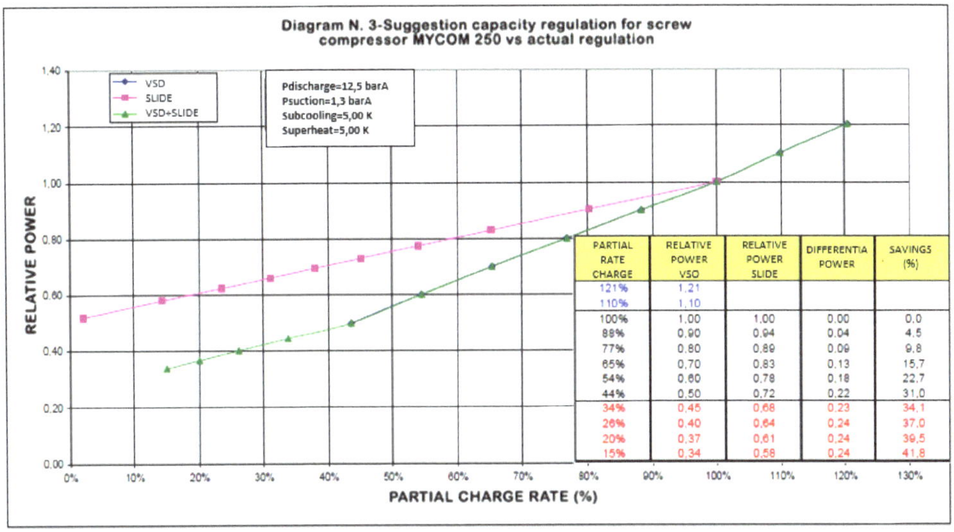

PARTIAL RATE CHARGE	RELATIVE POWER VSD	RELATIVE POWER SLIDE	DIFFERENTIA POWER	SAVINGS (%)
121%	1,21			
110%	1,10			
100%	1,00	1,00	0,00	0,0
88%	0,90	0,94	0,04	4,5
77%	0,80	0,89	0,09	9,8
65%	0,70	0,83	0,13	15,7
54%	0,60	0,78	0,18	22,7
44%	0,50	0,72	0,22	31,0
34%	0,45	0,68	0,23	34,1
26%	0,40	0,64	0,24	37,0
20%	0,37	0,61	0,24	39,5
15%	0,34	0,58	0,24	41,8

Figure 12. Integration graph of slide regulation with speed variation in compressor No. 6 and 9.

Another aspect to note is that with the speed variation option, the working range can be extended to reach the maximum rotation speed allowed by the manufacturer (mechanical considerations) of 3540 rpm, and in this way the compressors could expand their theoretical cooling capacity to 120% of the nominal. This possibility cannot be realized by means of mechanical slide control.

The variable speed drives will be those appropriate to the power of the compressor motor. It will involve integrating input/output signals to regulate the electronics of the compressors, PLC (Programmable Logic Controller) programming integrating the control ramp, as well as a new electrical installation to integrate cabinets with variable speed drives and new electrical wiring with the required technical condition for this type of facility.

In the case of the −40 °C line, for the savings calculation for the speed variation solution in compressor number 9, the histogram of Figure 9 is considered to determine the number of hours it operates at a determined degree of capacity, as well as the COP values for the same capacity percentages obtained from the manufacturer's computer application, the result is shown in Table 4. The percentage of time is based on the maximum hours that can be worked per year (8760 h).

The energy saving result for this compressor is 184,522 kWh (sum of all the savings obtained for each percentage of capacity). The total energy savings, considering the savings obtained by adding the savings of the rest of the compressors in the circuit with the measurements made in the A9 compressor regulating the capacity of the −40 °C circuit, is 201,707 kWh.

In the same way as with the A6 compressor of the −15 °C circuit, and according to the capacity analysis carried out, it was proposed to optimize the capacity management of the compressor A9 of the ammonia circuit at −40 °C, integrating a speed variator and managing the slide regulation in together with the speed variation as a function of the suction pressure (evaporation), so that the system can operate according to the graph in Figure 12.

Table 4. Energy saving results due to speed variation in compressor N°. 9.

COMP.A9 Capacity (%)	Time	Absorbed Power(kW)	Refrigeration Power (kW)	COP (no VSD)	Abs. Power VSD (kW)	COP (VSD)	Specific Saving	Estimated Saving (kWh)
95–100	13.53%	270.10	403.90	1.5	278.20	1.45	2.00%	6.404
90–95	4.98%	259.00	354.90	1.37	248.84	1.43	−3.92%	−4.429
85–90	4.31%	251.80	323.50	1.28	230.30	1.4	−8.54%	−8.125
80–85	3.89%	245.40	294.80	1.2	213.31	1.38	−13.08%	−10.929
75–80	3.40%	239.50	268.40	1.12	197.55	1.36	−17.52%	−12.513
70–75	2.86%	234.00	243.90	1.04	182.82	1.33	−21.87%	−12.809
65–70	2.44%	229.00	221.10	0.97	169.23	1.31	−26.10%	−12.765
60–65	2.84%	224.30	199.50	0.89	156.04	1.28	−30.43%	−16.979
55–60	2.84%	207.60	164.00	0.79	136.06	1.21	−34.46%	−17.787
50–55	7.07%	204.70	152.10	0.74	131.84	1.15	−35.59%	−45.119
45–50	0.71%							
40–45	0.75%							
35–40	0.74%							
30–35	0.86%	191.10	95.00	0.50	111.54	0.87	−42.86%	−49.471
25–30	0.86%							
20–25	0.87%							
15–20	0.84%							
10–15	1.27%							
0–10	44.95%	0.00	0.00	0.00	0.00	0.00	0.00	0.00

5. Discussion

In view of the study to improve the energy efficiency of the industrial refrigeration system for a meat processing factory, it has been observed that the combination of measures by means of capacity variation by mechanical slide of the refrigeration compressors together with speed variation actions produces better results that translate into a substantial improvement in combined energy efficiency.

The analysis should consider the whole of the entire refrigeration production system considering the total number of operating hours; this makes the best decision to be made, since it is not necessary to install speed regulation in all the electric motors of the compressors, but rather in those that make the regulation of the industrial refrigeration system more optimal to the joint capacity. This measure reduces the investment costs to be made for the total improve of the system.

These combined measures improved on energy saving close to 400 MWh per year, which entails an equivalent reduction of CO_2 emissions of 147.9 tons.

Likewise, it has been found that after a period of application of the measures, there are other advantages in addition to energy savings, such as improved maintainability due to a better distribution of capacity and preventing components from operating excessive hours and others few. In the same way, the reliability of the system has increased due to the reduction of failures due to the more efficient operation of the motors and the reduction of starts.

These collateral results are important in the operation and maintenance of this type of industrial plant [1,16], being one of the areas of concern to achieve efficient industries with respect to the environment [17–20]. With all this, the importance of combining different regulation systems in industrial refrigeration compressors is shown in order to improve energy efficiency, so that it can be understood by professionals in the sector and introduce new lines of research to improve the efficiency of these systems.

6. Conclusions

In large industrial refrigeration systems with capacity regulation by a mechanical slide, it was interesting to carry out the study in combination with speed regulation to optimize joint efficiency.

Obviously, given the complexity of the refrigeration facilities, the implementation of energy improvements does not always result in significant savings, and in some cases, they can have a direct impact on the reliability of the facilities, so carrying out technical and economic viability studies is essential.

It is important to note that, in addition to the economic savings and the use of variable speed drives that will entail as capacity control in screw compressors, there are other considerations that must be considered, such as the improvement in maintenance or the increase in the reliability of the system.

The reduction in capacity due to speed variation will reduce wear and damage to the compressor slide valves.

The stability of the suction pressures will be further optimized since the capacity control is direct.

Operating at reduced speed, if the load profile so requires, will reduce wear on the mechanical elements of the compressor.

At an electrical level, the operation of the installation and the motors will improve, since with the variable speed drives, the power factor will be constant closely to 1, so the reactive energy and power of the installation will be reduced.

The combined measures improved the refrigeration system with an energy saving close to 400 MWh per year.

The use of variable speed drives in standard motors should be studied in detail, since the speed reduction implies a linear reduction in power (constant torque), so at low rotor speeds, the level of cooling may not be adequate to extract the heat generated, causing overheating problems. This situation is easily solved with the inclusion of a forced ventilation accessory, control of the winding temperature (thermal probes) or control of motor overloads (drive control module).

Author Contributions: In this investigation, J.C.-C. and M.P.-G. conceived and designed the experiments; J.C.-C. and F.S.-V. performed the experiments; J.C.-C., F.S.-V. and M.P.-G. analyzed the data and contributed materials/analysis tools; J.C.-C. and F.S.-V. wrote the paper. All authors have read and agreed to the published version of the manuscript.

Funding: This research received no external funding.

Institutional Review Board Statement: Not applicable.

Informed Consent Statement: Not applicable.

Data Availability Statement: Not applicable.

Acknowledgments: This work has been possible thanks to the support shown by all the technical and managerial staff of the industry that is the object of this article, as well as the study carried out by the industrial group Schneider Electric España S.A. on energy efficiency in the factory subject of this case study.

Conflicts of Interest: The authors declare no conflict of interest.

References

1. Cárcel-Carrasco, J.; Roldan-Porta, C.; Grau-Carrion, J. La sinergia entre el diseño de planta industrial y mantenimiento-explotación eficiente. un ejemplo de éxito: El caso Martínez Loriente, S.A. *Dyna Ing. Ind.* **2013**, *89*, 159–164. [CrossRef]
2. The Birth of a European Distributed Energy Energy Partnership that will Help the Large-Scale Implementation of Distributed Energy Resources in Europe (EU-DEEP), the European Project Supported by the Sixth Framework Programme for Research and Technological Development. Available online: http://www.eu-deep.com (accessed on 8 September 2020).
3. Alfonso, D.; Pérez-Navarro, A.; Encina, N.; Álvarez, C.; Rodríguez, J.; Alcázar, M. Methodology for ranking of customer segments by their suitability for distributed energy resources applications. *Energy Convers. Manag.* **2007**, *48*, 1615–1623. [CrossRef]

4. Ramírez, C.A.; Patel, M.; Blok, K. How much energy to process one pound of meat? A comparison of energy use and specific energy consumption in the meat industry for European countries. *Energy* **2006**, *31*, 2047–2063. [CrossRef]
5. Álvarez, C.; Alcázar, M.; Escrivá, G.; Gabaldón, A. Technical and economical tools to assess customer demand response in the commercial sector. *Energy Convers. Manag.* **2009**, *50*, 2605–2612.
6. Álvarez, C.; Gabaldón, A.; Molina, A. Assessment and simulation of the responsive demand potential in end-user facilities: Application to a university customer. *IEEE Trans. Power Syst.* **2004**, *19*, 1223–1231. [CrossRef]
7. Alcázar-Ortega, M.; Álvarez-Bel, C.; Escrivá-Escrivá, G.; Domijan, A. Evaluation and assessment of demand response potential applied to the meat industry. *Appl. Energy* **2012**, *92*, 84–91.
8. Hakam, M.H.; Solvang, W.D. *On Flexibility and Sustainability in Container Ports*; IEEE International Conference on Service Operations, Logistics and Informatics: Chicago, IL, USA, 2009.
9. Zelenovic, D.M. Flexibility-a condition for effective production systems. *Int. J. Prod. Res.* **1982**, *20*, 319–337. [CrossRef]
10. Zahran, I.M.; Elmaghraby, A.S.; Shalaby, M.A. Evaluation of Flexibility in Manufacturing Systems. In Proceedings of the 1990 IEEE International Conference on Systems, Man, and Cybernetics Conference Proceedings, Los Angeles, CA, USA, 4–7 November 1990.
11. Calm, J.M. Emissions and environmental impacts from air-conditioning and refrigeration systems. *Int. J. Refrig.* **2002**, *25*, 293–305. [CrossRef]
12. Yang, L.; Zhang, C.-L. On subcooler design for integrated two-temperature supermarket refrigeration system. *Energy Build.* **2011**, *43*, 224–231. [CrossRef]
13. Aprea, C.; Maiorino, A.; Mastrullo, R. Change in energy performance as a result of a R422D retrofit: An experimental analysis for a vapor compression refrigeration plant for a walk-in cooler. *Appl. Energy* **2011**. [CrossRef]
14. Bell, I.H.; Groll, E.A. Air-side particulate fouling of microchannel heat exchangers: Experimental comparison of air-side pressure drop and heat transfer with plate-fin heat exchanger. *Appl. Therm. Eng.* **2010**, *31*, 742–749. [CrossRef]
15. Escrivá-Escrivá, G.; Segura-Heras, I.; Alcázar-Ortega, M. Application of an energy management and control system to assess the potential of different control strategies in HVAC systems. *Energy Build.* **2010**, *42*, 2258–2267. [CrossRef]
16. Cárcel-Carrasco, J.; Cárcel-Carrasco, J.-A. Analysis for the Knowledge Management Application in Maintenance Engineering: Perception from Maintenance Technicians. *Appl. Sci.* **2021**, *11*, 703. [CrossRef]
17. CARPL. *Centro de Actividad Regional Para la Producción Limpia (CAR/PL). Prevención de la Contaminación en la Industria Cárnica en la Región Mediterránea*; CAR/PL: Barcelona, Spain, 2006.
18. MMA, Ministerio de Medio Ambiente. *Guía de Mejores Técnicas Disponibles en España del Sector Cárnico*; Ministerio de Medio Ambiente-Spain: Madrid, Spain, 2005.
19. Jacques, S. Eficiencia Energética: Ventajas del uso de los Variadores de Velocidad en la Circulación de Fluidos. Cuaderno Técnico n° 214 de Schneider Electric. 2010. Available online: https://www.se.com/es/es/download/document/CT-214-Z002/ (accessed on 24 August 2020).
20. Web 1. Schneider Electric. Guía Práctica de Eficiencia Energética. Available online: https://www.se.com/es/es/download/document/ESMKT12001H10/ (accessed on 8 September 2020).

applied sciences

MDPI

Article

Analysis for the Knowledge Management Application in Maintenance Engineering: Perception from Maintenance Technicians

Javier Cárcel-Carrasco [1,*] and José-Antonio Cárcel-Carrasco [1,2]

1 Department of Architectural Constructions, Universitat Politècnica de València, Camino de Vera s/n, 46022 València, Spain; jacarcel@tecnatom.es
2 Tecnatom S.A., C.N. Cofrentes s/n, Cofrentes, 46625 València, Spain
* Correspondence: fracarc1@csa.upv.es; Tel.: +34-96-387-7000

Abstract: Knowledge based on personal experience (tacit knowledge) acquired in problem solving actions and in maintenance actions is the fundamental basis for maintenance technicians in companies with great physical assets. Generally, there is no proper policy for managing strategic knowledge and its capture. In this article, through qualitative studies (grounded theory) and surveys conducted with technicians, the aim was to obtain the perception of the maintenance technicians' part of the companies, in order to establish the characteristics of the relation between the strategic aspects and the engineering aspects of industrial maintenance, regarding knowledge management, as well as the enablers and barriers to its application. The results show how a high level of tacit knowledge is used in this activity, which requires more time for new staff. The values obtained from this survey show that the knowledge recorded by the companies (explicit) is 51.25%, compared to the personal knowledge (tacit) of maintenance technicians regarding reliability and breakdowns. In operational/exploitational actions it is 43.90%, for energy efficiency actions it is 49.61%, and in maintenance actions (preventive, predictive, and corrective) the value is 68.78%. This shows the significant gap between the perception of recorded knowledge (explicit), and the knowledge that maintenance technicians have (tacit knowledge). All this can affect the companies, as part of the strategic knowledge is lost when a maintenance technician leaves the company.

Keywords: knowledge management; industrial maintenance; tacit knowledge; large building maintenance; Industry 4.0

Citation: Cárcel-Carrasco, J.; Cárcel-Carrasco, J.-A. Analysis for the Knowledge Management Application in Maintenance Engineering: Perception from Maintenance Technicians. *Appl. Sci.* **2021**, *11*, 703. https://doi.org/10.3390/app11020703

Received: 6 December 2020
Accepted: 10 January 2021
Published: 13 January 2021

1. Introduction

In industrial maintenance engineering, in companies with large physical and productive assets or buildings, knowledge of facilities is essential, since it is key in transforming tacit knowledge into explicit strategic knowledge of the operational experiences of the maintenance technicians who operate them. An in-depth study of these measures will lead to energy efficiency and the improved reliability of the processes and production of the company.

The main indicators of maintenance efficiency are based on various pillars, such as reliability, availability, maintainability, safety, operating costs, and the human factor, which is a fundamental one for operability, and in which operates a high level of tacit knowledge that is generated by people in the performance of assigned maintenance tasks. The reliability of the machines or systems requires a high level of knowledge and experience, and is linked to the study of the failure process, including the aspects of reliability and quality of the work that maintenance departments carry out [1–6], and it is possible to establish new indicators that allow estimating the security level of these systems, and which describe the impact on infrastructure, and the associated risks [7,8].

The term "Industry 4.0" refers to the latest industrial revolution, which uses artificial intelligence to greatly change the way in which machines and facilities collect and interpret data and information [9,10]. However, a lot of information (often strategic) is in the hands of maintenance technicians, based on their own knowledge of the workplace (tacit knowledge), and which is lost when a technician leaves the company.

The technicians which work in the maintenance departments of companies, normally operate using their professional experience acquired from years of experience, which has a high component of tacit knowledge, and traditionally without knowledge management policies being applied in these companies [11,12].

The fundamental factors that determine operational reliability (human and technical) of maintenance departments must be taken into account [13–16]. Exploitation operations (also called facilities management) are the normal processes that happen during the production or service provided by the company, which involves utilization of the facilities, rearming of switches, start-up manoeuvres, stopping processes, etc.

An adequate maintenance management connects and transmits information between the departments of a company, but also between sensors, devices, or systems. It also transmits information between maintenance technicians themselves [17–20]. Through the data collected, the new devising of maintenance solutions can generate strong analyses, in order to support professionals in decision-making and troubleshooting actions [21,22].

The main advantage of predictive maintenance is the reduction in maintenance interventions or unscheduled breakdowns, based on data from sensors, equipment vibration, lubrication, noise, or increases in temperature. Once they are properly evaluated within the maintenance management solution, and analyzed with the available data on equipment (intervention history, management of spare part information), these measures can predict the occurrence of a breakdown, repair, or replacement [23–26].

Therefore, the solutions of Industry 4.0 should allow companies to implement preventive maintenance and actions to improve reliability, troubleshooting, energy efficiency, and the normal actions of facilities usage [26]. Maintenance technicians can improve cycles and adjust maintenance monitoring parameters over time, as well as add new parameters. The maintenance knowledge and instructions library can also be expanded and updated, which leads to a continuous improvement in repair and operating times.

The concepts of Industry 4.0, big data, and predictive maintenance (PdM) are closely related [19,27,28]. PdM relies on monitoring machinery in order to optimize maintenance tasks, so that repair tasks are only executed when it is strictly necessary. The main value of this system is the notification of abnormalities or failures in development, which allow the scheduling of maintenance tasks, and thus, the risks and costs of unexpected breakdowns are minimized, and the industry has a greater awareness of the possibility of optimizing the operation and maintenance of assets through data on their current status or forecasted status, on their operation, or the quality of the manufactured product.

The relation of Industry 4.0 with maintenance departments is important due to aspects such as updating, efficiency, and monitoring, being important tools to take into account, and also due to the interaction capacities, and the information exchanged, between humans, machines, and facilities; favouring the construction of capacities which allow companies to adapt to changes. All this involves proper knowledge management in the philosophy of Industry 4.0 [29–32].

For this reason, technicians who work in this activity need highly sophisticated technical and knowledge components, and the human factor is implicated highly in their performance, with a high degree of tacit knowledge; although in this activity, organizational factors, behaviour of materials, and failure, are also widely studied. Regarding the knowledge management process, it is not conducted in the same way, although in other departments of the company (marketing, commerce, development, communications, etc.), it has begun to be introduced, as another benchmark of competitiveness and innovation [33–37].

Therefore, the starting point is the visualization of knowledge management as a strategic resource [38], in any approach aimed at the efficient development of a maintenance department, and therefore of the company itself.

For this reason, it is important to define and extract the mechanisms of structural coordination, facilitators, and creation of knowledge [39], which occur in a maintenance organization, improving its operational processes [40,41], and transmitting an improvement throughout the company, and with it a management of human resources towards the achievement of knowledge management [42] with a useful and productive purpose. This knowledge management process must be taken into account within the maintenance activity, which consists basically in the generation, coding, transfer, and use of knowledge [43–45].

This article shows the result of a study carried out among maintenance technicians of large companies, using qualitative research techniques, whose main objective was to define the relationship of knowledge management within maintenance activities, from the perspective of the technicians which operate in these departments, visualizing the strategic aspects that make knowledge management important in maintenance engineering, as well as extracting the fundamental barriers and facilitators that these technicians consider for the creation, transmission, and use of this strategic knowledge. The maintenance technicians, who participated in the study, work in different types of industries in Spain, although the activities done by technicians are similar (maintenance actions, troubleshooting, maneuvering actions for machines and industrial process equipment, and energy efficiency actions, etc.). All this can affect the companies, since part of the strategic knowledge is lost when a maintenance technician leaves the company. From the study an important difference was observed between the perception of registered knowledge (explicit), and the knowledge that maintenance technicians have (tacit knowledge).

2. Instruments and Methods

A working hypothesis was made on which this research was based, in order to analyze the implementation of knowledge management in maintenance engineering among maintenance technicians, based on their perception while working in the companies:

(a) Maintenance technicians operate with a high component of knowledge that is not registered in the company (tacit knowledge).
(b) Tacit knowledge directly impacts on the resolution of critical actions and breakdowns in a shorter time (especially the non-cyclical ones), operational maneuver time reduction, and improvement in knowledge for detecting and improving energy efficiency.
(c) The use of knowledge management techniques in maintenance activities, consequently, induces a reduction in the operational familiarization time of new staff.
(d) Proper knowledge management by the maintenance organization positively influences the operability of the company and the connection of work teams.

The instruments used in this research were: the semi-structured interview and analysis done by grounded theory; direct observation and company documents related to the study phenomenon were the main data collection methods in this investigation. Collecting information from various sources, people, or sites, using a variety of methods reduces the risk that the conclusions reflect only the predispositions or limitations of a specific method, allowing a better evaluation of the validation and generalization of the results [46].

All this is focused on showing the different barriers and facilitators to implementing knowledge management techniques in industrial maintenance activities, and detecting their involvement in the essential factors considered in maintenance (reliability, performance in operational activities, maintainability, and energy efficiency). For this, qualitative research techniques were used in various phases, in order to detect the evolution in companies, the evolution in maintenance departments, and the involvement of people.

In qualitative research [47], being objective does not mean controlling variables but being open, having the will to listen and "give voice" to the interviewees, whether they are individuals or organizations. It means listening to what others have to say and seeing what others are doing, and representing them as accurately as possible.

Grounded theory (within qualitative techniques) [48,49] was in the analysis of the research data. For this, the process indicated by Charmaz [48] was followed: (a) data collection through theoretical sampling; (b) initial coding; (c) oriented coding; (d) elevation of the codes to provisional categories by theoretical coding; and (e) writing the obtained results.

The main characteristic of grounded theory research is theoretical sampling, selecting cases according to their potential in order to obtain new points of view and a refinement of the object of study [50].

With the implementation of grounded theory, the following results should be obtained [51,52]: (a) Obtain an exposition of the main variables that explain how the studied group solve their problems; (b) The results obtained identify the basic processes that people use to solve key problems; (c) It is not enough to describe the phenomenon. It is necessary to go one step further, and to interpret and explain what is happening.

Through observation techniques, an attentive examination of the different aspects of a phenomenon is carried out in order to study its characteristics and behaviour within the environment in which it develops. It is a technique that consists of observing the fact or case, taking information, and recording it for subsequent analysis.

Direct observation of phenomena helps to provide appropriate methods for the problem to be studied. In addition, among other advantages, it allows for the global presentation of research, including plans and the use of techniques and tools.

The phases of this research for achieving the defined objectives can be summarized with the following characteristics:

(a) First, using grounded theory, 76 people belonging to operational maintenance staff from different sections were interviewed (Table 1). At the same time, the direct observation technique was used, during this research phase, with access to the facilities, documentation, and equipment of the factory by the researcher, the real characteristics of the work carried out in maintenance, the study of their internal relations, and the characteristics of the information used by the maintenance teams were recorded, giving the researcher an insight into the phenomena in the research environment. With this, an attentive examination of the different aspects of a phenomenon is achieved, in order to study its characteristics and behaviour within the environment in which it develops.

Table 1. Features of the sample of the technicians interviewed.

Labor Category	Work Experience (under 5 Years)	Work Experience (between 10 and 15 Years)	Work Experience (>15 Years)
operational maintenance technicians. (mechanics)	6	12	9
operational maintenance technicians. (electrical-systems)	10	9	7
operational maintenance technicians. (production)	9	8	6
subtotal	25	29	22
total		76	

The format of the questions for the individual interview (and also used in other qualitative techniques) was based on 22 basic questions, the basic script of the interview being as follows:

- Based on your experience in the field of industrial maintenance engineering, it is intended to study the strategic factors of maintenance, and their relationships and evolution with knowledge management processes, answer the following questions:
- Q01. What do you consider to be the strategic activities of the maintenance activity that most affects the company?

- Q02. To what extent do you think the experience of technical maintenance personnel affects these strategic activities?
- Q03. What level of information/knowledge do you manage at your own level regarding maintenance activities (tacit knowledge, not registered), and what is precisely documented in the company (explicit knowledge)? Can you give an example?
- Q04. From the explicit information that the company may have access to in order to carry out their work (computer programs, machinery, and equipment manuals, planimetry, work orders, etc.), to what extent is it useful and what deficiencies do you observe?
- Q05. How do you document or transmit your daily jobs/experiences in your maintenance job, and how much time do you spend on it?
- Q06. What is the usual way in which you capture the (important) operational experiences of your colleagues (through meetings, informal conversations, etc.), so that you can resolve such an action when it happens to you (example: operational manoeuvres in the event of breakdowns) or perform this task (example: maintenance work)?
- Q07. Has a knowledge management program been implemented in your organization involving tactical maintenance actions? If yes, what opinion do you have of it?
- Q08. What information/knowledge should be captured or made explicit, that helps you in the performance of your duties?
- Q09. How should such information/knowledge, its accessibility (to share it), and its maintenance (how to collect and update it) be structured, so that it is easily usable and accessible to you?
- Q10. What would be the benefit of the capture and conversion of tacit to explicit knowledge, personally and at the company level?
- Q11. What would facilitate, in your opinion, the capture and conversion of tacit to explicit knowledge? How should such knowledge capture be done?
- Q12. Which barriers do you consider most important for the implementation of a knowledge management program in the maintenance activity?
- Q13. What would motivate you to support and be interested in capturing and recording your tacit knowledge and that of your colleagues, and which could improve the work of your colleagues and help improve the productivity and efficiency of the company?
- Q14. What type of actions/experiences should be documented that affect the tactical actions of maintenance engineering, such as: Reliability of equipment and systems, Operation/operation of facilities, Energy efficiency, Maintainability?
- Q15. How do you think it would affect the familiarization time of new personnel, and the action times of all maintenance technicians, if information structuring the and capture of tactical actions, as well as operational experiences lived, were based on experience?
- Q16. What factors should be controlled quantitatively (measured), to see what affects the improvement of knowledge management in the tactical actions of maintenance?
- Q17. Before a new installation, machinery, reform, etc., would it be convenient to introduce in the gant/pert diagrams of the duration of the works, a new activity in which the registration and collection of practical and useful knowledge is found, reflecting the actions or relevant information that would help in future installations?
- Q18. What tools/techniques, means, etc., do you think would help you capture the important tactical and strategic information in your maintenance activity?
- Q19. In your opinion, what consideration does the company's management and maintenance clients (production, other areas of the company, etc.) have of the activities and missions of the maintenance department?
- Q20. Do you need to know more about these topics, with reference to knowledge management of maintenance activity? What knowledge gaps do you have on these topics, which prevent you from getting more out of it?

- Q21. What type of training would it be appropriate to receive, to what degree and in what way, that could improve your work efficiency?
- Q22. Enter below any information or suggestions that you consider relevant, and that have not been addressed in the questionnaire.

The data was analyzed with the help of the Atlas.ti 5.0 application from ResearchTalk Inc. (Berlin, Germany) [53–55].

(b) With the aim of deepening the perception of knowledge management on the strategic actions of industrial maintenance engineering (reliability, operations, energy efficiency and maintainability) by maintenance technicians, and allowing much greater access to it, by a greater number of members of the operative staff (Table 2), a survey (Figure 1) was passed to all the operative personnel (174 technicians) in order to identify and quantify their perceptions, about the personal knowledge they use (tacit) and the knowledge that they perceive as documented in a useful and precise way by the organization (explicit), regarding factors involved in the performance of their functions.

Table 2. Survey population based on years of maintenance experience.

Technicians' Work Experience (years)	Number of Surveys
<3	34
3 to 5	48
>5	92
total	174

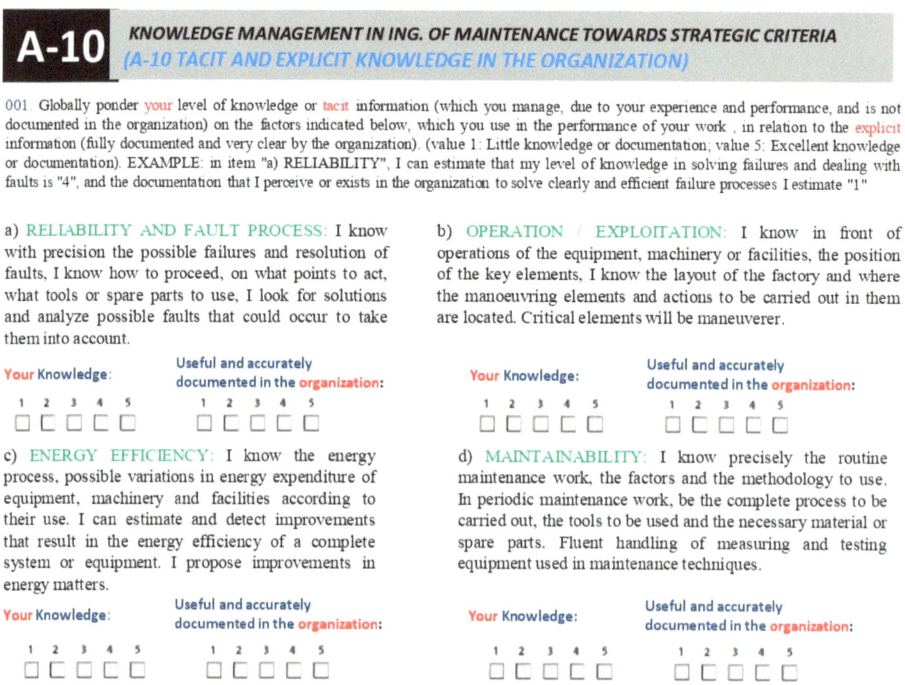
A-10 KNOWLEDGE MANAGEMENT IN ING. OF MAINTENANCE TOWARDS STRATEGIC CRITERIA
(A-10 TACIT AND EXPLICIT KNOWLEDGE IN THE ORGANIZATION)

001. Globally ponder your level of knowledge or tacit information (which you manage, due to your experience and performance, and is not documented in the organization) on the factors indicated below, which you use in the performance of your work , in relation to the explicit information (fully documented and very clear by the organization). (value 1: Little knowledge or documentation; value 5: Excellent knowledge or documentation). EXAMPLE: in item "a) RELIABILITY", I can estimate that my level of knowledge in solving failures and dealing with faults is "4", and the documentation that I perceive or exists in the organization to solve clearly and efficient failure processes I estimate "1"

a) RELIABILITY AND FAULT PROCESS: I know with precision the possible failures and resolution of faults, I know how to proceed, on what points to act, what tools or spare parts to use, I look for solutions and analyze possible faults that could occur to take them into account.

Your Knowledge: 1 2 3 4 5 — Useful and accurately documented in the organization: 1 2 3 4 5

b) OPERATION / EXPLOITATION: I know in front of operations of the equipment, machinery or facilities, the position of the key elements, I know the layout of the factory and where the manoeuvring elements and actions to be carried out in them are located. Critical elements will be maneuverer.

Your Knowledge: 1 2 3 4 5 — Useful and accurately documented in the organization: 1 2 3 4 5

c) ENERGY EFFICIENCY: I know the energy process, possible variations in energy expenditure of equipment, machinery and facilities according to their use. I can estimate and detect improvements that result in the energy efficiency of a complete system or equipment. I propose improvements in energy matters.

Your Knowledge: 1 2 3 4 5 — Useful and accurately documented in the organization: 1 2 3 4 5

d) MAINTAINABILITY: I know precisely the routine maintenance work, the factors and the methodology to use. In periodic maintenance work, be the complete process to be carried out, the tools to be used and the necessary material or spare parts. Fluent handling of measuring and testing equipment used in maintenance techniques.

Your Knowledge: 1 2 3 4 5 — Useful and accurately documented in the organization: 1 2 3 4 5

Figure 1. Survey format, among technical maintenance members.

3. Results

As results, the different elements detected in the qualitative research from the perception of the maintenance technicians in relation to knowledge management in the activity of the company, in relation to strategic maintenance actions, and the involvement of operators, are listed, as well as the quantitative perception that they had between their own knowledge (tacit), and the explicit knowledge which is registered in the company.

Sections 3.1 and 3.2 show the results obtained, based on the usage of the grounded theory technique through semi-structured interviews held with 76 maintenance technicians.

3.1. Relation between Knowledge Management and Fundamental Strategic Maintenance Actions

From the qualitative study based on interviews with maintenance operators, it was confirmed that proper knowledge management would greatly affect strategic activities, improving the following actions:

- Capture of the tacit strategic knowledge from operational maintenance technicians.
- Resolution of critical failures in a short period of time (especially non-cyclical ones).
- Reduction of operating manoeuvre times.
- Facilitate the change of area or staff substitutions.
- Decrease in the engagement times of new staff members.
- Information capture and transfer of subcontracted companies.
- Sharing of employee knowledge that can be used by others, who can detect new opportunities for improvement.
- Improved knowledge of the reliability of equipment and facilities.
- Improvement of knowledge for the detection and improvement of energy efficiency actions.
- Time optimization, with effects on knowledge management and the reduction of maintenance costs.

Maintainability: This affects all the equipment and infrastructure of the company. It considers the great variety of processes and actions needed to perform efficient maintenance of each of the elements, and which requires a great deal of experience and knowledge. The adaptation of employees to carry out maintenance work occurs through knowledge of the environment, where the facilities are located, and normally acquiring the necessary knowledge by observing and commenting on employees with more experience, until being fully autonomous in those activities. They consider employees with less work experience a delay in carrying out the processes to be executed. All consider that, even though it is the activity that is most documented within the organization, due to the use of computerized maintenance management programs, the actions to be carried out and useful anecdotal experiences are not usually documented, and should be experienced by the operators that have not been through a particular situation.

Reliability: The main demand of the production departments is to avoid stoppages of the equipment and the dependent facilities for production. Interviewees considered that the knowledge of cyclical and non-cyclical failures has been acquired based on experience in their performance, with greater security in the prevention and resolution of breakdowns by more experienced employees. The resolution procedures are not usually documented, the learning process having been carried out based on a trial and error process and informal comments from other employees who have experienced these situations previously. Diagrams of criticality and failure processes are not usually made, and regarding critical non-cyclical actions, a very significant loss of time occurs in the resolution of a breakdown, which economically affects the company, due to the excess time spent in the replacement of the service.

Energy efficiency: This is of high economic relevance as it affects the final price of the products made by the company at an economic level. It is the most controlled strategic action by the company's management, and effort is normally focused on the quantification of general energy consumption or the revision of rates by the supply companies. All considered that there are numerous actions for energy efficiency of little expense (review of useless consumption, valve closures, control of consumption of machinery in the process of production stoppage, etc.), however they are not usually documented. Many of the low impact actions are carried out directly by the operators using their good know-how, due to their own experience in the factory, and knowing the characteristics of the manufacturing processes. Employees take time to adapt, and only when you have consolidated experience in the factory do you gain enough knowledge to make useful decisions in that direction. It is recognized that with the adoption of many of these small energy efficiency actions, it is possible to achieve significant savings, as well as anticipating and planning new actions that will result in their improvement.

Operation/exploitation: These are normal actions in a factory's operating cycle (facilities manoeuvres, machinery shutdown and rearming processes, actuation of automatic switches by a trip, etc.). Operational actions directly affect the efficiency of the processes or services provided. All the interviewees agreed that with a change in maintenance personnel or the subcontracted company there is a loss during the first months of operation from the new personnel is recruited, until there is a greater knowledge of the facilities and characteristics demanded by the company. It is recognized that the subcontracting processes means that the subcontractor company manages a strategic knowledge of the company itself, which is normally lost due to a change or substitution of such a company.

3.2. Operators Involvement

The involvement of operators is another key facilitator of sustainability in a maintenance knowledge management project, which was verified in this research. They must be fully involved as a fundamental source of strategic knowledge and of the improvements developed, as well as the base of ideas and part of the improvement process. Without the participation and involvement of operators, the knowledge management project is doomed to failure, since it must, as a principle, involve all members of the organization. To get the involvement of the operators requires training, support, and explicit recognition by the company management and maintenance managers.

A large part of the interviewees reacted very positively to the introduction of material incentives based on improvements achieved by the work, both in groups and individually. When starting a knowledge management (KM) project, it is advisable to have incentives, but once the culture has been assimilated, they consider the express recognition of the company as sufficient.

At an individual level, personal motivation, and the opportunity to learn and facilitate the generation of knowledge, which is shared with other members of the group, gives rise to organizational knowledge.

3.3. The Quantitative Perception of Maintenance Operators

In order to estimate the differences in perception between the knowledge based on maintenance operators own experience, in relation to the knowledge that they perceive to be explicit in the organization, a questionnaire (Figure 1) was passed to all the maintenance operating personnel of the organization (174 maintenance technicians), comprising four items, subdivided between the two perceptions. Based on a maximum knowledge index valued of five, the following means were obtained based on the different strategic activities (Figure 2), and the seniority of the operators.

It was observed that the knowledge that the operators use to carry out their daily actions, is based mainly on their own (tacit) knowledge, considering that many of these actions are not included in the explicit knowledge of the company. This was observed to a greater extent among older operators, where such contrast was much higher.

Figure 3 shows a radar-type graph, where it can be seen, according to the study, the comparison between the operators own strategic knowledge, in contrast to that explicitly provided by the maintenance organization from the perception of maintenance technicians. Although it was based on a subjective view on the part of the operators, for all of them a higher level of perception of their own knowledge, as a mechanism for the performance of their fundamental missions was reported. There were higher levels of knowledge on the part of the operators and in the organization on maintenance actions. This may be mostly due to the fact that it is where the bulk of the maintenance department's information and procedures are normally concentrated (maintenance management programs, maintenance estimate tables, etc.). In the same way, it can be concluded that the level of tacit knowledge compared to the explicit knowledge of the organization, increased due to the increase in seniority, and therefore the experience, of the operators.

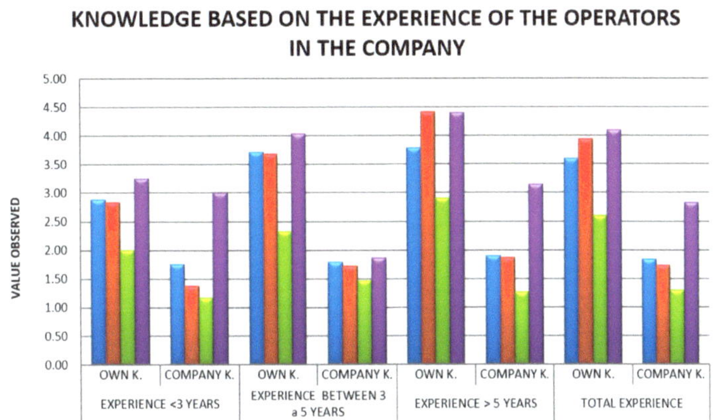

	OWN K.	COMPANY K.	OWN K.	COMPANY K.	OWN K.	COMPANY K.	OWN K.	COMPANY K.
	EXPERIENCE <3 YEARS		EXPERIENCE BETWEEN 3 a 5 YEARS		EXPERIENCE > 5 YEARS		TOTAL EXPERIENCE	
RELIABILITY	2.88	1.75	3.71	1.79	3.78	1.89	3.59	1.84
OPERATION	2.83	1.38	3.68	1.71	4.42	1.86	3.94	1.73
ENERGY EFFIC.	2.00	1.17	2.32	1.46	2.90	1.26	2.60	1.29
MAINTAINABILITY	3.25	3.00	4.04	1.86	4.40	3.14	4.10	2.82

Figure 2. Graph of the relation between own knowledge vs. knowledge registered in the company regarding the strategic activities of the maintenance department (reliability, operation, energy efficiency, and maintainability) according to the perception of maintenance technicians.

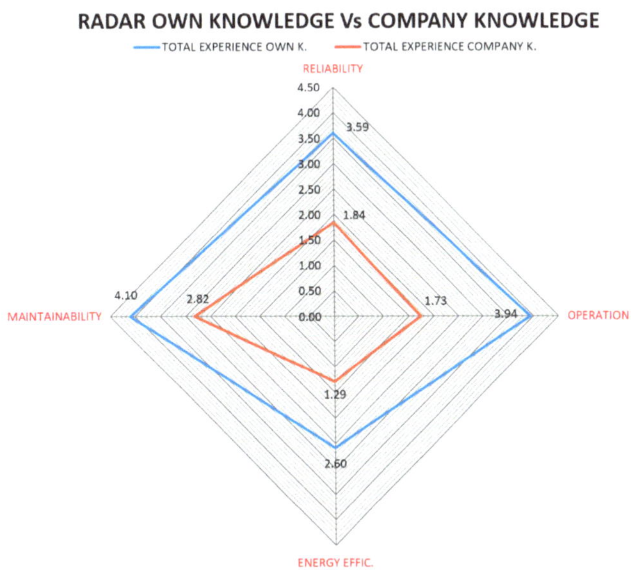

Figure 3. Graph of the relation between the technicians' own knowledge in relation to the explicit knowledge in the company.

4. Discussion

The fundamental barriers found by this study for the adequate management of knowledge in the maintenance activity according to the perception of the technicians of these departments, were the short period of time available to properly document some important actions, and the existence of cultural barriers, with a culture based on "own knowledge", that is not shared, especially by the operational technicians, as well as getting the full involvement of the staff. This study confirms some processes and actions that must be taken into account when proposing a knowledge management model in this activity:

- ➢ The normal process of familiarizing the staff to the conditions of the company relies on tacit knowledge (learning based on experience in the environment), which involves this internal knowledge and the way to make it explicit to the entire organization [12]
- ➢ This coupling is necessary before operators with experience in the company change their environment (by changing to another headquarters or changing the work section).
- ➢ The usual process is to learn the actions based on others experience in carrying them out over time.
- ➢ There is a unit of operators with more experience and knowledge of the facilities and equipment.
- ➢ Knowledge about solving breakdowns is critical since it strongly affects the production of the company or the service it provides.
- ➢ Experience in the resolution of non-cyclical breakdowns is not usually documented, and the resolution process starts from scratch when it happens to an operator who has not gone through such an experience.
- ➢ There is usually no critical study of reliability or a crisis knowledge map. In-depth knowledge of key processes is required.
- ➢ The knowledge process in the routine actions of operation, is characteristic of the facilities of each company and involves a familiarization time for maintenance technicians.
- ➢ Such operational actions directly affect the efficiency of the processes or services provided.
- ➢ In-depth knowledge of the facilities and equipment is necessary to determine the best energy efficiency options.
- ➢ Many of the energy efficiency options are observed during the operation of the facilities, with simple implementations, which are not normally realized or executed, due to factors related to the poor transfer of information, or the knowledge of the operators who observe them.

Similarly, the massive use of informal knowledge transfer mechanisms has been identified, with information being found on "islands" within the organization itself. A large volume of tacit knowledge handled by the operators is present, and which is the fundamental way of operating, compared to the explicit information or knowledge of the organization (Figure 4).

As can be observed in Figure 4, the perception of maintenance technicians, between their own knowledge (tacit knowledge) and that registered in the company (explicit), varies considerably. If we compare the values obtained for this perception, it is seen that the knowledge registered in the company (explicit) is 51.25% (1.84/3.59) compared to the knowledge of the maintenance technicians (tacit), for the case of reliability and fault resolution. In the case of operation/exploitation share, it is 43.9% (1.73/3.94); for energy efficiency actions 49.61% (1.29/2.60); and in maintenance actions (preventive, predictive, and corrective) 68.78% (2.82/4.10). This shows the significant gap between the perception of the registered information (explicit) and that maintenance technicians have by themselves (tacit knowledge), based on their experiences accumulated over years in their jobs. All this can affect the company regarding staff modifications (layoffs, staff reductions, sick leave, staff retirements) since part of the strategic knowledge is lost when a technician leaves the company.

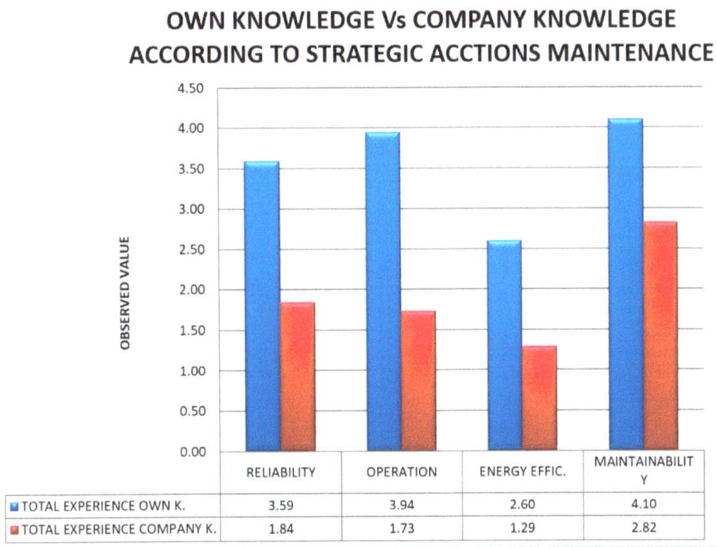

	RELIABILITY	OPERATION	ENERGY EFFIC.	MAINTAINABILIT Y
TOTAL EXPERIENCE OWN K.	3.59	3.94	2.60	4.10
TOTAL EXPERIENCE COMPANY K.	1.84	1.73	1.29	2.82

Figure 4. Graph of the relation between the technicians' own knowledge compared to the explicit knowledge in the company based on the strategic aspects of maintenance.

The perception of the maintenance technicians was that they have limited awareness in knowledge management techniques, normally functioning as "islands of knowledge" within the department. Moreover, the vast majority consider that they need training applied more to the company's own facilities (they suggest that the training they normally receive is very generic on technical aspects of electricity, mechanics, etc.).

This study confirms the importance that adequate knowledge management can have on the fundamental strategic maintenance activities, and which was confirmed by all the interviewed personnel (reliability, maintainability, energy efficiency, and operation/exploitation). In Figure 5, the main observed characteristics are extracted based on the processes and strategic aspects of maintenance, and the result on the efficiency of the company's activity.

It is recognized that an improvement in the management of information and knowledge, positively effects all these actions, and especially in the resolution of major breakdowns, or non-cyclical failures spaced in time, and with their performance not normally recorded.

As for the tools that can be used for the strategic information capture that helps improve knowledge management, they are normally little used in all maintenance environments. There is a low use of audits on internal actions, information and knowledge maps are recognized, and criticality diagrams are made only for certain facilities or equipment essential for the company's activity.

A greater use of informal meetings was suggested as a means of generating and transferring knowledge, especially among groups of operational technicians, who have less of an organizational culture than the managers or maintenance managers.

The participants in the analysis considered that applying their own knowledge on organizational activities would motivate them in fields such as self-learning, learning new tools, and creating new ways of executing different activities. When this personal motivation is reinforced by knowing that their opinions and suggestions to acquire external knowledge will be taken into account, the processes of knowledge transfer and usage are enhanced.

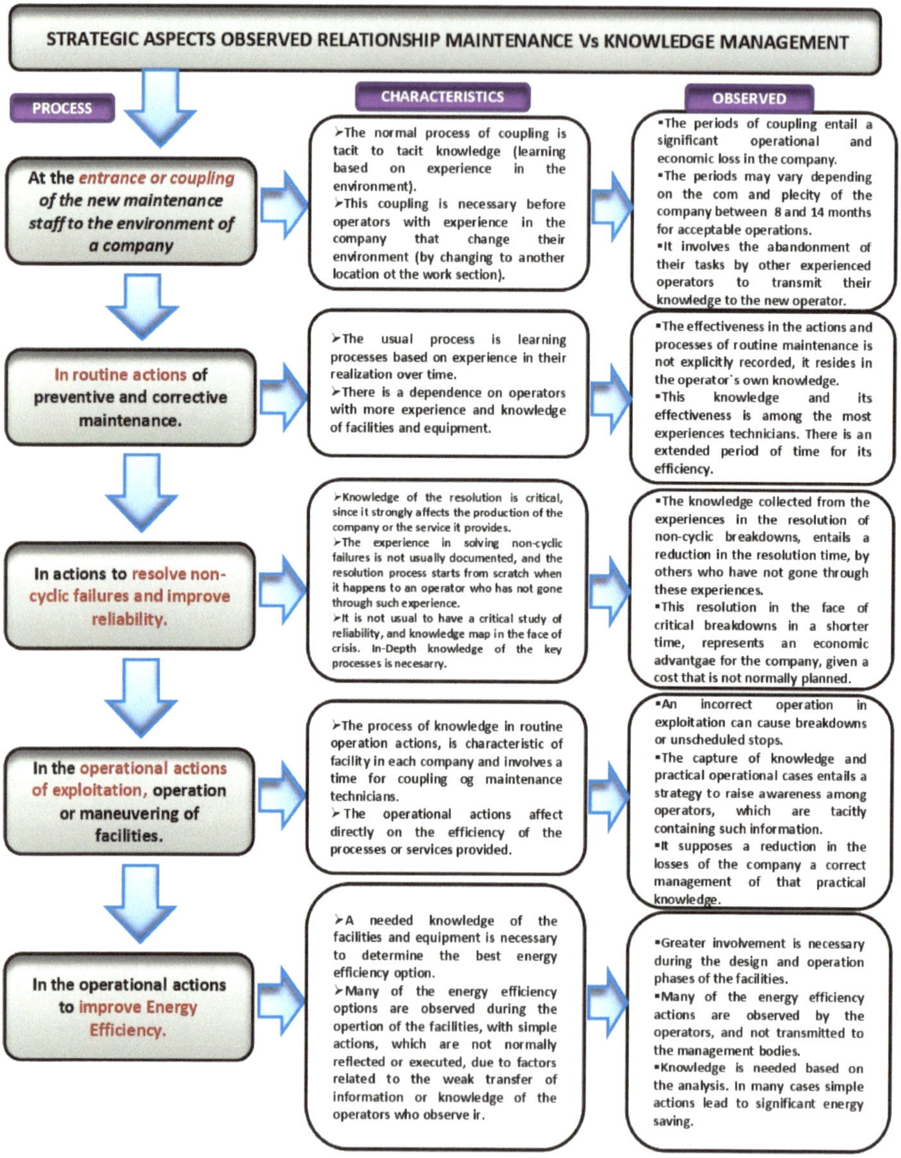

Figure 5. Strategic aspects of maintenance and its relationship with knowledge management.

Table 3 shows a summary of the main characteristics that have been identified in the present study in terms of tools, barriers and facilitators, and implications in the knowledge management processes, according to the perception of maintenance technicians.

According to maintenance technicians, personal motivation and learning opportunities promote the creation of knowledge, when the knowledge is shared with other members of the company organizational knowledge is created, and the open culture of the organization enhances the organizational knowledge. The research participants thought that being able to use their knowledge in organized activities would motivate them in self-learning, to learn new tools, and create new ways of doing activities. This would improve knowledge transfer, when this personal motivation increases by knowing that their opinions and suggestions about acquiring external knowledge will be considered.

Table 3. Tools, barriers, and facilitators in the K.M. in maintenance activity according to the perception of maintenance technicians.

Category of the Studied Phenomenon	Maintenance Operational Technicians
Tools for knowledge management	Information and knowledge maps. Agile and simple systems to capture experiences. Mobile computing tools to capture images, videos, and experiences.
Barriers in Knowledge Management	Little time available to properly document important actions. Cultural barriers. Culture based on "own knowledge", not shared. Staff involvement. Greater use of informal knowledge transfer mechanisms.
Facilitators in Knowledge Management	Open and flexible proactive organizational culture. Participatory style of management. Employee's personal motivation. Opportunity to learn. Organizational culture of the maintenance area. Management style. Media. Use of a manager of own knowledge of the maintenance activity.
Observations	A lot of strategic information, collected in a handwritten form disaggregated in personal notes and notebooks, annotations on plans, not shared with the rest of the organization, which hinders the transmission and use of the knowledge by the rest of the organization.
Results of Proper Management of Knowledge in Maintenance Activity	Capture of the tacit strategic knowledge of operational maintenance technicians. Resolving critical failures in less time (especially non-cyclical ones). Reduction of operating manoeuvre times. Facilitate the change of area or personnel substitutions. Reduction of familiarization times for new personnel. Capture of information and transfer of subcontractor companies. Sharing the knowledge of employees that can be used by others who can detect new opportunities for improvement. Improved knowledge of the reliability of equipment and facilities. Improvement of knowledge for the detection and improvement of energy efficiency actions. Optimization of time, which again results in knowledge management and the reduction of maintenance costs. Improvement in self-learning in order to solve problems in the factory itself.

5. Conclusions

Industrial maintenance engineering requires a staff with deep technical knowledge, and who have a high component of tacit knowledge, acquired through years of work experience.

The main contributions of the research that are presented in this article, and that allow the extension of understanding on knowledge management in maintenance activities, are:

- The main facilitators/barriers detected are summarized based on the qualitative research carried out.
- The main strategic aspects of maintenance that can increase its efficiency are confirmed by adopting a knowledge management model.

- The high level of tacit knowledge used in this activity is confirmed, normally based on the high experience level required of the operators, and which requires high engagement times with new personnel.
- The knowledge directly impacts on, the resolution of critical actions and breakdowns in a shorter time, operational maneuver time reduction, and the improvement in knowledge for detecting and improving energy efficiency.

When addressing a knowledge management model for this activity, it is necessary to have an impact on the fundamental strategic maintenance activities confirmed by all the interviewed staff (reliability, maintainability, energy efficiency, and operation/exploitation), and which can improve the efficiency of all facilities.

From the survey of the perceptions of maintenance technicians (Figure 2), it is concluded that the perception that they have on their own knowledge (tacit) and the knowledge actually registered in the company (explicit), vary considerably, being approximately 50% recorded knowledge (explicit), as compared to their own knowledge (tacit). This is a problem when a workman leaves the company since their knowledge, with a high strategic component, is lost.

Participants in the study thought that the possibility of applying their knowledge to the organization's activities would stimulates their motivation to learn by themselves, learn new tools, and create new ways to carry out activities.

A greater use of informal meetings was suggested as a means of generating and transferring knowledge, especially among groups of operational technicians, with less of an organizational culture than the managers or maintenance managers.

A great use of informal knowledge transfer mechanisms was identified, which means that the information is found on "islands" within the organization itself.

The fundamental barriers located by this study were the limited availability of time to adequately document important actions, the cultural barriers, with a culture based on "own knowledge", not shared, especially among operational technicians, as well as achieving full involvement of the staff.

With this study, we have attempted to identify some key elements in order to improve the information and knowledge capture programs in the maintenance areas, according to the perception of the technicians who operate in these departments and facilitate its extension to all areas of the company.

This research through qualitative study should serve as a basis for future research, in order to design a knowledge management model applied to the industrial maintenance departments of companies, which reduces the gap between technicians' tacit knowledge compared to the knowledge registered in the company. It would also be suitable for making new studies with more companies from different countries in order to expand and confirm the data collected.

Author Contributions: J.C.-C. developed the methodology J.C.-C. and J.-A.C.-C. prepared the conceptualization and data curation; J.C.-C. gathered and analyzed the data. J.C.-C. and J.-A.C.-C. wrote the paper. All authors have read and agreed to the published version of the manuscript.

Funding: This work has been conducted within the framework of "Sistemas de mantenimiento industrial" project (Ref. 20140432) and the CONDAP project "Digital skills for workplace mentors in construction sector apprenticeships" funded by the European Commission within the Key Action 2: Cooperation for innovation and the exchange of good practices, reference number 2018-1-UK01-KA202-048122.

Institutional Review Board Statement: Not applicable.

Informed Consent Statement: Not applicable.

Data Availability Statement: Not applicable.

Acknowledgments: The authors deeply thank the participating companies and also the participation of all members involved in this research, consisting mainly of maintenance engineering professionals.

Conflicts of Interest: The authors declare no conflict of interest.

References

1. IEEE Std 493-2007. *IEEE Recommended Practice for the Design of Reliable Industrial and Commercial Power Systems*; IEEE-SA Standards Board: Washington, DC, USA, 2007.
2. Koval, D.; Zhang, X.; Prost, J.; Coyle, T.; Arno, R.; Hale, R. Reliability Methodologies Applied to the IEEE Gold Book Standard Network. *IEEE Ind. Appl. Mag.* **2003**, *9*, 32–41. [CrossRef]
3. Wang, W.; Loman, J.; Arno, R.; Vassiliou, P.; Furlong, E.; Ogden, D. Reliability block diagram simulation techniques applied to the IEEE std. 493 Standard Network. *IEEE Trans. Ind. Appl.* **2004**, *40*, 887–895. [CrossRef]
4. Lu, Y. Industry 4.0: A survey on technologies, applications, and open research issues. *J. Ind. Inf. Integr.* **2017**, *6*, 1–10. [CrossRef]
5. Cacique, J. *Diseño de un Programa Para Calcula la Confiabilidad en un Sistema de Distribución de Energía Eléctrica*; UNEXPO: Bolívar, Venezuela, 2007; p. 138.
6. Baeza, G.; Rodríguez, P.; Hernández, J. *Evaluación de Confiabilidad de Sistemas de Distribución Eléctrica en Desregulación*; Revista Facultad de Ingeniería: Santiago, Chile, 2003; Volume 11, pp. 33–39.
7. McGranaghan, M. Quantifying Reliability and Service Quality for Distribution Systems. *IEEE Trans. Ind. Appl.* **2007**, *43*, 188–195. [CrossRef]
8. Sexto, L. Confiabilidad integral del activo. In Proceedings of the Seminario Internacional de Mantenimiento Celebrado en Perú-Arequipa-Tecsup, Arequipa, Perú, 23–25 February 2005.
9. Wilkesmann, M.; Wilkesmann, U. Industry 4.0—Organizing routines or innovations? *VINE J. Inf. Knowl. Manag. Syst.* **2018**, *48*, 238–254. [CrossRef]
10. Cárcel-Carrasco, J.; Gómez-Gómez, C. Qualitative Analysis of the Perception of Company Managers in Knowledge Management in the Maintenance Activity in the Era of Industry 4.0. *Processes* **2021**, *9*, 121. [CrossRef]
11. Isworowati, D.; Muhammad, F.; Kurniawati, A.; Kurniawan, M. Knowledge Management System for Maintenance Activity: Case Study at the Maintenance Department of XYZ Corporation. In Proceedings of the IEEE International Conference on Industrial Engineering and Engineering Management (IEEM), Macao, China, 15–18 December 2019; pp. 139–143.
12. Carcel-Carrasco, F.J.; Rodriguez-Mendez, M. Industrial maintenance and tacit knowledge: An introduction about its incidence. *DYNA Manag.* **2013**, *1*, 15. [CrossRef]
13. Altmann, C. El Análisis de Causa Raíz, como herramienta en la mejora de la Confiabilidad. In Proceedings of the 2nd Congreso Uruguayo de Mantenimiento, Gestión de Activos y Confiabilidad, Montevideo, Uruguay, 16–18 August 2006.
14. Amendola, L. *Modelos Mixtos de Confiabilidad Projet Managament*; PMM Institute for Learning: València, Spain, 2016; ISBN 978-84-943897-3-3.
15. Amendola, L. Estrategias y Técnicas en la Dirección y Gestión de Proyectos. In *ProjetManagament*; Universitat Politècnica de València: València, Spain, 2004; ISBN 84-9705-522-5.
16. Tavares, L. *Administración Moderna de Mantenimiento*; Novo Polo Publicacoes: Rio de Janeiro, Brasil, 2004.
17. Ruiz-Sarmiento, J.; Monroy, J.; Moreno, F.-A.; Galindo, C.; Bonelo, J.-M.; Gonzalez-Jimenez, J. A predictive model for the maintenance of industrial machinery in the context of industry 4.0. *Eng. Appl. Artif. Intell.* **2020**, *87*, 103289. [CrossRef]
18. Ghobakhloo, M.; Fathi, M. Corporate survival in Industry 4.0 era: The enabling role of lean-digitized manufacturing. *J. Manuf. Technol. Manag.* **2019**, *31*, 1–30. [CrossRef]
19. Çınar, Z.M.; Nuhu, A.A.; Zeeshan, Q.; Korhan, O.; Asmael, M.; Safaei, B. Machine Learning in Predictive Maintenance towards Sustainable Smart Manufacturing in Industry 4.0. *Sustainability* **2020**, *12*, 8211. [CrossRef]
20. Jasiulewicz-Kaczmarek, M.; Legutko, S.; Kluk, P. Maintenance 4.0 Technologies—New Opportunities for Sustainability Driven Maintenance. *Manag. Prod. Eng. Rev.* **2020**, *11*, 74–87.
21. Zonta, T.; da Costa, C.A.; da Rosa Righi, R.; de Lima, M.J.; da Trindade, E.S.; Li, G.P. Predictive Maintenance in the Industry 4.0: A Systematic Literature Review. *Comput. Ind. Eng.* **2020**, *150*, 106889. [CrossRef]
22. Ceruti, A.; Marzocca, P.; Liverani, A.; Bil, C. Maintenance in aeronautics in an Industry 4.0 context: The role of Augmented Reality and Additive Manufacturing. *J. Comput. Des. Eng.* **2019**, *6*, 516–526. [CrossRef]
23. Schumacher, A.; Nemeth, T.; Sihn, W. Roadmapping towards industrial digitalization based on an Industry 4.0 maturity model for manufacturing enterprises. *Procedia CIRP* **2019**, *79*, 409–414. [CrossRef]
24. Bousdekis, A.; Lepenioti, K.; Apostolou, D.; Mentzas, G. Decision Making in Predictive Maintenance: Literature Review and Research Agenda for Industry 4.0. *IFAC-PapersOnLine* **2019**, *52*, 607–612. [CrossRef]
25. Lee, S.M.; Lee, D.; Kim, Y.S. The quality management ecosystem for predictive maintenance in the Industry 4.0 era. *Int. J. Qual. Innov.* **2019**, *5*, 4. [CrossRef]
26. Patalas-Maliszewska, J.; Kłos, S. An Approach to Supporting the Selection of Maintenance Experts in the Context of Industry 4.0. *Appl. Sci.* **2019**, *9*, 1848. [CrossRef]
27. Sang, G.M.; Xu, L.; de Vrieze, P.T.; Bai, Y.; Pan, F. Predictive Maintenance in Industry 4.0. In Proceedings of the ICIST 2020: 10th International Conference on Information Systems and Technologies, Lecce, Italy, 4–5 June 2020.
28. Bag, S.; Pretorius, J.H.C. Relationships between industry 4.0, sustainable manufacturing and circular economy: Proposal of a research framework. *Int. J. Organ. Anal.* **2020**. [CrossRef]

29. Manesh, M.F.; Pellegrini, M.M.; Marzi, G.; Dabić, M. Knowledge Management in the Fourth Industrial Revolution: Mapping the Literature and Scoping Future Avenues. *IEEE Trans. Eng. Manag.* **2021**, *68*, 289–300. [CrossRef]
30. Pokojski, J.; Oleksiński, K.; Pruszyński, J. Conceptual and detailed design knowledge management in customized production—Industrial perspective. *J. Comput. Des. Eng.* **2019**, *6*, 479–506. [CrossRef]
31. Emmanouilidis, C.; Pistofidis, P.; Bertoncelj, L.; Katsouros, V.; Fournaris, A.P.; Koulamas, C.; Ruiz-Carcel, C. Enabling the human in the loop: Linked data and knowledge in industrial cyber-physical systems. *Annu. Rev. Control* **2019**, *47*, 249–265. [CrossRef]
32. Wan, S.; Li, D.; Gao, J.; Li, J. A knowledge based machine tool maintenance planning system using case-based reasoning techniques. *Robot. Comput. Manuf.* **2019**, *58*, 80–96. [CrossRef]
33. Bravo-ibarra, E.; Herrera, L. Capacidad de innovación y configuración de recursos organizativos. *Intang. Cap.* **2009**, *5*, 301–320.
34. Griffiths, P.; Remenyi, D. Aligning Knowledge Management with Competitive Strategy: A Framework. *Electron. J. Knowl. Manag.* **2008**, *6*, 125–134.
35. Argote, L.; Ingram, P. Knowledge trasnfer: A basis for competitive advantage in firms. *Organ. Behav. Hum. Decis. Process.* **2000**, *82*, 150–169. [CrossRef]
36. Claver, E.; Zaragoza, P. La dirección de recursos humanos en las organizaciones inteligentes. Una evidencia empírica desde la dirección del conocimiento. *Investigaciones Europeas de Dirección y Economía de la Empresa* **2007**, *13*, 55–73.
37. Cárcel-Carrasco, F.-J.; Cárcel-Carrasco, J.-A.; Peñalvo-López, E. Factors in the Relationship between Maintenance Engineering and Knowledge Management. *Appl. Sci.* **2020**, *10*, 2810. [CrossRef]
38. Pawlowski, J.; Bick, M. The Global Knowledge Management Framework: Towards a Theory for Knowledge Management in Globally Distributed Settings. *Electron. J. Knowl. Manag.* **2012**, *10*, 92–108.
39. Pauleen, D. Personal knowledge management: Putting the "person" back into the knowledge equation. *Online Inf. Rev.* **2009**, *33*, 221–224. [CrossRef]
40. Hoelzle, K.; Gemuenden, H.G. Cultural vs. structural aspects of Open Innovation—How to implement Open Innovation. In *TUHH User and Open Innovation Workshop 2009*; Workshop 2009 Hamburg, Germany, 3 June 2009; University of Hamburg: Hamburg, Germany, 2009.
41. Cárcel Carrasco, F.J. La gestión del conocimiento en la ingeniería del mantenimiento industrial: Investigación sobre la incidencia en sus actividades estratégicas. *Omnia Sci.* **2014**. [CrossRef]
42. De Andrade, F.G.; Araújo, C.K.C.; Basso, S.; Przybysz, A.L.; De Francisco, A.C. The Perception of Knowledge Management by Small and Medium-Sized IT Companies, Campos Gerais, Brazil. In *Proceedings of the International Joint Conference on Industrial Engineering and Operations Management, Novi Sad, Serbia, 21 March 2020*; Springer: Cham, Switzerland, 2019; pp. 522–531.
43. Martins, V.W.B.; Rampasso, I.S.; Anholon, R.; Quelhas, O.L.G.; Filho, W.L. Knowledge management in the context of sustainability: Literature review and opportunities for future research. *J. Clean. Prod.* **2019**, *229*, 489–500. [CrossRef]
44. Nonaka, I.; Takeuchi, H. *The Knowledge-Creating Company: How Japanese Companies Create the Dynamics of Innovation*; Oxford University Press: New York, NY, USA, 1995.
45. Nisar, T.M.; Prabhakar, G.P.; Strakova, L. Social media information benefits, knowledge management and smart organizations. *J. Bus. Res.* **2019**, *94*, 264–272. [CrossRef]
46. Maxwell, J.A. *Qualitative Research Design: An Interactive Approach*; Sage Publications: Szende Oaks, CA, USA, 1996.
47. Corbin, J.; Strauss, A. *Basics of Qualitative Research: Techniques and Procedures for Developing Grounded Theory*; Sage Publications: London, UK, 2015; ISBN 978-1-4129-9746-1.
48. Charmaz, K. *Constructing Grounded Theory: A Practical Guide through Qualitative Analysis*; SAGE: London, UK, 2006.
49. Glaser, B.G.; Strauss, A.L. *The Discovery of Grounded Theory*; Aldine deGruyter: New York, NY, USA, 1967.
50. Pace, S. A grounded theory of the flow experiences of Web users. *Int. J. Hum. Comput. Stud.* **2004**, *60*, 327–363. [CrossRef]
51. Cutcliffe, J. Adapt or Adopt: Developing and Transgressing the Methodological Boundaries of Grounded Theory. *J. Adv. Nurs.* **2005**, *21*, 42. [CrossRef] [PubMed]
52. Passmore, J.; McGoldrick, S. Super-vision, Extra-vision, or Blind Faith? A Grounded Theory Study of the Efficacy of Coaching Supervision. In *Coaching Researched: A Coaching Psychology Reader*; Wiley Online Library: Hackensack, NJ, USA, 2020; pp. 145–165.
53. Hwang, S. Utilizing qualitative data analysis software: A review of ATLAS.ti. *Soc. Sci. Comput. Rev.* **2007**, *26*, 519–527. [CrossRef]
54. Gibbs, G.R. Media review: ATLAS.ti software to assist with the qualitative analysis of data. *J. Mix. Methods Res.* **2007**, *1*, 103–104. [CrossRef]
55. Lewis, R.B. NVivo 2.0 and ATLAS.ti 5.0: A comparative review of two popular qualitative data-analysis programs. *Field Methods* **2004**, *16*, 439–464. [CrossRef]

applied sciences

MDPI

Article

Digital Twins in Product Lifecycle for Sustainability in Manufacturing and Maintenance

Izabela Rojek [1,*], Dariusz Mikołajewski [1] and Ewa Dostatni [2]

1 Institute of Computer Science, Kazimierz Wielki University, 85-064 Bydgoszcz, Poland; dmikolaj@ukw.edu.pl
2 Faculty of Mechanical Engineering, Poznań University of Technology, 60-965 Poznań, Poland; ewa.dostatni@put.poznan.pl
* Correspondence: izarojek@ukw.edu.pl

Abstract: A "digital twin" is a dynamic, digital replica of a technical object (e.g., a physical system, device, machine or production process) or a living organism. Using this type of solution has become an integral part of Industry 4.0, offering businesses tangible benefits, in addition to being particularly effective within the context of sustainable production and maintenance. The purpose of this paper is to present the results of research on the development of digital twins of technical objects, which involved data acquisition and their conversion into knowledge, the use of physical models to simulate tasks and processes, and the use of simulation models to improve the physical tasks and processes. In addition, monitoring processes and process parameters allow for the continued improvement of existing processes as regards intelligent eco-designing and planning and monitoring production processes while taking into account sustainable production and maintenance.

Keywords: digital twins; sustainability; manufacturing; maintenance; computational intelligence

Citation: Rojek, I.; Mikołajewski, D.; Dostatni, E. Digital Twins in Product Lifecycle for Sustainability in Manufacturing and Maintenance. *Appl. Sci.* **2021**, *11*, 31. https://dx.doi.org/app11010031

Received: 2 December 2020
Accepted: 21 December 2020
Published: 23 December 2020

1. Introduction

The term digital twin (DT) stems from the concept of a virtual equivalent to a physical phenomenon, developed in 2002 at the University of Michigan. A dynamic, digital replica of a physical system, device, machine, production process or a living organism is more than just a model [1]. Any changes to which the physical object is subjected are detected by sensors and reflected in its digital replica. This offers a more in-depth insight into the processes which occur when using the physical object, rendering it possible to predict events, enabling effective remote management, early malfunction detection, element wear monitoring and predictive responding to such situations. Conclusions drawn from an analysis of a digital twin can subsequently be applied to the original object with reduced risk and increased return on investment. In addition to being implemented in new and already established solutions (e.g., production lines that are planned to operate for many more years), the digital twin technology is also used by businesses to test new products before committing to serial production and use, enabling the introduction of improvements and further development. A key requirement for developing a virtual recreation of the physical world is real-time access to a complete spectrum of appropriate-quality data and that the virtual world can "learn" the behaviors of the physical world. Currently, available dedicated solutions meet this requirement. This is rendered possible as a result of the development of sensors and the Internet of Things, which offers the ability to continuously acquire data and transfer it in large volumes (currently in exabytes). The other end of the process of developing a digital twin involves analytical tools, machine learning and artificial intelligence instruments which make use of this data. Their dynamic development enables the effective use of data to build knowledge about a physical object, its behaviors and reactions to changes in its environment, and continuous verification of the recreation, and in some cases also creating variations for the purpose of analyzing different scenarios. The ability to visualize a technical object (e.g., a machine) and to simulate its operation (both functionally and structurally) is expanded to include communication methods that enable teamwork. Digital twins, due to their features, revolutionize industrial and business activities thanks to the results which are possible to achieve with their help, in particular their predictive potential, product monitoring throughout the entire product life cycle and the ability to utilize this knowledge to optimize and develop products, including developing new and improved generations. The availability of technological solutions enabling the implementation of the digital twin concept, combined with the results it generates, make its implementation a key part of modern digital transformations. In annual reports published by Gartner on major strategic technological trends entitled "Top 10 Strategic Technology Trends for 2019", "Top 10 Strategic Technology Trends for 2020", "Top 10 Strategic Technology Trends for 2021", digital twins and related technologies are frequently in the top ten. According to the report, half of all large manufacturing companies will be using digital twins by 2021, resulting in a 10% increase in effectiveness. In the new generation of the industry—Industry 4.0, digital twins are of key importance. Depending on the industry, DT offers improvements at every stage of the production process, in particular: shorter solution modeling time (up to 20%), shorter construction time (up to 25%), a marked increase in the quality of documentation and reduced amount of design errors (up to 20%) [2–4]. In the middle and long term, it also increases sales and market share, as the enhancement of the above-mentioned processes results in businesses being able to participate in a higher number of projects than in the past, and for specific industries, it also means larger, more complex projects, possibly featuring more advanced technologies and offering a higher return on investment. Monitoring product life cycles renders exchanging information between various departments more effective, which may give rise to completely new possibilities regarding analysis, prediction and creative applications, including faster product line replacement and simultaneous demand creation across multiple markets previously unavailable due to scale of production or logistical limitations, for example. As a result, the reduction in production, service/helpdesk time and/or cost is a natural consequence. The digital twin concept becomes even more attractive in sustainable manufacturing, in both subtractive and additive machining.

Sustainable development is one of the most important issues for current and future generations. The assumption that natural resources are infinite, i.e., that the ability of the environment to regenerate is sufficient to repair all damage caused by the entirety of human activity, is changing. Thus, the idea of sustainable development now influences all organizational aspects of human life, including economic, political, social and environmental. Manufacturing, being one of the foundations of the civilized lifestyle, will be profoundly affected by it and will have an important role to play in creating a sustainable path towards development. Despite this rising awareness, nearly all current production models are based on the old, outdated paradigm. The technology on which manufacturing is largely based must, along with culture and economics, provide tools and opportunities to create new solutions to bring us closer to sustainable production [5,6]. New technologies, business models and lifestyles will become a milestone marking the advent of a new, sustainable world, and nowhere will this be truer than in the manufacturing sector. The inevitable restrictions and increased requirements will impact the entire industrial sector, as well as education and science, as we move closer to sustainable development [7,8]. Research and development will be of key significance, their purpose of enabling society to adequately satisfy the above-mentioned needs in the form of properly trained staff and innovative technologies [9,10]. The main research challenges related to sustainable production have been presented by the authors involved in the international project "IMS2020: Supporting Global Research for IMS2020 Vision", promoted by the European Commission for the purpose of developing a roadmap for future (2020) production [5,6].

The purpose of this paper is to present the results of the authors' research on the development of digital twins, which involved data acquisition and their conversion into knowledge, the use of physical models to simulate tasks and processes, and the use of simulation models to improve physical tasks and processes. In addition, monitoring processes and process parameters allow for the continued improvement of existing processes as regards intelligent eco-designing and planning and monitoring production processes, including sustainable production and maintenance.

The novelty proposed by the authors, which is missing from the literature, is presented in the following chapters: Materials and methods, Procedures and Results, are the result of the authors' own research, simulations and calculations. The industry uses DT technology to create benchmarks for predictive analysis of asset performance. As an advanced type of process model simulation, digital twins provide real-time data, and operators can apply them in a variety of ways throughout planning, production and supply. In our study, various support systems are being created, which we have included in the article. As for the digital twins that are being created, they are the research of different researchers, not patterns.

The Introduction section defines the concept of the digital twin in the context of Industry 4.0, sustainable production and maintenance. This part also defines research objectives and methods to achieve them and presents the structure of the article. The rest of this paper is structured as follows: Section two provides a review of the literature related to the use of DTs in enterprises and highlights the importance of machine-learning methods in the context of their use in digitizing data in DTs. The Materials and Methods section presents the own method used to create a digital twin. The Procedures section demonstrates the application of the DT in eco-designing, planning and monitoring of manufacturing processes for sustainability in manufacturing and maintenance. The Results section contains an assessment of the solution, an example application of digital twins and a system architecture incorporating DTs. The final two sections contain Conclusions and References.

2. Literature Review

A comprehensive literature review was carried out using Whittemore and Knafel's approach. This integrative method of study is the only approach that combines different methodologies (including experimental and non-experimental research) and has the potential to play a greater role in practice. Literature research was conducted in scientific journals and monographs. The most basic criterion for inclusion was the relevance for this group of technologies, both for their theory, simulation and experimental research, and practical implementation in the economy.

The authors of article [7] identify eight DT development prospects, including modular DTs, ensuring modeling coherency and accuracy, incorporating big data analytics into DT models, improving DT simulations, integrating virtual and augmented reality systems with DT, expanding the applications of DT, effective mapping of cyber-physical data and integration with cloud/edge computing. The DT technology is an effective tool that meets the requirements of smart manufacturing by way of recreating physical processes in virtual space [8,9]. This is performed within a broad-spectrum of CPS (cyber-physical systems). The DT paradigm is well suited to a lifecycle-based paradigm [10]. Technology is increasingly considered to be of paramount importance to improving and evolving global production, including its globalization. Because DTs are now being used in new sectors to increase productivity, efficiency and competitiveness, a variety of tools and methods must be used, which include tools for managing data and connectivity, tools for representing and storing data, machine learning tools and analytical methods.

Data acquisition and transmission are of key significance to DT as the technology requires real-time connectivity and data transfer. As an example, Freeman [11] proposes a data stream processing system in which data are analyzed and queried continuously. The above-mentioned data acquisition systems are of key significance to the implementation of DT in production environments containing data collected using temporary (due to the hardware used, type of materials, etc.) and permanent data storage processes (processing systems utilizing real-time sensors) [12]. Heterogeneous data and industry-specific knowledge acquired from production processes require modeling and integration with production systems. The most important knowledge representation tools for developing DTs include NoSQL ontologies and databases [13–17], relational databases [18,19], MySQL [20], SQLite 3 [21], real-time databases for storing various UML data structures [22], transactional graph databases [23] and databases handling data transformation and predictive queries [24].

In order to process data into knowledge, it is necessary to utilize machine learning and data exploration methods. The following methods are used in the case of digital twins:

- Neural networks, used in quality prediction and operation control in metallurgy [25], smart control in the paint and finishing works industry [26,27], error prediction and diagnostics in CNC machining [28];
- Deep neural networks, used in autonomous production in smart production plants [29] and error diagnostics [30];
- Dynamic Bayesian networks, used by DTs to monitor the condition of aircraft wings [3
- NSGA-II genetic algorithm used to optimize the efficiency of a machine [19,26,28,32].

Businesses utilizing the DT technology also make use of what is referred to as microservices. These are programming tools developed as a set of loosely connected services. Thönes [12] defines them as functions that enable the development of an application as a set of relatively small, coherent, isolated and autonomous services that perform specific tasks. Rojko [33] analyzed virtualization tools in modern production systems, which enable the monitoring and tracking of resource services in a production plant for the purpose of automatic conflict resolution and increasing efficiency by way of facilitating decision-making and control.

Potential applications of DTs in the product lifecycle management process from creation to disposal (PLM) are analyzed within the general competitive process framework proposed in Casadesus-Masanell and Ricart [34], which outlines the structure and process of implementing DTs. It should be noted that introducing a DT renders it possible to limit the number of samples in product assessment drastically and enables businesses to mitigate production risks related to production line imperfections.

From an innovation perspective, process innovation involving a marked increase in the effective utilization of the creative potential of R&D and engineering staff compared to traditional methods translates into product innovations in the long term (increased quality of new product and/or service families). Moreover, the process of training the staff is expedited thanks to increased productivity and the number of projects completed, in addition to better error identification, including among younger, less experienced employees. This is due to the increased effectiveness of DT, including in modeling and analyzing designed objects while accounting for their statics, kinematics and material properties, space optimization thanks to analyzing large-scale designs and multilayered blueprints (electricity, piping, etc.), or creating technical documents or visualizations which take into consideration the technologies used to manufacture the objects (subtractive, additive or hybrid).

Gharaei et al. proposed a systems engineering approach to define DT requirements to formalize the DT concept from a systemic perspective, including the conceptual architecture of DT, which is defined based on ISO 42,010 standard. DT architectures have been identified by capturing formalized requirements using the EARS approach assessed based on a case from the IMPULSE project [35].

Some UE projects are also proposed for this topic, such as FACTLOG—energy-aware factory analytics for process industries, QU4LITY—digital reality in zero-defect manufacturing and the Innosuisse, IMPULSE project on digital twins, etc. We strongly believe that their outcomes will enrich and fuel the future of DT technology and its environment.

3. Materials and Methods

In line with the DT paradigms, research was conducted, which involved data acquisition and the development of the principles according to which these data would be transferred. The data were then processed into knowledge. In our approach, we try to extract the knowledge and experience of the company's employees, which are hidden in the developed examples included in databases. Data-mining methods extract knowledge from data. Based on decision trees, decision rules are created and then included in the applications used by employees. You could say that knowledge has been extracted and incorporated into computer applications.

The actual data come from several manufacturing companies in the area of eco-design, process planning and process supervision. These companies were characterized mainly by unit and small batch production, for which the knowledge gained for them is very important. In order to develop models in the form of decision trees with designers, technologists and machine operators, important criteria for solving individual problems were established, and predicators (conditional attributes) and decision classes were defined, which were used to file learning examples periodically. To make the data useful for exploration, they were cleaned and transformed. Cleaning the data consisted of standardizing the record, completing the missing data and identifying remote points, then converting the data into normalization and coding.

Physical models were used to simulate tasks and processes. Simulation models were subsequently developed for the purpose of improving physical tasks and processes. Monitoring processes and process parameters render it possible to continually improve physical processes from the perspective of intelligent eco-designing, planning and monitoring manufacturing processes in line with sustainable manufacturing and maintenance principles. Digital twins were created for the above-mentioned areas. The digital twins were developed in stages:

- Analysis of physical processes and input data collected from real processes using sensors and measuring devices;
- Development of artificial intelligence models in the form of neural networks, decision trees and rules, and fuzzy logic;
- AI model assessment;
- Creation of a digital replica of the physical process or machine;
- Simulation of a physical process or machine operation;
- AI model improvement;
- Application of the AI models in existing systems;
- Upgrading and teaching AI models (Figure 1).

We considered individual objects: in eco-design, this is a choice of 3 ways of material. We broke down the planning of the technological process into simpler elements. We divided process supervision into the supervision of selected processes. As far as the size of the teaching files was concerned:

- for eco-design, the files had about 200 examples of learners;
- for design and supervision, there were about 300 examples of learners.

Concerning eco-designing, the focus was primarily on material selection. The goal of the study was to develop material selection methods and models utilizing AI and simulate the implementation of these methods to achieve optimal selection from the perspective of eco-construction. Material selection focused on a multicriteria analysis taking into consideration realistic factors from industry, i.e., determining the compatibility of materials, selecting additional materials while accounting for a set compatibility level and selecting methods of connecting materials. An eco-constructor selects materials and their fastenings in a way that minimizes their negative impact on the environment and renders recycling easier. For example, the compatibility of the materials was determined by the compatibility matrix (Figure 2).

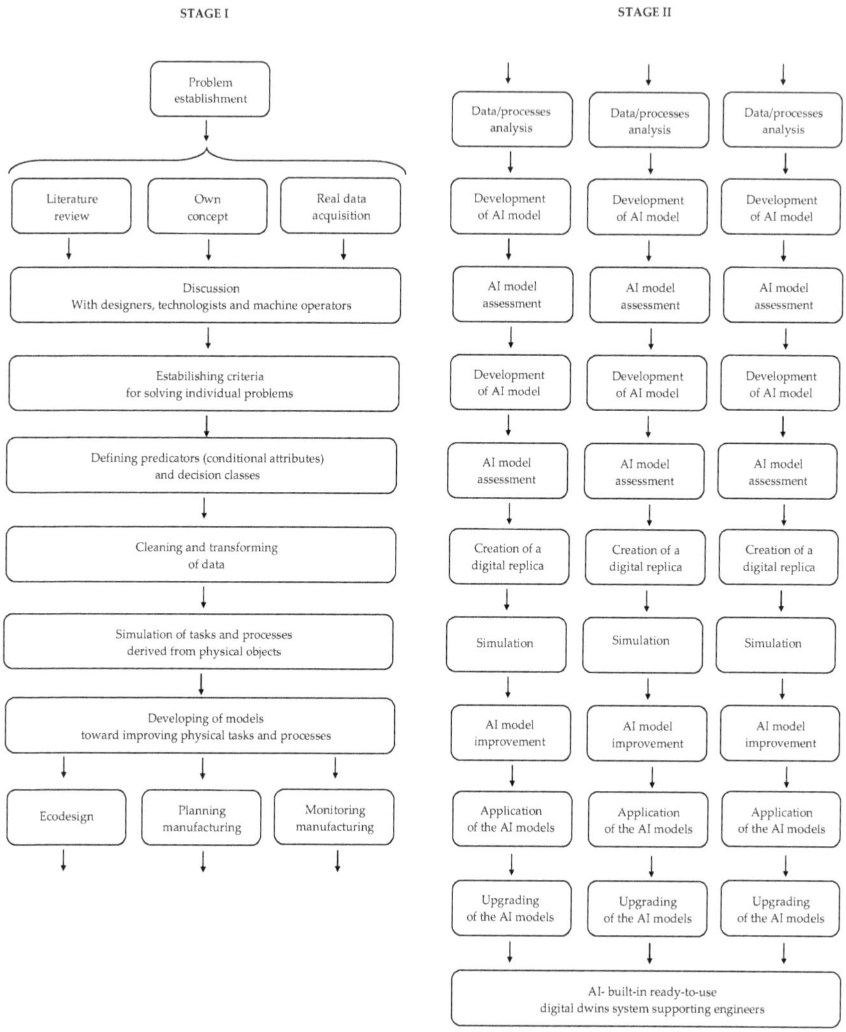

Figure 1. Main phases of the proposed method.

As regards the planning of manufacturing processes, special emphasis was put on technological process design. The purpose of the study was to develop technological process design methods and models utilizing AI and to simulate the operation of these processes to achieve optimal results from the perspective of a particular business. Decision rules were used to determine the structure of the technological process, i.e., the order of technological operations and procedures. Neural networks rendered it possible to construct models for selecting semi-finished products, instruments, machine tools, tools, tool holders and machining parameters, and decision trees—models for selecting machine tools, tools and tooling. Example selections used to create models were processed, i.e., normalized using fuzzy logic. Subsequently, the models of selecting semi-finished products, tooling, machine tools, tools, tool holders and machining parameters were implemented in the form of a prototype expert system for designing technological processes. The system is dedicated to technologists who do not possess sufficient experience in designing technological processes or have only just begun working at a given production company and thus are not yet familiar with the machines or other means of production.

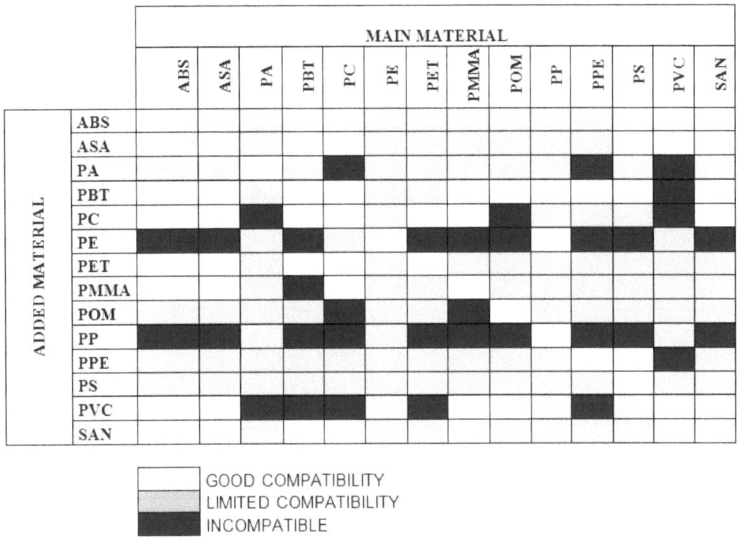

Figure 2. Matrix of compatible materials [36].

As regards the monitoring of manufacturing processes, special emphasis was put on machining process monitoring. Simply designing a technological process was not sufficient. The product is manufactured under the planned process; however, various disruptions may occur during this process, which impacts the quality of the final product. Another research goal was to eliminate this type of hazard using our methods and models of monitoring the technological process of machining. The monitoring encompassed:

- Machining process disruptions; a simulation of such disruptions was conducted—when a particular disruption occurs, the operator selects its cause and is provided with a solution to the issue. The knowledge assumes the form of decision rules;
- Machining process stability—control charts were used to develop process stability disruption patterns, and a neural network was used to simulate the monitoring of such a process. After improving the example scenarios, the neural network was learned to react to instabilities properly. This model was then implemented in the system. The network is learned and improved based on actual actions;
- Ra and Rz surface roughness value—using a system consisting of models made up of neural networks and decision trees; the system informs the operator about norms being exceeded, i.e., it signals that the machining parameters require correcting;
- The CNC machine via controlling and compensating for the thermal deformation of ball screws—a system was developed which incorporates neural network-based models for monitoring current speed and load values and predicting the elongation of the CNC machine ball screw based on these values;
- Monitoring and predicting to predict blade wear based on various input data (cutting forces, acoustic emission and mechanical vibration). Selected measurements of physical parameters were made during the processing, and a simulation of blade wear was conducted based on these values. The neural networks were learned to predict the wear level of the blade.

A new element introduced in this study was developing DT as a new approach to selected issues while taking into account maintenance in sustainable manufacturing. Following the paradigm of sustainable production, this change in approach must be implemented on three levels: product, process and system. At the product level, the traditional 3R (reduce, reuse, recycle) concept was transformed into a more sustainable 6R approach (reduce, reuse, recycle, recover, redesign, remanufacture) [37,38], which is being gradually implemented in eco-designing and manufacturing process planning. At the process level an attempt was made to optimize technological processes to reduce resource use, the amount of waste generated, and hazards related to the work and working environment which finds implementation in the designing and monitoring of machining processes.

According to the EN 13306: 2010 standard, maintenance is a "combination of all technical, administrative and managerial actions during the life cycle of an item intended to retain it in, or restore it to, a state in which it can perform the required function." From the perspective of a business, this means that maintenance operations should focus on e.g., ensuring the required level of reliability and availability of machines and devices and their efficiency, optimal use of the capital invested, ensuring the required level of safety for users and technical operators, monitoring the environmental aspects of machine operation and operation processes, modernizations ensuring the economic efficiency of the objects used, cooperation with providers of machines, replacement parts and services regular improvement of technical service employee skills, etc. Therefore, selecting the correct combination of operational strategies for every technical object (corrective, preventive, predictive) must take into account not only economic and technical factors but also environmental and security-related ones and how they relate to the business strategy of the company [39,40]. It is thus evident that maintenance can have a profound impact on the effectiveness of sustainable production. Such major maintenance factors, which are of significance to the development of sustainable production processes, include replacement part and consumable management, cooperation with a machine, device and repair service providers, machine and device modernization, cooperation with design and product development departments, cooperation with production and quality departments, cooperation with OHS and environmental departments, employee competences, use of preventive and predictive operational strategies and systems for collecting and processing operational data [41–45]. Operational data, as analyzed from the perspective of sustainable production, find use in all of the analyzed areas: eco-designing, planning manufacturing and monitoring manufacturing processes. The application of DT in optimizing the physical processes of a business is presented in Figure 3.

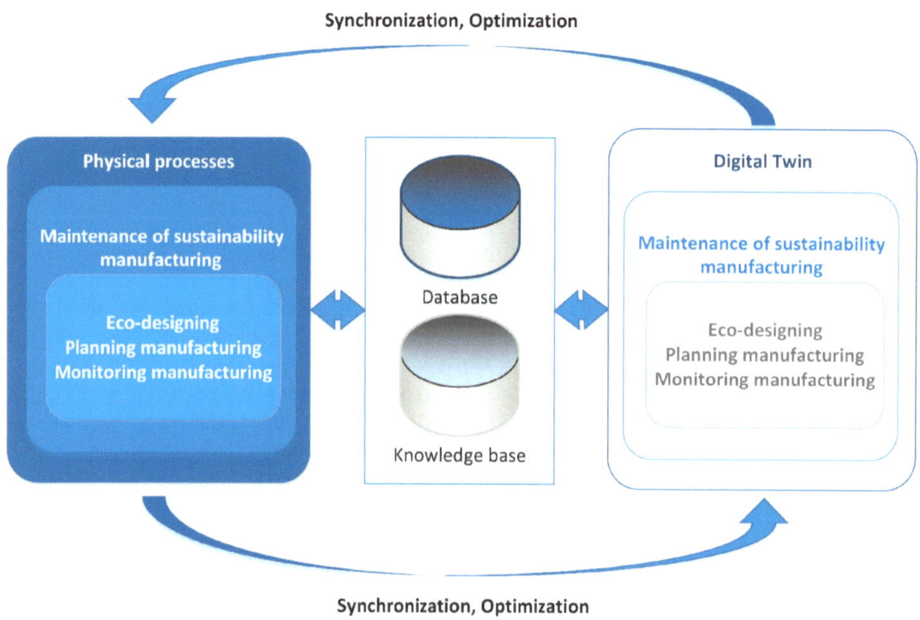

Figure 3. Physical process optimization using digital twin (DT).

4. Procedures

The digital twins developed have a profound impact on certain parts of the business in which they were implemented. At the construction level, data acquired from users who test and use products in their everyday life were used. Construction changes were simulated in a digital twin to improve product design via correct material selection. At the production planning and monitoring stage, a DT rendered the process more efficient, reliable and flexible thanks to a wide range of innovative research. This was particularly significant at the technological process design and production process monitoring stage. DTs can be used to digitize process models to better react to consumer trends. In order to smartly facilitate tasks in a business, digital twins make use of a range of classification models that predict particular parameters used in manufacturing goods. The models were developed in the form of classification trees: C4.5, C&RT, CHAID, reinforced decision trees and random forests.

DT and eco-designing

In the field of eco-designing, the digital twin offers a series of classification models for selecting materials that take into consideration maintenance in sustainable manufacturing.

Recycling-focused eco-designing primarily involves selecting appropriate construction materials and methods of fastening them. Products should be designed to incorporate the maximum possible amount of normalized and recyclable materials. This has a positive impact on the environment in the final stages of the product's life cycle, including conservation and decommissioning. When selecting materials for use in products, their compatibility should also be taken into consideration: the materials used should be recyclable at the end of the product's life cycle without the need to separate them. Additional data were also collected about the materials, based on which they can be recovered, redesigned and remanufactured. The authors' previous papers on the development of neural network classifiers can be found in [46].

Materials can be selected in two ways. The first method involves selecting added materials based on the primary material and the required compatibility level between the two. In the second method, the user defines the primary and added materials of a new element, and the system informs them about the compatibility between these materials. The fastening of materials is another important issue. Material fastening methods must ensure quick and smooth disassembly, especially in cases where the use of incompatible or hazardous materials is necessary due to functional reasons [47]. If the compatibility is high, temporary or permanent fastening can be used. However, if the compatibility is low or if the materials are incompatible, only temporary fasteners should be used.

Files containing examples used to develop the classifiers were created based on an analysis of material properties, including material name (e.g., PVC), a tensile strength in megapascals (e.g., 35.5), density in grams per cubic centimeter (e.g., 7.88), processing temperature in Celsius (e.g., 20.8), elongation at yield (Re) in percentage (e.g., 5.5), Young's modulus E in gigapascals (e.g., 4.61), dielectric constant (e.g., 2.0), dielectric strength in kilowatts per millimeter (e.g., 22.0), water absorptivity in percentage (e.g., 22.55), recycling cost in PLN per kilogram (1 PLN = 0.23 euros) (e.g., 4.25); a positive value denotes profit from selling the material, while a negative value denotes disposal cost, negative impact on the environment (e.g., true) and the name of the added material (e.g., ABS). The material parameters were established with designers.

For the first problem, i.e., material selection where the user defines a material selected for an element (part) of the product and the required level of compatibility, and the system recommended which material to add, a file was created containing examples with the conditional attributes "main material" and "compatibility", and the decision class "added material". For the second problem, i.e., material selection where the user defines the main and added material of a new element, and the system informs the user about the compatibility between these materials, a file containing examples was created with the conditional attributes "main material" and "added material", and the decision class "compatibility". For the third problem, i.e., selecting material connections, a file containing examples was created with the following conditional attributes: "main material", "added material" and "compatibility", and the decision class "connection type". Example classifiers are presented in Figure 4 (in the StatSoft Statistica DataMiner software). The root of the tree shown in Figure 4a,b was the main material attribute, which provided the highest information gain (the division criterion of examples).

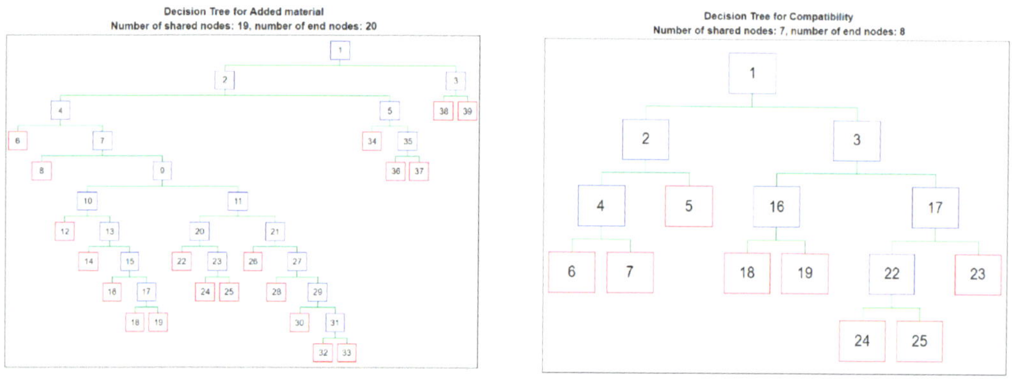

(a) (b)

Figure 4. Example classifiers: (**a**) for selecting a second product material, (**b**) for determining the compatibility of two materials.

DT and planning manufacturing

In the field of planning manufacturing, the digital twin contains many classification models for technological process design while taking into account maintenance in sustainable production. The selection models are based on the following criteria matching the technologies used in a given manufacturing company and related to particular selections:

- Semi-finished products: this selection takes into account production scale, part quality, part shape, semi-finished product availability and type of material;
- Technological process structure: this selection determines the order of technological operations and procedures, taking into account production scale, type of semi-finished product, requirements and shape of the workpiece;
- Workpiece equipment: taking into account production scale, type of semi-finished product, machining method, part shape; also, the technologist assesses the speed of mounting and positioning repeatability;
- Machine tools for technological operations and procedures: taking into account machining accuracy, workpiece dimensions, expected load, production efficiency and hourly cost of machine tool operation;
- Machining tools for technological operations and procedures: taking into account the shape of the machined surface, machining method, scale of production, type and accuracy of the machining, kind of machine tool and type of workpiece material;
- Tooling: taking into account the dimensions of the tooling compatible with the tool and machine;
- Machining parameters are taking into account the type of machine tool, part material, surface quality requirements and type of blade material.

Classification models were developed for all of these selections. The authors' early work demonstrates certain selection aspects whose classification models were based on neural networks [48]. As an example, an optimal classifier for tool holder selection using decision trees was presented. Tooling selection models were built in a similar fashion to the other classification models. Figure 5 depicts a decision tree, i.e., the optimal classifier for selecting a tool holder. This selection occurred when the tool was not compatible with the machine. The root of the tree shown in Figure 5 was the milling tool attribute, which provided the highest information gain (the division criterion of examples).

DT and monitoring manufacturing

In the field of monitoring manufacturing, the digital twin comprises a range of classification and predictive models for monitoring production processes, which take into consideration maintenance in sustainable manufacturing. This module is constantly being developed. Models were developed which facilitate the elimination of machining process disruptions (simulations were conducted of these disruptions and the methods of their repair), maintaining process stability based on control charts, signaling the need for machining parameter correction based on workpiece surface roughness assessment, CNC machine control including compensating for the thermal deformation of ball screws, predicting blade wear based on various input data, in particular cutting forces, acoustic emission and mechanical vibrations. The authors' earlier work was related to constructing models in the form of neural networks [49–51] and other authors [52]. This paper demonstrates classification models in the form of decision trees for predicting VBc - tool wear (Figure 6) based on cutting force. The root of the tree shown in Figure 5 was the time of cutting force attribute, which provided the highest information gain (the division criterion of examples).

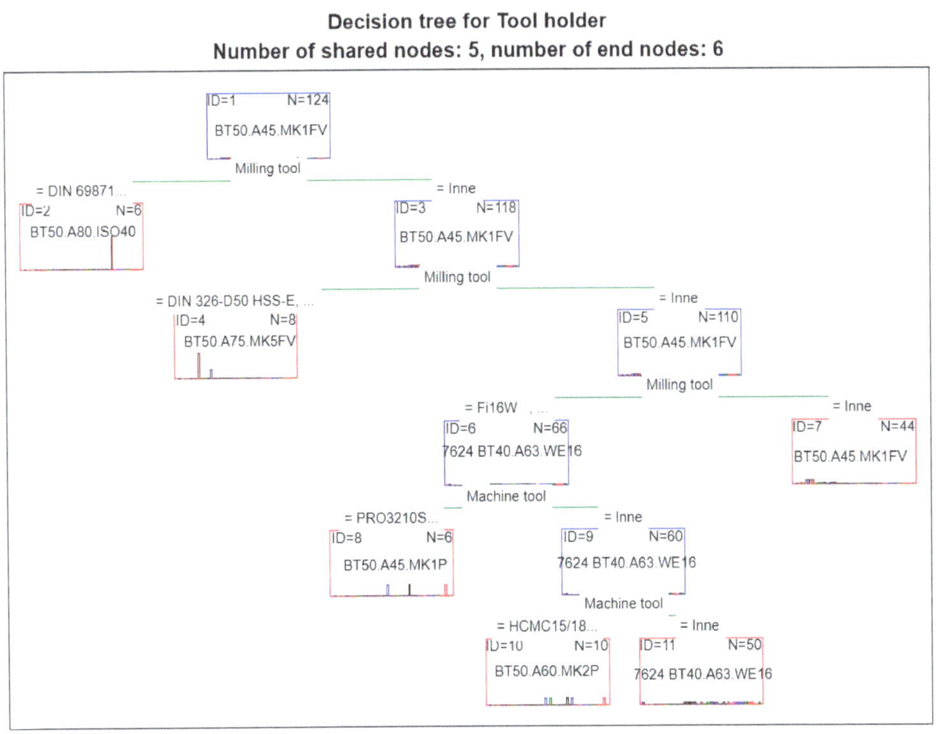

Figure 5. Optimal classifier for tool holder selection—the tree with strictly defined nodes.

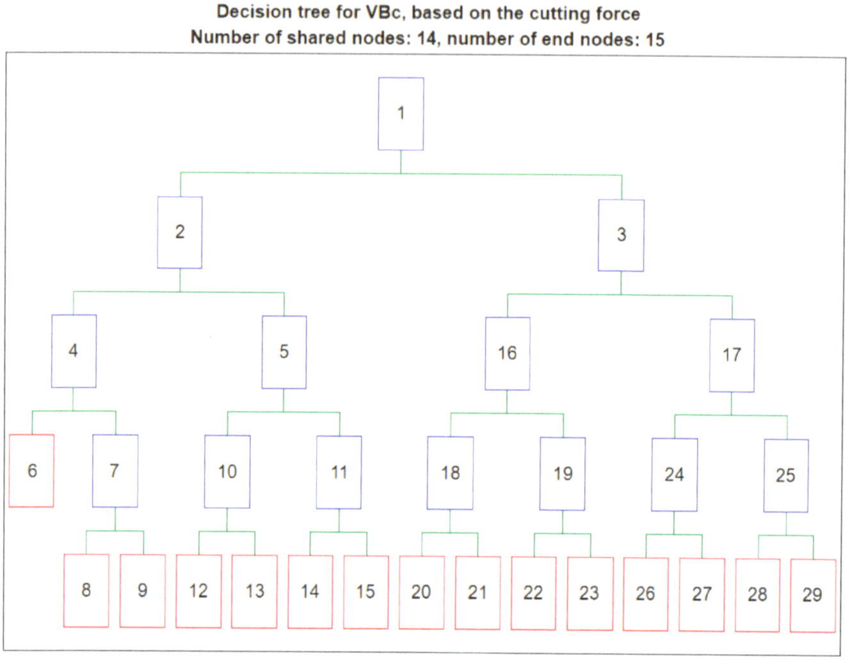

Figure 6. Tool wear (VBc) level prediction classifier utilizing cutting force analysis.

5. Results

The decision tree models were verified and assessed for the purpose of selecting the most effective models. The models were developed in the form of classification trees: C&RT, CHAID, reinforced decision trees and random forests. Model parameters were changed for every type of tree. The following classification parameters were set for the C&RT model: the cost of erroneous classification, the goodness of fit and a priori probability. The stop criterion incorporated the stop rule: cut according to variance and the parameter of a minimum number of examples per node. For the CHAID model, the cost of erroneous classification was set. The stop criterion required a minimum number of examples per node. For the reinforced tree model, the following classification parameters were set: cost of an erroneous classification and a priori probability. The stop criterion incorporated the parameter of the minimum number of examples per node. For the random forest model, the following classification parameters were set: cost of an erroneous classification and a priori probability. The stop criterion incorporated the parameter of the minimum number of examples per node.

The cost of erroneous classification relates to the distribution of examples across classes. Cost minimization is equivalent to minimizing the overall proportion of erroneously classified cases when a priori probabilities are proportional to class size, and the cost of erroneous classification is equal for every class [53].

The goodness of fit involves finding the split for every predictor, which offers the highest increase in the goodness of fit. How does one define the goodness of fit increase? The goodness of fit can be measured in three ways. The Gini impurity measure reaches zero when a given node contains only a single class (with a priori probabilities estimated based on the size of classes and equal cost of erroneous classification, the Gini measure is calculated as the sum of the products of all class proportion pairs in a given node; and reaches the maximum when the number of classes in a given node is equal). The Gini index was the preferred measure of goodness of fit for the developers of C&RT [53]. A perfect fit means perfect classification.

A priori probabilities determine the probability that a given case or object fits a given class without any prior knowledge about the predictive variables in the model. They are used in cost minimization and may impact case or object classification. Cost minimization is equivalent to minimizing the general proportion of erroneously classified cases when a priori probabilities are proportional to class size (and the cost of erroneous classification is equal for every class), as predicting should be more effective for larger classes and yield a generally lower rate of erroneous classifications [53].

As for the attributes: attached is a different form of the tree, where you can see what the root is, how these attributes are distributed. Classification matrices demonstrate the effectiveness of the classifiers. These were developed and analyzed for every decision tree to choose the most optimal one. An example classification matrix is presented in Figure 7. The material added to the main material constitutes the observed and predicted class. Table 1 presents classes of added materials, which we can connect to main materials.

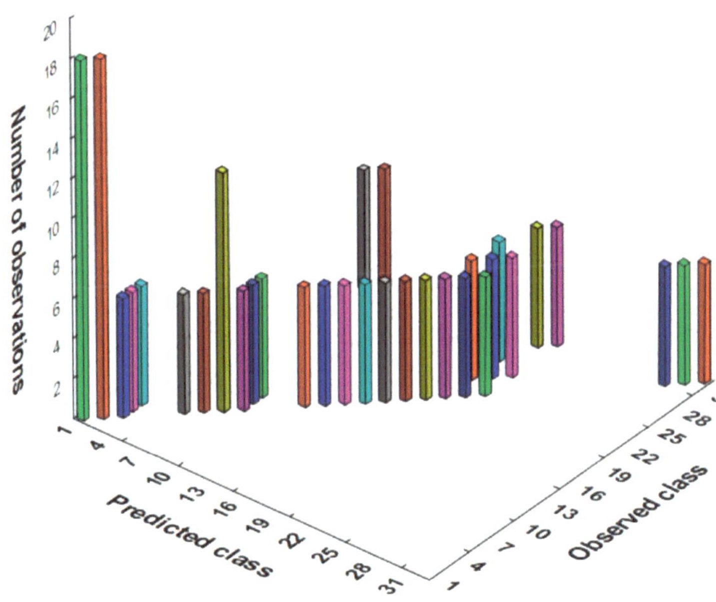

Figure 7. Decision tree classification matrix for selecting materials based on the added material (the observed and predicted values of the additional material).

Table 1. Classes of added materials.

Class	Added Materials
1	SAN, PVC, PMMA, PC, PBT, ASA, ABS
2	PS, PPE, PP, POM, PET, PE, PA
3	PA
4	SAN, PS, PP, POM, PMMA, PET, PE, PBT, ASA, ABS
5	PVC, PPE, PC
6	SAN, PC, PBT, ASA, ABS
7	PS, PPE, PP, POM, PMMA, PET, PE, PA
8	PVC
9	SAN, PMMA, PET, PC, PBT, ASA, ABS
10	PS, PPE, PP, PE
11	PVC, POM, PA
12	PP, PE
13	PA, PVC, PC
14	SAN, PS, PPE, POM, PMMA, PET, PBT, ASA, ABS
15	PET, PC, PBT, ASA, ABS
16	SAN, PA, PVC, PE, PS, PMMA, PPE, POM, PP
17	PC, ABS, PMMA, ASA
18	PA, SAN, PE, PVC, PET, PS, POM, PP, PPE
19	PBT
20	POM
21	ABS, PVC, ASA, SAN, PA, PS, PBT, PE, PET, PPE, PP
22	PMMA, PC
23	PP
24	PVC, PE, PA
25	ABS, SAN, ASA, PS, PBT, PPE, PC, POM, PET, PMMA
26	PS, PPE
27	ABS, ASA, SAN, PA, PP, PBT, POM, PC, PMMA, PET, PE
28	PS
29	ABS, ASA, PA, POM, PBT, PP, PC, PPE, PE, PVC, PET, SAN, PMMA
30	SAN, ABS, PVC, ASA, POM, PMMA
31	PS, PP, PE
32	PET, PPE, PA, PC, PBT

The costs of misclassifications concern the distribution of examples between classes. The minimization of costs corresponds to minimizing the proportion of misclassified cases where probabilities are taken a priori proportional to the size of the classes, and the costs of misclassification equal for each class. In order to assess the models, a cross-match test based on a test sample was used, as well as 10-fold cross-validation. Table 2 compares the cost, risk assessment and standard error of the decision trees.

Table 2. Comparison of classifiers of material selection based on the added material.

Classifier Type	Classifier Assessment	
	Risk Cost/Assessment	Standard Error
C&RT	0.035217	0.110509
CHAID	0.201187	0.016060
Reinforced decision trees	0.000001	0.000001
Random forest	0.021846	0.011653

Reinforced decision trees and random forest proved to be the most effective models. Correct classification rates were 100% and 97.64%, respectively. Decision rules were developed based on the trees, which were then implemented in the system. Example rules developed based on the second decision tree are:

1. *if added material = "ABS" and main material = "ABS" then compatibility = "good"*
2. *if added material = "ABS" and main material = "ASA" then compatibility = "good"*

3. *if added material = "ABS" and main material = "PA" then compatibility = "limited"*
4. *if added material = "ABS" and main material = "PE" then compatibility = "limited"*

The rules were imported into an existing module for determining material compatibility: main and added product (Figure 8).

Material selection - result from the network - selected compability

Main material	PVC
Density	1.43
Tensile_stress	40
Yield_point_elongation	35
Processing_temperature	170
Dielectric_constant	2
Modules_of_elasticity	470
Water_absorptivity	0.02
Recycling_cost	80
Added_material	ASA
Compability	good

Figure 8. Module for determining material compatibility.

The result is a system that facilitates a company's operations in the following areas: eco-designing, planning manufacturing and monitoring manufacturing while accounting for maintenance in sustainable manufacturing.

Figure 9 depicts a system architecture that incorporates digital twins in areas of importance to the company for which it was developed by us. It provides a dynamic virtual representation of a physical object/process at selected stages of its life cycle. Our DT system includes physical processes chosen area: eco-designing, manufacturing planning and monitoring. In databases, we have descriptions of these physical processes. Systems can both present observations concerning the current state of the system and answer the questions "what-if". Many decision trees have been defined for each decision problem in these three areas and in their specific tasks. Some trees from various areas are shown in Figures 4–6. Figure 7 shows the evaluation of an exemplary tree. All trees were assessed in this way, and the best ones were selected. The best decision trees are in the knowledge base. Decision rules were created from the trees, which are included in computer applications. The computer programs aid engineers in creating new solutions. Figure 9 shows the architecture of a DT system in the company.

The DT system architecture combines:

- Physical processes that are performed in the departments of the company;
- Digital mapping of physical processes, which is transferred to the database in the form of data;
- Based on standardized and coded data, AI models are created, which are then stored in the knowledge base;
- The system control module supervises all these activities.

When the customer's order arrives, the control module divides the tasks into individual departments of the company. The order is being processed. New physical processes are the basis for expanding the data and knowledge in the system. This is how the system learns all the time. Data that is relevant to sustainable production and maintenance are collected during the operation of the system.

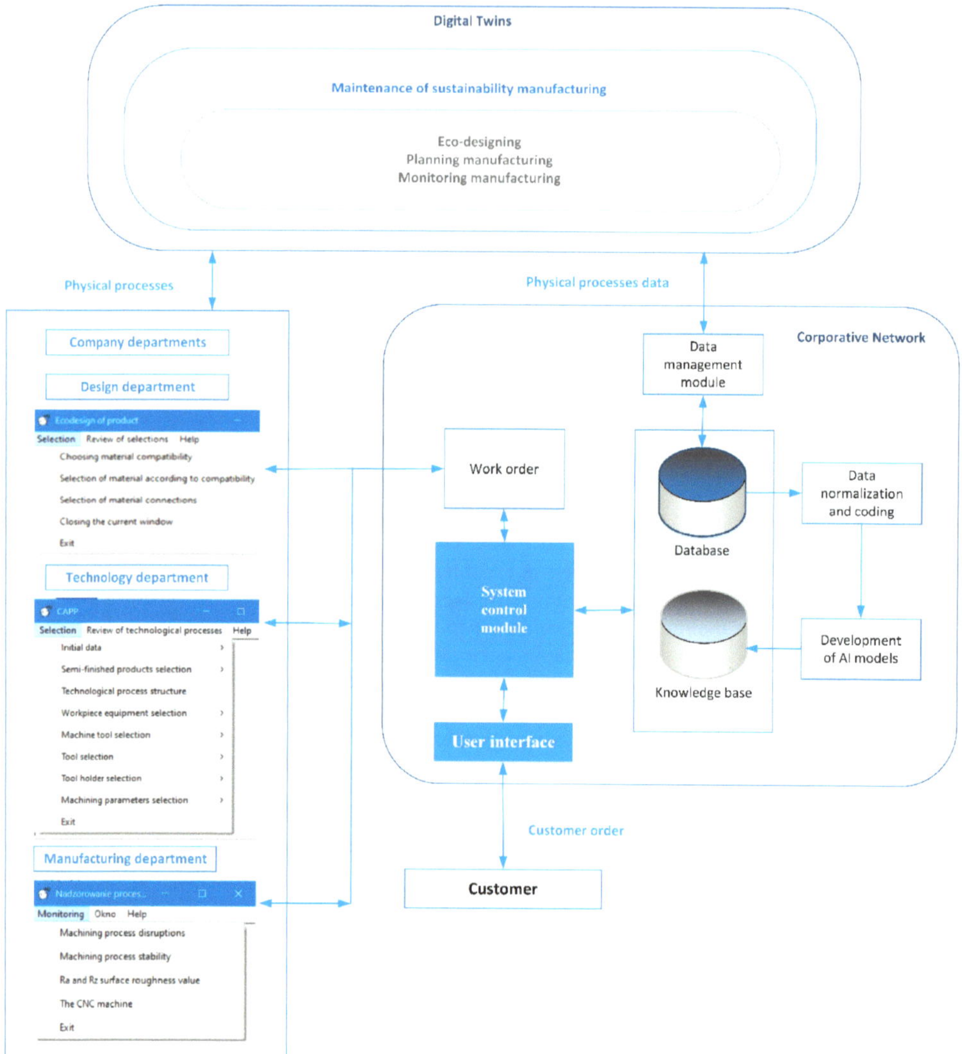

Figure 9. The architecture of a DT system.

6. Conclusions

The system supports selected areas: eco-designing, planning manufacturing and monitoring manufacturing, for the purpose of manufacturing goods, and may be of help in sustainable manufacturing. This is of particular utility in situations where green design and manufacturing requirements are difficult to formalize, uncertain, incomplete or unavailable.

Machine-learning proved a precious tool for acquiring knowledge in facilitating operations in companies. It was discovered that designers and technologists frequently make decisions based on intuition and are unable to formulate with precision the rules behind their choices. As they acquire knowledge, machine-learning methods can be used to create classic rules used in smart systems automatically. In the case of sustainable product design, decision tree induction must be used as the classification method due to the large volumes of input data presented in symbolic form. Rules generated based on decision trees are more concise, and the time required to draw conclusions is significantly reduced. Our research demonstrates that machine learning can be effectively applied in DT. Thanks to DT, the system gains improved connectivity and flexibility, in addition to increased intelligence as a result of introducing knowledge and experience to a computer program. Moreover, manufacturing efficiency is increased, as well as the quality of the goods produced, reducing costs and increasing profits. In addition to improving manufacturing processes, DT renders it possible to transition into individualized production. It also offers the possibility of conducting simulations for the purpose of amending manufacturing process irregularities and accounting for maintenance in sustainable manufacturing. All data are saved in a database, and the knowledge and experience in the system knowledge base. Utilizing data and analyses generated by sensors built into smart products and instruments, it is possible to streamline operations, reconfigurations and maintenance processes. DT facilitates decision-making in multidimensional processes, strategic planning and process prediction with the help of knowledge recycling and experience. DT expands conventional engineering analyses to include information integration to create digital product lifecycles. As monitoring and simulating in real time enables predicting tool and machine damage, all data necessary for this type of prediction are saved in the system database.

Comparing research results with solutions developed by other authors poses difficulties due to the fact that no identical product has been identified on the market, which would include all set elements from the same industry. In addition, the boundary conditions for the application of the solution in question could potentially differ to the degree that would significantly impact the final numerical result, rendering a direct comparison impossible. This indicates the necessity to develop a comparative research standard pertaining to DT solutions, including a division into industries according to their specificity and the level of complexity of the solutions themselves.

Further research will involve integrating the system as a smart module with a system responsible for managing an entire manufacturing company. This will not only enable further verification of current assumptions with the use of large sets of real manufacturing data but also contribute to gaining a more thorough understanding of how to implement DT in an existing business.

From a scientific point of view, we should also monitor the activity of research groups and their progress as published and delivered at conferences and through initiatives integrating the research community into larger umbrella projects. This will allow us to know how our research results are linked to other research, i.e., how they enrich the research carried out by other scientists and also how they affect industrial implementation. This will allow us to better target our own research, including optimizing the effect, speeding up the development of research-based on the experience of other teams, avoiding their mistakes, but also avoiding the risks of even a lack of sustainability.

Author Contributions: Conceptualization, I.R., D.M. and E.D.; methodology, I.R., D.M. and E.D.; software, I.R.; validation, I.R. and D.M.; formal analysis, I.R., D.M. and E.D.; investigation, I.R., D.M. and E.D.; resources, I.R., D.M. and E.D.; data curation, I.R. and D.M.; writing—original draft preparation, I.R., D.M. and E.D.; writing—review and editing, I.R., D.M. and E.D.; visualization, I.R., D.M. and E.D.; supervision, I.R., D.M. and E.D.; project administration, I.R. and D.M. All authors have read and agreed to the published version of the manuscript.

Funding: The work presented in the paper was co-financed under the 0613/SBAD/4677 grant for Poznan University of Technology and a grant to maintain the research potential of Kazimierz Wielki University.

Data Availability Statement: Please refer to suggested Data Availability Statements in section "MDPI Research Data Policies" at https://www.mdpi.com/ethics.

Conflicts of Interest: The authors declare no conflict of interest.

References

1. Barricelli, B.R.; Casiraghi, E.; Fogli, D. A survey on digital twin: Definitions, characteristics, applications, and design implications. *IEEE Access* **2019**, *7*, 167653–167671. [CrossRef]
2. Qi, Q.; Tao, F.; Hu, T.; Anwer, N.; Liu, A.; Wei, Y.; Wang, L.; Nee, A. Enabling technologies and tools for digital twin *J. Manuf. Syst.* **2020**. [CrossRef]
3. Rasheed, A.; San, O.; Kvamsdal, T. Digital twin: Values, challenges and enablers from a modeling perspective. *IEEE Access* **2020**, *8*, 21980–22012. [CrossRef]
4. Bevilacqua, M.; Bottani, E.; Ciarapica, F.E.; Costantino, F.; Di Donato, L.; Ferraro, A.; Mazzuto, G.; Monteriù, A.; Nardini, G.; Ortenzi, M.; et al. Digital twin reference model development to prevent operators' risk in process plants. *Sustainability* **2020**, *12*, 1088. [CrossRef]
5. Garetti, M.; Taisch, M. Sustainable manufacturing: Trends and research challenges. *Prod. Plan. Control* **2012**, *23*, 83–104. [CrossRef]
6. Garretson, I.C.; Mani, M.; Leong, S.; Lyons, K.W.; Haapala, K.H. Terminology to support manufacturing process characterization and assessment for sustainable production. *J. Clean. Prod.* **2016**, *139*, 986–1000. [CrossRef]
7. Kendrik, Y.; Hong, L.; Pai, Z.; Chun-Hsien, C. A state-of-the-art survey of digital twin: Techniques, engineering product lifecycle management and business innovation perspectives. *J. Intell. Manuf.* **2020**, *31*, 1313–1337. [CrossRef]
8. Tao, F.; Qi, Q.; Liu, A.; Kusiak, A. Data-driven smart manufacturing. *J. Manuf. Syst.* **2018**, *48*, 157–169. [CrossRef]
9. Zheng, Y.; Yang, S.; Cheng, H. An application framework of digital twin and its case study. *J. Ambient Intell. Humaniz. Comput.* **2018**, *10*, 1–13. [CrossRef]
10. Schneider, G.F.; Wicaksono, H.; Ovtcharova, J. Virtual engineering of cyber-physical automation systems: The case of control logic. *Adv. Eng. Inform.* **2019**, *39*, 127–143. [CrossRef]
11. Freeman, H. Streaming Analytics 101: The What, Why, and How. Available online: http://www.dataversity.net/streaming-analytics-101/ (accessed on 22 September 2020).
12. Thönes, J. Microservices. *IEEE Softw.* **2015**, *32*, 116–118. [CrossRef]
13. Uhlemann, T.H.J.; Schock, C.; Lehmann, C.; Freiberger, S.; Steinhilper, R. The digital twin: Demonstrating the potential of real time data acquisition in production systems. *Procedia Manuf.* **2017**, *9*, 113–120. [CrossRef]
14. Angrish, A.; Starly, B.; Lee, Y.S.; Cohen, P.H. A flexible data schema and system architecture for the virtualization of manufacturing machines (VMM). *J. Manuf. Syst.* **2017**, *45*, 236–247. [CrossRef]
15. Arafsha, F.; Laamarti, F.; El Saddik, A. Cyber-physical system framework for measurement and analysis of physical activities. *Electronics* **2019**, *8*, 248. [CrossRef]
16. Lu, Y.; Xu, X. Resource virtualization: A core technology for developing cyber-physical production systems. *J. Manuf. Syst.* **2018**, *47*, 128–140. [CrossRef]
17. Lu, Y.; Xu, X. Cloud-based manufacturing equipment and big data analytics to enable on-demand manufacturing services. *Robot. Comput. Integr. Manuf.* **2019**, *57*, 92–102. [CrossRef]
18. Nikolakis, N.; Alexopoulos, K.; Xanthakis, E.; Chryssolouris, G. The digital twin implementation for linking the virtual representation of human-based production tasks to their physical counterpart in the factory-floor. *Int. J. Comput. Integr. Manuf.* **2019**, *32*, 1–12. [CrossRef]
19. Rodič, B. Industry 4.0 and the new simulation modelling paradigm. *Organizacija* **2017**, *50*, 193–207. [CrossRef]
20. Liu, Q.; Zhang, H.; Leng, J.; Chen, X. Digital twin-driven rapid individualised designing of automated flow-shop manufacturing system. *Int. J. Prod. Res.* **2018**, *7543*, 1–17. [CrossRef]
21. Lovas, R.; Farkas, A.; Marosi, A.C.; Ács, S.; Kovács, J.; Szalóki, Á.; Kádár, B. Orchestrated platform for cyber-physical systems. *Complexity* **2018**, 8281079. [CrossRef]
22. Zheng, P.; Lin, T.J.; Chen, C.H.; Xu, X. A systematic design approach for service innovation of smart product-service systems. *J. Clean. Prod.* **2018**, *201*, 657–667. [CrossRef]
23. Schluse, M.; Priggemeyer, M.; Atorf, L.; Rossmann, J. Experimentable digital twins-streamlining simulation-based systems engineering for industry 4.0. *IEEE Trans. Ind. Inform.* **2018**, *14*, 1722–1731. [CrossRef]
24. Abramovici, M.; Göbel, J.C.; Dang, H.B. Semantic data management for the development and continuous reconfiguration of smart products and systems. *CIRP Ann. Manuf. Technol.* **2016**, *65*, 185–188. [CrossRef]
25. Damjanovic-Behrendt, V.; Behrendt, W. An open source approach to the design and implementation of digital twins for smart manufacturing. *Int. J. Comput. Integr. Manuf.* **2019**, *32*, 1–19. [CrossRef]
26. Lee, J.H.; Do Noh, S.; Kim, H.J.; Kang, Y.S. Implementation of cyber-physical production systems for quality prediction and operation control in metal casting. *Sensors* **2018**, 1428. [CrossRef]
27. Park, K.T.; Im, S.J.; Kang, Y.S.; Noh, S.D.; Kang, Y.T.; Yang, S.G. Service-oriented platform for smart operation of dyeing and finishing industry. *Int. J. Comput. Integr. Manuf.* **2019**, *32*, 307–326. [CrossRef]
28. Luo, W.; Hu, T.; Zhang, C.; Wei, Y. Digital twin for CNC machine tool: Modeling and using strategy. *J. Ambient Intell. Humaniz. Comput.* **2018**, *10*, 1129–1140. [CrossRef]
29. Ding, K.; Chan, F.T.S.; Zhang, X.; Zhou, G.; Zhang, F. Defining a digital twin-based cyber-physical production system for autonomous manufacturing in smart shop floors. *Int. J. Prod. Res.* **2019**, *61*. [CrossRef]
30. Xu, Y.; Sun, Y.; Liu, X.; Zheng, Y. A digital-twin-assisted fault diagnosis using deep transfer learning. *IEEE Access* **2019**, *7*, 66. [CrossRef]

31. Li, C.; Mahadevan, S.; Ling, Y.; Choze, S.; Wang, L. Dynamic Bayesian network for aircraft wing health monitoring digital twin. *AIAA J.* **2017**, *55*, 930–941. [CrossRef]
32. Wang, W.; Zhang, Y.; Zhong, R.Y. A proactive material handling method for CPS enabled shop-floor. *Robot. Comput. Integr. Manuf.* **2020**, *61*, 101849. [CrossRef]
33. Rojko, A. Industry 4.0 concept: Background and overview. *Int. J. Interact. Mob. Technol.* **2017**, *11*, 77–90. [CrossRef]
34. Casadesus-Masanell, R.; Ricart, J.E. From strategy to business models and onto tactics. *Long Range Plan.* **2010**, *43*, 195–215. [CrossRef]
35. Gharaei, A.; Jinzhi, L.; Stoll, O.; Zheng, X.; West, S.; Kiritsis, D. Systems engineering approach to identify requirements for digital twins development. In *Advances in Production Management Systems. The Path to Digital Transformation and Innovation of Production Management Systems*; Lalic, B., Majstorovic, V., Marjanovic, U., von Cieminski, G., Romero, D., Eds.; Springer: Cham, Switzerland, 2020; pp. 82–90.
36. Ecodesign Guide of WEEE Compliance Schemes. Available online: http://eco3e.eu/en/ (accessed on 10 January 2019).
37. Jayal, A.D.; Badurdeen, F.; Dillon, O.W., Jr.; Jawahir, I.S. Sustainable manufacturing: Modelling and optimization challenges at the product, process and system levers. *CIRP J. Manuf. Sci. Technol.* **2010**, *2*, 144–152. [CrossRef]
38. Jasiulewicz-Kaczmarek, M.; Żywica, P. The concept of maintenance sustainability performance assessment by integrating balanced scorecard with non-additive fuzzy integral. *Eksploatacja i Niezawodnosc Maint. Reliab.* **2018**, *20*, 650–661. [CrossRef]
39. Aivaliotis, P.; Georgoulias, K.; Alexopoulos, K. Using digital twin for maintenance applications in manufacturing: State of the Art and Gap analysis. In Proceedings of the IEEE International Conference on Engineering, Technology and Innovation (ICE/ITMC), Valbonne Sophia-Antipolis, France, 17–19 June 2019.
40. Aivaliotis, P.; Georgoulias, K.; Chryssolouris, G. The use of Digital Twin for predictive maintenance in manufacturing. *Int. J. Comput. Integr. Manuf.* **2019**, *32*, 1067–1080. [CrossRef]
41. Zhang, Y.; Ren, S.; Liu, Y.; Si, S. A big data analytics architecture for cleaner manufacturing and maintenance processes of complex products. *J. Clean. Prod.* **2017**, *142*, 626–641. [CrossRef]
42. Mwanza, B.G.; Mbohwa, C. Safety in maintenance: An improvement framework. *Procedia Manuf.* **2017**, *8*, 657–664. [CrossRef]
43. Roy, R.; Stark, R.; Tracht, K.; Takata, S.; Mori, M. Continuous maintenance and the future–Foundations and technological challenges. *CIRP Ann. Manuf. Technol.* **2016**, *65*, 667–688. [CrossRef]
44. Jasiulewicz-Kaczmarek, M.; Legutko, S.; Kluk, P. Maintenance 4.0 technologies—New opportunities for sustainability driven maintenance. *Manag. Prod. Eng. Rev.* **2020**, *11*, 74–87.
45. Daily, J.; Peterson, J. Predictive maintenance: How big data analysis can improve maintenance. In *Supply Chain Integration Challenges in Commercial Aerospace*; Richter, K., Walther, J., Eds.; Springer: Cham, Germany, 2017; pp. 267–278.
46. Rojek, I.; Dostatni, E. Machine learning methods for optimal compatibility of materials in eco-design. *Bull. Pol. Ac. Tech.* **2020**, *68*, 199–206.
47. Dostatni, E.; Rojek, I.; Hamrol, A. The use of machine learning method in concurrent ecodesign of products and technological processes. In *Advances in Manufacturing. Lecture Notes in Mechanical Engineering*; Hamrol, A., Ciszak, O., Legutko, S., Jurczyk, M., Eds.; Springer: Cham, Germany, 2018; pp. 321–330.
48. Rojek, I. Neural networks as performance improvement models in intelligent CAPP systems. *Control Cybern.* **2010**, *39*, 55–68.
49. Rojek, I.; Kujawińska, A.; Hamrol, A.; Rogalewicz, M. Artificial neural networks as a means for making process control charts user friendly. In *Intelligent Systems in Production Engineering and Maintenance—ISPEM 2017, Proceedings of the International Conference on Intelligent Systems in Production Engineering and Maintenance, Wroclaw, Poland, 17–18 September*; Burduk, A., Mazurkiewicz, D., Eds.; Advances in Intelligent Systems and Computing; Springer: Cham, Germany, 2018; Volume 637, pp. 168–178.
50. Rojek, I.; Kowal, M.; Stoic, A. Predictive compensation of thermal deformations of ball screws in CNC machines using neural networks. *Tehnicki Vjesnik* **2017**, *24*, 1697–1703.
51. Twardowski, P.; Rojek, I.; Znojkiewicz, N. Application of artificial neural networks for the prediction of tool wear during turning of hardened steel. *Int. J. Mod. Manuf. Technol.* **2020**, *12*, 196–205.
52. Twardowski, P.; Wiciak-Pikuła, M. Prediction of tool wear using artificial neural networks during turning of hardened steel. *Materials* **2019**, *12*, 3091. [CrossRef] [PubMed]
53. StatSoft Statistica Internetowy Podręcznik Statystyki. Available online: http://www.statsoft.pl/textbook/stathome_stat.html?http%3A%2F%2Fwww.statsoft.pl%2Ftextbook%2Fstclatre.html (accessed on 10 September 2020).

applied sciences

MDPI

Review

Digital Twin Lean Intralogistics: Research Implications

Pawel Pawlewski, Monika Kosacka-Olejnik * and Karolina Werner-Lewandowska

Faculty of Engineering Management, Poznan University of Technology, 60-965 Poznan, Poland; pawel.pawlewski@put.poznan.pl (P.P.); karolina.werner@put.poznan.pl (K.W.-L.)
* Correspondence: monika.kosacka@put.poznan.pl; Tel.: +48-616653414

Abstract: This article presents research implications related to the analysis of current trends occurring in the industry and resulting from the analysis of trends in literature. A new trend is noticeable in the range of computer simulations using digital twin technologies in the optimization of intralogistics processes, the implementation of which is based on Lean philosophy. This article shows the connection of Industry 4.0 with Lean in the context of Digital Twin (simulation) in the area of intralogistics. A three-step methodology of literature research was developed and described. In accordance with the adopted research methodology, research questions were indicated and a detailed list of selection criteria was developed. The research methods included brainstorming and statistical analysis. The research results are presented in three sections: the results of the trend analysis, the results of the quantitative literature research, and the results of the complementary research. The research results confirm the existence of a new trend and form the basis for formulating objectives for further research.

Keywords: digital twin; lean; intralogistics; digital twin lean intralogistics

Citation: Pawlewski, P.; Kosacka-Olejnik, M.; Werner-Lewandowska, K. Digital Twin Lean Intralogistics: Research Implications. *Appl. Sci.* **2021**, *11*, 1495. https://doi.org/10.3390/app11041495

Academic Editor: Piera Centobelli
Received: 14 January 2021
Accepted: 3 February 2021
Published: 7 February 2021

1. Introduction

This article presents the effect of research carried out in recent years by the team of authors. The result of the research is the authors' thoughts on new trends and changes in the industry. On the one hand, there is a mechanistic approach by Elon Musk in the production of the Tesla Model 3, which was expressed by excessive automation (too many robots in the final assembly), something that Elon Musk assessed as a mistake [1] "Yes, excessive automation at Tesla was a mistake. To be precise, my mistake. Humans are underrated" On the other hand, there is the Toyota approach, expressed in the book [2]; today's car complexity requires a methodology such as the one developed by Toyota through many decades. Employees are the central focus as they have the best knowledge in the range of production processes. Automation is secondary and should be used to an extent not greater than absolutely necessary. Toyota first tries to understand the production process very well, and only then does it introduce robots into it. In turn, 2011 saw the emergence of an initiative of the German government called Industry 4.0 (shortened to I4.0).

The aim of the article is to present research implications related to the analysis of current trends observed in the industry and resulting from the analysis of trends in the literature. A new trend is noticeable in the range of computer simulations using digital twin technologies in the optimization of intralogistics processes, the implementation of which is based on Lean philosophy.

The main contribution of this paper is to demonstrate the research implications of the new trend referred to by the authors as Digital Twin Lean Intralogistics and justify the need to define it based on the quantitative analysis of literature.

The paper is divided into five sections. Following the introduction, the second section concerns the demonstration of the background of the presented research. It explains the rationale for undertaking research and provides arguments concerning the proposal to introduce the concept of Digital Twin Lean Intralogistics into the terminology. The second section presents the existing implications between the concepts of Digital Twin, Lean, and Intralogistics. The third section describes the adopted research methodology for the quantitative review of the literature. The results of the conducted quantitative literature research are presented in section four. The last section presents conclusions and outlines further research.

2. Related Work

In learning about complex systems (complexity of details and dynamic complexity), simulation, due to its ability to manipulate space-time, is the only tool that allows us to grasp and understand cause-and-effect relationships distant in time and space and related by frequent feedback (dynamic complexity) [3].

Supply chains and factories are complex and dynamic systems. On the one hand, complexity results from the intricacy of manufactured products and manufacturing technologies. On the other hand, complexity is influenced by the structure of processes implemented in the factory. Dynamics are the effects of changes in the market (requiring changes in the products offered) and internal changes resulting from changes in processes implemented inside factories caused by, for example, changes in the organization of material flows, the introduction of new products, the termination of processes related to products withdrawn from the offer, etc. Figure 1 presents four levels of complexity for the supply chains and factories and the impact of external (market) and internal changes (changes in processes, introduction of new and termination of old processes) on the dynamics of the entire system.

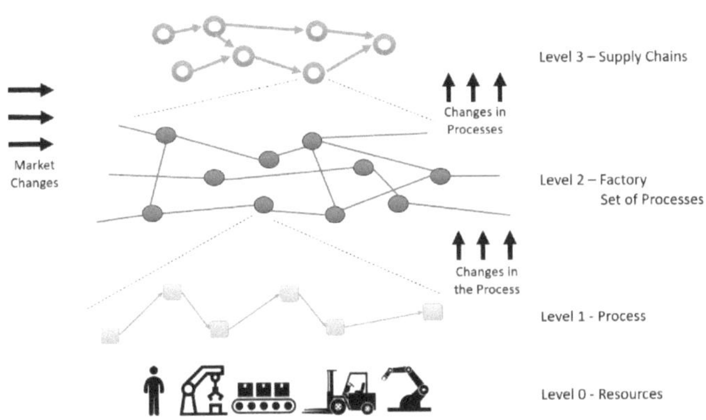

Figure 1. Levels of complexity in supply chains and factories and the impact of external and internal changes on the dynamics of the entire system.

We are now at a time referred to as the industrial revolution 4.0. Industry 4.0., along with the associated involvement of information and communication technologies, is a time for manufacturing companies to think about adapting and changing their current production systems. Industry 4.0 is a general term for various digital concepts such as: the Internet of Things (IoT), the Cyber-Physical System (CPS), Big Data, Data Analytics, Digital Twin, Digital Shadow, Human-Robot Collaboration (HRC), etc. [4]. These concepts promise new potential for production planning and control.

The term "Industry 4.0" was used for the first time in 2011 at the Hannover trade fair in Germany. At this event, the German government presented the I4.0 initiative plan for the first time to protect the long-term competitiveness of the domestic manufacturing industry [5]. A group called the "Industry 4.0 Working Group" was created, chaired by Siegfried Dais (Robert Bosch GmbH) and Henning Kagermann (Acatech). The Industry 4.0 concept defines a new organization of factories (called smart factories), enabling better customer service through great flexibility and optimization of resources.

The key principles for I4.0 are [6]:

1. The factory becomes digital and flexible, which means continuous and immediate communication between various workstations and tools integrated with production lines and supply chains;

2. The use of simulation and data processing tools to collect and analyze data from assembly lines which are used for modelling and testing, which is of great value for employees who want to better understand industrial conditions and processes;

3. Factories become energy-and resource-efficient by using communication networks for the continuous and immediate exchange of information to coordinate needs and availability.

This approach is characterized by a strong link between processes, products, and services represented by the Internet of Things. This concept was widely discussed by scientists and organizations, and its high level of integration creates a working network that connects the physical space and the virtual world via the Cyber-Physical System [7]. In this sense, the I4.0 concept can be interpreted as a strategy to increase competitiveness in the future scenario. It focuses on value-chain optimization because of dynamic and autonomously controlled production [7]. As a result, it facilitates fundamental improvements in industrial processes related to production, material use, supply chain, and life-cycle management Smart factories, which are already beginning to appear, are adopting a new, productive approach. Intelligent products are uniquely identified. They can be located at any stage of the process and one can learn about their history, current status, and alternative ways to achieve the goal [7].

At the same time, a new term has recently appeared: "Intralogistics". This term is particularly popular in German industries; there are many scientific articles in this area from recognized German universities [8–10]. This term was defined by the Intralogistics Forum Verband Deutscher Maschinen- und Anlagenbau (VDMA) [11] (p. 132) as: "The organisation, control, implementation and optimisation of the internal flow of materials, the flow of information and the handling of goods in industry, retail and public facilities." Other definitions can also be found, e.g., [12]: "Every dimension of logistics within the four walls related to implementing, managing, monitoring and optimizing materials handling and information flows."

Together with the product, process, and layout, intralogistics creates one coherent system in which each element depends on the other. The layout as the central point in this system plays a special role. The layout is a floor plan of the plant that locates equipment according to its functions. It is the integration of the physical arrangement of departments, workstations, machines, equipment, materials, common areas, etc. within the existing or planned enterprise [13].

A change in the product is, for example, a change in the product design, which may result in changing the technology. This, in turn, affects the process structure. The introduction of a new product involves the design of a new process, and the implementation of a new process in the same factory results in changes in layout as well as changes in supporting intralogistic processes. Figure 2 shows three instances of how the situation in the factory has changed: first, three processes P1, P2, P3 were implemented; then, process P4 was introduced in place of P2 and P3, and then process P5 instead of process P4, which does not share common positions with process P1 but uses the same transport infrastructure. The described situation took place in a large factory producing parts for the automotive industry and is characteristic for short series of products.

Another situation is a change in the organization of intralogistic processes based on, for example, forklifts into processes based on logistics trains; the main process remains unchanged while the processes of supplying workstations and receiving finished products change completely.

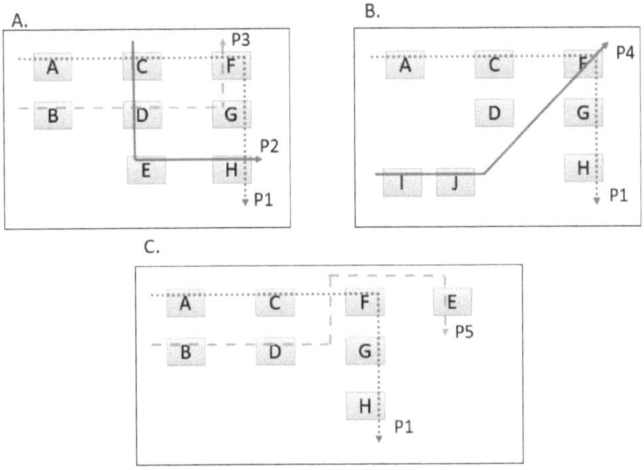

Figure 2. Factory layout dynamics.

Data availability (thanks to the IoT and Digital Twin) makes it possible to increase the efficiency of production planning and control [14]. Lean manufacturing system managers keep asking themselves how to integrate these new opportunities with the existing philosophy and optimization projects. It is currently uncertain whether the I4.0 approach will replace or revive Lean Manufacturing [4]. There are articles whose authors give examples that Lean and I4.0 should be treated complementarily [9,15].

The connection of Lean with I4.0 and the location of simulation regarding intralogistics processes is shown in Figure 3, which is inspired by the siemens.com website [16].

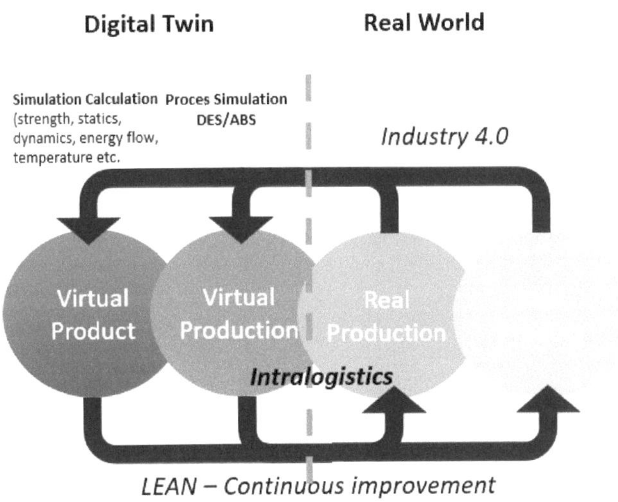

Figure 3. Industry 4.0 and Lean connection in the context of Digital Twin—based on [16].

3. Methodology

In this study, the following research methodology was chosen (Figure 4) considering works [17,18].

Referring to Figure 4, a three-stage study was conducted to demonstrate research implications in the range of Digital Twin Lean Intralogistics.

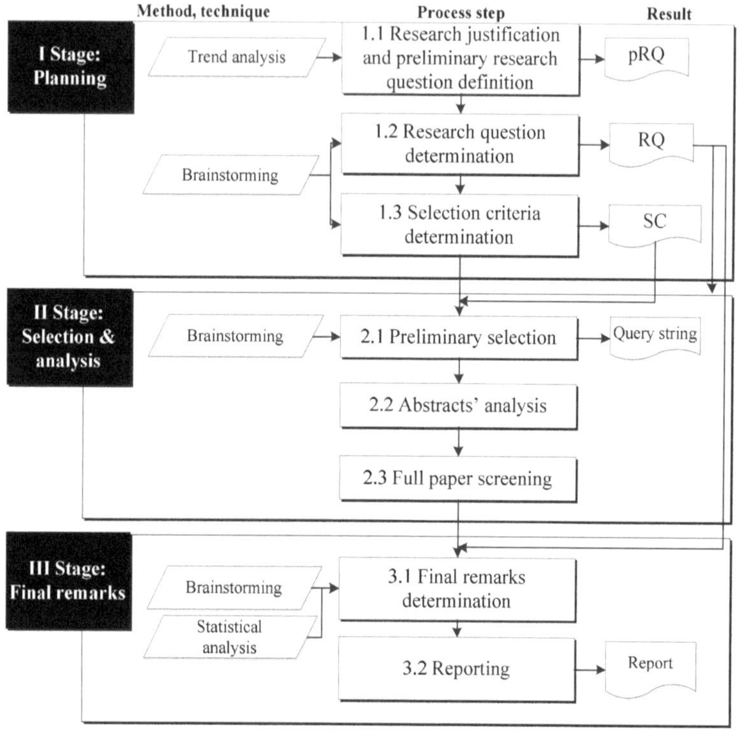

Figure 4. Research methodology.

Stage I: Planning. At this stage, studies were planned that concerned the justification of research on the presented topic (step 1.1.), the determination of research questions (step 1.2.), and the specification of detailed selection criteria (step 1.3.).

In order to justify undertaking research on the implication of the Digital Twin, Lean, and Intralogistics concepts, in accordance with the adopted methodology, a trend analysis was performed. The authors used Google Trends, which allows the analysis of words and phrases entered by users into the Google search engine. The results show the popularity of a search term within a given period and in a selected region. Data in Google Trends are normalized and presented in the range of 0 to 100. Normalization consists of finding the day on which Internet users in a given region and at a given time most often ask for the selected term in the Google search engine (compared to all searches in a given area and time). The obtained result is marked as 100. The following days are given numbers from 0 to 100 in proportion to the highest score.

In the study related to the analysis of trends, the following preliminary research question was posed (hereinafter referred to as pRQ):

- pRQ1: What is the extent of interest in terms such as: "digital twin", "intralogistics", "lean intralogistics", or "digital twin lean intralogistics"?
- pRQ2: Has the digital revolution/Industry 4.0 influenced an increase in interest in digital twin?

The research question pRQ1 results directly from the purpose of the research, while pRQ2 is a kind of derivative question indicating a relationship between I4.0 and Digital Twin technology, which was discussed in Section 2 of the article.

The results of the study of trend analysis allowed the definition of the main research questions according to step 1.2. of the adopted research cycle methodology. It is therefore held that it is justified to conduct a literature analysis related to concepts such as Digital

Twin, Lean intralogistics, and Digital Twin Lean Intralogistics. After justifying the need for literature research, research questions (RQ) were developed:

- RQ1: Is scientific research related to digital twin being undertaken?
- RQ2: Is research on intralogistics being undertaken?
- RQ2: Is research on lean intralogistics being undertaken?
- RQ3: Is the application of digital twin in intralogistics based on lean philosophy being studied in the range of research on digital twin?

In accordance with the adopted research methodology, in response to the indicated research questions (RQ1–RQ3), a detailed summary of the selection criteria used (hereinafter SC) was developed (step 1.3.). It is presented in Table 1.

Table 1. Selection criteria definition (SC).

No	Selection Criteria		
1	Keyword	ID	Description
		H1	Digital twin, digital–twin, digital twean, digital-twean, digital twins, digital-twins
		H2	Intralogistics, intralogistics
		H3	Lean intralogistics, lean intralogistic
		H4	H1 + H3
2	Boolean operators		AND, OR
3	Search range	Z1	Title, abstract, keywords
		Z2	Title
4	Time		without limitation
5	Language		English
6	Publication type		without limitation
7	Research area		without limitation
8	Databases		WoS, Scopus

The research used several keywords (H1, H2, H3, H4) which combined issues such as digital twin, intralogistics, and lean, in accordance with the needs expressed in RQ. The keywords were presented in different notation variants. The literature research was planned to be carried out using logical operators AND, OR, making the appropriate combinations of keywords in the title, abstract, or keywords (Z1), or just in the title (Z2). There were no restrictions on the publication date, type of publication, or research area. However, due to the language of publication, searches were limited to English, which stemmed from the assumption that the relevant studies are mainly written in this language.

The abstract and citation databases used in the research included Web of Science and Scopus. The selected databases are mainly used internationally in management studies and are also chosen by online libraries of major universities. In the authors' opinion, these databases proved to be the best in terms of the collection of scientific publications, and provided a great combination of variables, useful for performing the systematic literature review.

Stage II: Selection and Analysis. Three steps were taken at this stage. In the first step, specified as the preliminary selection (step 2.1), query strings for each database were defined considering the required keywords and search range in the frame of the research questions (RQ). The analysis began with a broad approach to the topic (digital twin) in accordance with RQ1. Then, in order to obtain an answer to RQ2, an analysis of publications on the issue of "intralogistics" and RQ 4 "lean intralogistics" was conducted. In accordance with RQ4, publications including the issues of "digital twin lean intralogistics" were also examined. The publications, which were selected in step 2.1, were subjected to an abstract analysis. In the next step (step 2.3), the full texts of publications were examined, which after analyzing the abstracts, were considered to have the potential to obtain answers to RQ2 and RQ3.

Stage III: Final remarks. At the final stage of research, final remarks were defined (step 3.1), which should be associated with research questions (RQ), as shown in Table 2.

Table 2. Quantitative literature review—Research results (as of 12.01.2021).

RQ	Keyword	Search Range	WoS	Scopus
RQ1	H1	Z1 Z2	1432 711	2854 1387
RQ2	H2	Z1 Z2	114 46	210 124
RQ3	H3	Z1 Z2	0 0	17 0
RQ4	H4	Z1 Z2	0 0	0 0

In the literature research, the indicated terms (H1, H2, H3, H4) were analyzed in a broad context (expressed in Z1) and then narrowed to Z2. Due to the broader context, searching in the Z1 range resulted in better quantitative results (e.g., searching in the Z1 range in the Scopus database gathered 206% more results than in the Z2 range). However, it was recognized that for issues such as H1 and H2, which are general concepts, searches limited to the title of the publication (Z2) garnered better results from the perspective of answering the research questions; these concepts should be included if the publication concerns them. Despite the limitation of the scope of the search for concepts H1 and H2, the number of publications is significant, and that is in spite of the novel character of these concepts. In the case of complex issues expressed in terms H3 and H4, it was considered worth searching for results in a wider way. After all, for the H3 term, which combined "lean" and "intralogistics", only 17 publications in Scopus were identified and analyzed.

The test results were prepared in the form of a report (step 3.2) in accordance with the research methodology (Figure 4). They are described in Section 4 of the article: Results.

4. Results

4.1. Trend Analysis Results

In reference to pRQ1, Google Trends was searched to verify global interest in the term "digital twin". The research concerned the period from 1 January 2010 to 12 January 2021. The scope of the study covered the whole world. The results are shown in Figure 5.

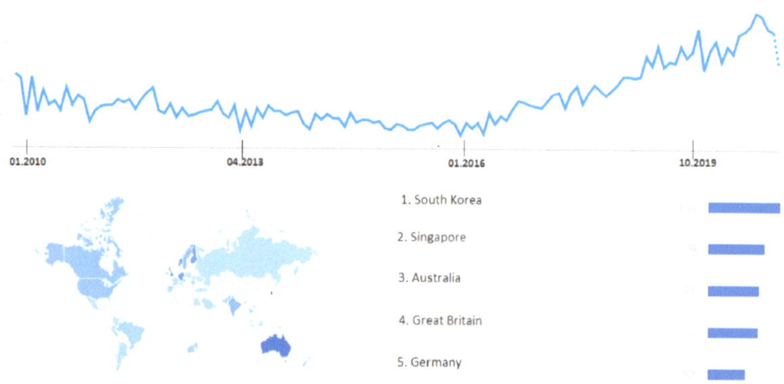

Figure 5. Global extent of interest in the term "digital twin" (concerns the period 1 Jan 2010–12 Jan 2021).

According to the data in Figure 5, it is noted that interest in digital twin has clearly increased since 2016 and the growing trend may continue. The greatest interest in the term "digital twin", according to Google Trends, occurred in Asia and Europe. The authors link the growing interest in digital twin among Internet users to the birth of the industrial revolution, called the era of the digital revolution Industry 4.0, whose element is digital twin, as mentioned in Section 2 of this article.

The authors expressed this relationship in a research question, pRQ2. In order to find answers to pRQ2, the authors examined how trends in interest in the terms "Industry 4.0" and "digital twin" are evolving. Figure 6 shows the results.

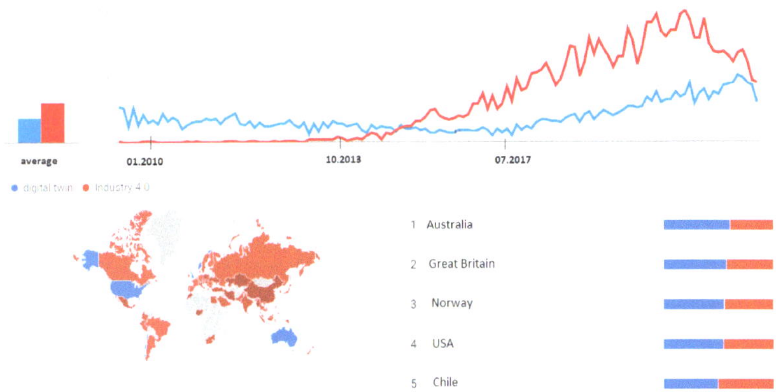

Figure 6. Trends in interest in the terms "Industry 4.0" (red line) and "digital twin" (blue line) (concerns the period from 1 Jan 2010–12 Jan 2021).

The data in Figure 6 clearly indicate that, along with the emergence of interest in Industry 4.0, interest in the term "digital twin" also increased, with the highest level of interest recorded in Australia, Europe and the USA. It is also worth noticing that until 2014 interest in digital twin was higher than interest in I4.0. A detailed explanation of this phenomenon may be the subject of other studies. During the last year, the pandemic situation has changed the global trend, so the interest rate in digital twin as well as in I4.0 has been decreasing. However, when it comes to the relationship between digital twin and I4.0, the authors relate growing interest in digital twin, influenced by the development of I4.0, to the fact that the digital revolution somehow "promoted" or continues to promote IT solutions that are integral to its development. This may be demonstrated by the growing interest of Internet users in the concept of the IoT, the idea of which lies at the heart of the digital age; digital twin solutions are one of the IoT technologies. Figure 7 shows how interest in these terms is shaped.

According to the data in Figure 7, since 2010, the greatest interest among the analyzed trends was related to the Internet of Things, or IoT (these terms mean the same thing, but can be entered in the search engine differently). The greatest interest on a global scale is noted in mostly European countries, as well as in Australia. It is also worth noting that at the time when the IoT reached maximum values according to the Google Trends method, interest in digital twin began to increase. This proves, in the authors' opinion, that Internet users are looking for IT tools and technologies generally called Industry 4.0. Both Digital Twin and IoT are digital technologies that are part of I4.0, but Digital Twin can use IoT (see Figure 3). An increase in IoT possibilities results in an increase in possibilities for monitoring the on-line production process, i.e., an increase in the possibility of reading (collecting) data online and using them directly in Digital Twin. Therefore, in the authors' opinion, one can expect an increased interest in Digital Twin in the coming years as a result of the expansion of the IoT concept.

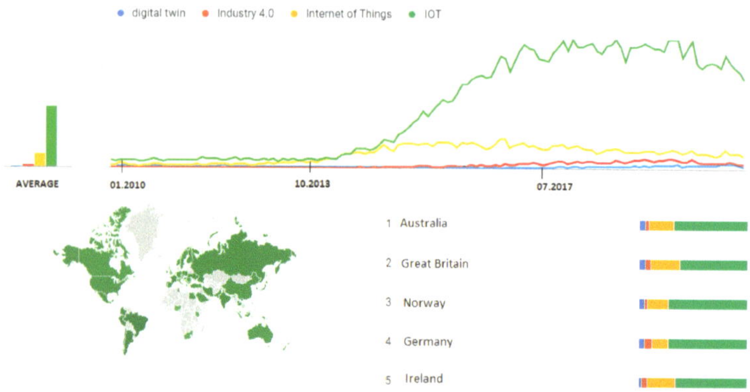

Figure 7. Trends in interest in the terms "Industry 4.0" (red line), "digital twin" (blue line), "Internet of Things" (yellow line), and "IoT" (green line) (concerns the period 1 Jan 2010–12 Jan 2021).

Similar to the study of the term "digital twin", an analysis of the interest-related trend concerning issues such intralogistics, lean intralogistics and digital twin lean intralogistics was made, with reference to pRQ1. Figure 8 shows the global extent of interest in the term "intralogistics".

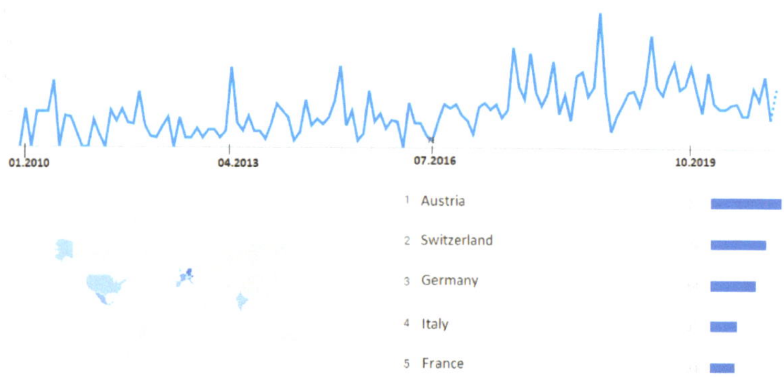

Figure 8. Global trend of interest in the term "intralogistics" (concerns the period 1 Jan 2010–12 Jan 2021).

The data presented in Figure 8 indicate that, since 2010, the trend of interest in the term "intralogistics" has been significantly dynamic, jumping and falling with a simultaneous growing trend that, in the authors' opinion, will continue. It is also worth noting that the term "intralogistics" is most often searched for in Western European countries, including Germany, where originates from, as previously mentioned in Section 2 of the article.

The analysis of trends related to interest in the term "lean intralogistics" showed that, according to Google Trend data since 2010, no one has searched for this term globally in the Google search engine.

After the individual analysis of the trends of each of the analyzed terms, a collective comparison of three terms was made: "digital twin", "lean", and "intralogistics", and the term that is the subject of the article, "digital twin lean intralogistics". The results of the analysis of these terms treated individually are presented in Figure 9.

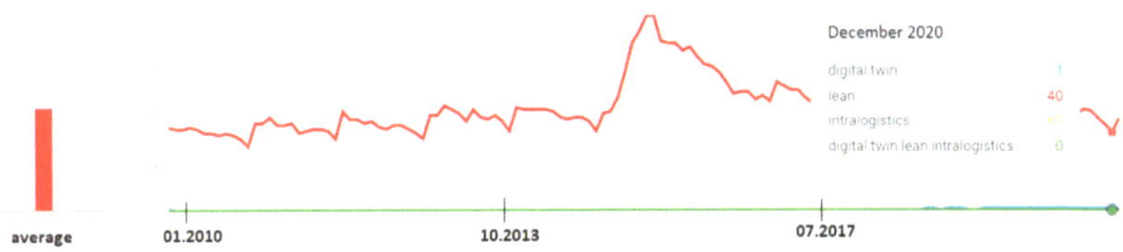

Figure 9. Global interest in the terms: digital twin + lean + intralogistics + digital twin lean intralogistics (concerns the period 1 Jan 2010–12 Jan 2021).

In relation to the whole phrase "Digital Twin Lean Intralogistics", according to Google Trend data, since 2004 such a wording has not been searched for globally on Google.

The compensation of trends in one chart allowed the formulation of answers to pRQ1. The greatest interest among Google users applies to the term "Digital Twin" in comparison with the term "Intralogistics". For both terms, high dynamics and an upward trend are noticeable, which indicates a progressive interest in these areas. Analyzing trends using Google Trend also showed that, until December 2020, the terms "Lean intralogistics" and "Digital Twin Lean Intralogistics" had not been searched for, indicating a research gap.

The conclusion resulting from this part of the research became a premise for continuing studies in the form of quantitative literature research on the issue of digital twin lean intralogistics.

4.2. Results of the Quantitative Literature Research

According to the adopted research methodology (Figure 4), quantitative literature studies were performed after the trend analysis. The research was conducted in January 2021. The results of the quantitative literature research related to terms such as "Digital twin", "digital–twin", "digital twin", "digital-twin", "digital twins", "digital-twins" (H1), "Intralogistics", "intralogistics" (H2), "Lean intralogistics", and "lean intralogistics" (H3), and H1 and H3 combined are presented in Table 3. The data in Table 3 refer to the number of publications for which, according to Table 2, Z2 was adopted for the search range.

Table 3. Summary of the literature review (12.01.2021).

Keyword	**H1**		**H2**		**H3**		**H4**	
Database	**WoS [1]**	**S [2]**	**WoS [1]**	**S [2]**	**WoS [1]**	**S [2]**	**WoS [1]**	**S [2]**
till 2014	5	17	8	48	0	0	0	0
2015	1	1	1	3	0	0	0	0
2016	5	8	3	9	0	0	0	0
2017	32	48	5	9	0	0	0	0
2018	129	155	8	15	0	0	0	0
2019	268	417	17	24	0	0	0	0
2020	264	701	4	16	0	0	0	0
2021	7	40	0	0	0	0	0	0
TOTAL	**711**	**1387**	**46**	**124**	**0**	**0**	**0**	**0**
first publication	1973	1973	2005	2004				
max. number of papers per year	268	701	17	24				
% share of papers (2019–2020)	74.82%	80.61%	45.65%	32.26%				

[1] Web of Science, [2] Scopus.

Despite the fact that in both WoS and Scopus databases publications whose titles concerned digital twin (H1) or intralogistics (H2) were identified, in the case of "lean intralogistics" or "lean intralogistic", no work meeting the indicated selection criteria was found.

The term Digital Twin Lean Intralogistics was not identified in either of the databases regardless of the scope used (Z1 or Z2), which again proves the existence of a research gap in this area.

4.3. Results of Complementary Research

Having conducted quantitative literature research related to publications on the concepts studied and done in accordance with the adopted methodology, complementary studies were carried out. Firstly, based on the data from Table 3, tendency in publishing was assessed (see Figure 10). Complementary studies supplement the analysis of trends carried out in the first stage of the research cycle (Planning).

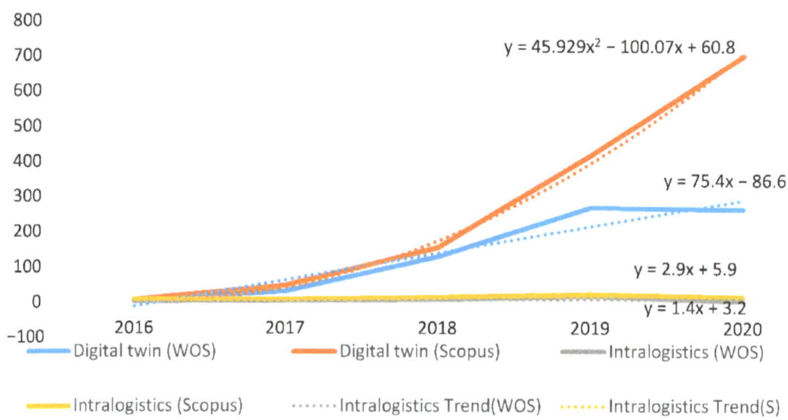

Figure 10. Trends in publications related to terms in H1 and H2.

As the results from Figure 10 show, in each case, the function reflecting the trend lines is growing (a > 0), which confirms the growing number of publications related to the research subject. Considering the results of analysis from the last five years (2016–2021) and the trend functions, it is expected that the total number of publications on digital twin will grow to the value of 1114 in 2021 in Scopus and to the value of 366 in WOS. The interest rate on intralogistics should be growing, too, but the growth rate is rather small. As a consequence, it should be noted that digital twin is becoming a highly explored research topic.

Next, correlation between the number of publications related to Digital Twin (H1) and Intralogistics (H2) was examined. To demonstrate the relationship, the Pearson's correlation coefficient rxy was used and the coefficient of determination R2 was calculated. The number of publications related to Digital Twin was adopted as variable x and the number of publications on Intralogistics was adopted as y. The analyses were made separately for publications identified in the Web of Science and Scopus databases. The results are presented in Table 4.

Table 4. Pearson's coefficient and coefficient of determination for the analyzed population of publications.

Correlation Coefficients	WoS	Scopus
r_{xy}	0.82	0.80
R^2	67%	64%

As the data in Table 4 show, there is a positive correlation between publications related to Digital Twin and Intralogistics. Along with an increase in the number of publications in the Digital Twin area, the number of publications on Intralogistics also increased. The analysis did not include 2021 as this is too short a period so far.

5. Discussion and Conclusions

The main goal of the article was to demonstrate the research implications of a new trend in computer simulations using digital twin technologies to optimize intralogistics processes, the implementation of which is based on Lean philosophy, named Digital Twin Lean Intralogistics by the authors. According to the authors, this trend is a consequence of current trends noticeable in the industry, as presented in Section 2.

In order to demonstrate the research implications, quantitative literature studies were carried out in accordance with the adopted methodology (Section 3) and the results of these studies (Section 4) indicated a research gap in the area of research on the use of digital twin technology in the optimization of intralogistics processes, the implementation of which is based on Lean philosophy.

Considering the results of individual stages of research, in accordance with the adopted research methodology, as well as the results of complementary studies, the authors conclude that:

1. There is an implication between Digital Twin and Intralogistics, as shown by the results of the trend analysis and the values of the Pearson's coefficient and coefficient of determination. Changes in the number of publications related to intralogistics were 67% for data from WoS and 64% for data from Scopus, conditioned by changes in the number of publications related to digital twin. The remaining part of the change results from other factors that were not analyzed. However, it should be remembered that both examined correlation coefficients only focus on the strength and direction of the relationship between the analyzed variables. They do not indicate a cause-and-effect relationship. Such a relationship can only be demonstrated through an in-depth qualitative analysis of selected publications. Such analyses are the subject of the authors' further scientific research.
2. Interest in the terms "Digital Twin" and "Intralogistics" is increasing (pRQ1). Both of these issues are also the subject of scientific research, as evidenced by the growing number of publications related to them (RQ2 and RQ3).
3. Growing interest in digital twin technology is determined by the emergence of the I4.0 concept (pRQ2) and related IoT solutions, as confirmed by the trend analysis in Google Trends.
4. There is a research gap in the range of Lean Intralogistics (RQ3), as evidenced by the lack of publications on this subject in the WoS and Scopus databases, undertaking the context of optimizing intralogistics in the spirit of lean philosophy.
5. No publications regarding research on the use of digital twin in intralogistics based on lean philosophy (RQ4) were identified and until 2021 the term "Digital Twin Lean Intralogistics" had not been searched for in the Google search engine.

On the basis of the obtained research results, the authors define a further two-step research direction. The first stage consists of conducting in-depth qualitative literature research on the issue of Digital Twin Lean Intralogistics in order to fully define the term "Digital Twin Lean Intralogistics", extending the scope of research using databases such as IEEE (Institute of Electrical and Electronics Engineers) Explore and IFAC (International Federation of Automatic Control). The second stage consists of developing theoretical assumptions for the Digital Twin Lean Intralogistics concept with a view to its subsequent implementation in a simulation environment and ultimately in industrial reality.

Author Contributions: Conceptualization, P.P. and K.W.-L.; methodology, M.K.-O.; validation, P.P., K.W.-L., M.K.-O.; formal analysis and investigation, P.P., K.W.-L., M.K.-O.; writing—original draft preparation, P.P.; writing—review and editing, M.K.-O., K.W.-L.; visualization, P.P., M.K.-O.;

supervision, K.W.-L.; funding acquisition, P.P., K.W.-L. and M.K.-O. All authors have read and agreed to the published version of the manuscript.

Funding: This research was funded by the Ministry of Science and Higher Education/Poznan University of Technology, Project ID: 0812/SBAD/4180.

Institutional Review Board Statement: Not applicable.

Informed Consent Statement: Not applicable.

Conflicts of Interest: The authors declare no conflict of interest.

References

1. Twitter. Available online: https://twitter.com/elonmusk/status/984882630947753984 (accessed on 13 April 2018).
2. Liker, J.K. *The Toyota Way: 14 Management Principles from the World's Greatest Manufacturer*; McGraw-Hill Education: New York NY, USA, 2004; ISBN 9780071392310.
3. Cempel, C. *Nowoczesne Zagadnienia Metodologii i Filozofii Badań (Eng. Modern Issues of Methodology and Research Philosophy)*; ITEi Radom, Poland, 2005.
4. Bauer, H.; Brandl, F.; Lock, C.; Reinhart, G. Integration of Industrie 4.0 in lean manufacturing learning factories. *Procedia Manuf* **2018**, *23*, 147–152. [CrossRef]
5. Hermann, M.; Pentek, T.; Otto, B. Design Principles for Industrie 4.0 Scenarios. In Proceedings of the 49th Hawaii International Conference on System Sciences HICSS, Koloa, HI, USA, 5–8 January 2016; pp. 3928–3937. [CrossRef]
6. Schwab, K. *The Fourth Industrial Revolution*; Crown Business: New York, NY, USA, 2017; p. 192. ISBN 978-1-5247-5886-8.
7. Kagermann, H.; Helbig, J.; Hellinger, A.; Wahlster, W. *Recommendations for Implementing the Strategic Initiative INDUSTRIE 4.0: Securing the Future of German Manufacturing Industry*; Final Report of the Industrie 4.0 Working Group; Forschungsunio: Berlin, Germany, 2013.
8. Kartnig, G.; Grosel, B.; Zrnic, N. Past, state-of-the-art and future of intralogistics in relation to megatrends. *FME Trans.* **2012**, *40*, 193–200.
9. Schuhmaher, J.; Hummel, V. Self-organization of changeable intralogistics systems at the ESB Logistics Learning Factory. *Procedia Manuf.* **2019**, *31*, 194–199. [CrossRef]
10. Trott, M.; Auf der Landwehr, M.; von Viebahn, C. A new Simulation Approach for Scheduling Consolidation Activities in Intralogistics—Optimising Material Flow Processes in Industrial Practice. In *Simulation in Produktion Und Logistik herausgegeben von Matthias Putz and Andreas Schlegel*; Verlag Wissenschaftliche Scripten: Auerbach, Germany, 2019; pp. 325–334.
11. Hompel, M.; Heidenblut, V. *Taschen-Lexikon Logistik*; Springer: Berlin/Heidelberg, Germany, 2008; ISBN 978-3-540-28581-6.
12. Intralogistics. Available online: wynright.com/intralogistics (accessed on 21 April 2020).
13. Muther, R.; Hales, L. *Systematic Layout Planning*, 4th ed.; Management & Industrial Research Publications: Marietta, GA, USA, 2015; ISBN 978-0-933684-06-5.
14. Oztemel, E.; Gursev, S. Literature review of Industry 4.0 and related technologies. *J. Intell. Manuf.* **2020**, *31*, 127–182. [CrossRef]
15. Sanders, A.; Elangeswaran, C.; Wulfsberg, J. Industry 4.0 Implies Lean Manufacturing: Research Activities in Industry 4.0 Function as Enablers for Lean Manufacturing. *J. Ind. Eng. Manag.* **2016**, *9*, 811–833. [CrossRef]
16. Drive Optimized Decision-Making: The Closed-Loop Digital Twin with IoT. Available online: https://blogs.sw.siemens.com/mindsphere/drive-optimized-decision-making-the-closed-loop-digital-twin-with-iot/ (accessed on 21 April 2020).
17. Denyer, D.; Tranfield, D. Producing a systematic review. In *The Sage Handbook of Organizational Research Methods*; Buchanan, D.A., Bryman, A., Eds.; Sage Publications Ltd.: Thousand Oaks, CA, USA, 2009; pp. 671–689.
18. Kitchenham, B. *Procedures for Performing Systematic Reviews*; Keele University Technical Report TR/SE-0401; Keele University: Keele, UK, 2004; pp. 1–26.

applied sciences

MDPI

Article

Dynamic Planning of Mobile Service Teams' Mission Subject to Orders Uncertainty Constraints

Grzegorz Bocewicz [1,2,*], Peter Nielsen [2], Małgorzata Jasiulewicz-Kaczmarek [3] and Zbigniew Banaszak [1]

1 Faculty of Electronics and Computer Science, Koszalin University of Technology, 75-453 Koszalin, Poland; zbigniew.banaszak@tu.koszalin.pl
2 Department of Materials and Production, Aalborg University, DK-9100 Aalborg, Denmark; peter@mp.aau.dk
3 Faculty of Engineering Management, Poznan University of Technology, 60-965 Poznań, Poland; malgorzata.jasiulewicz-kaczmarek@put.poznan.pl
* Correspondence: grzegorz.bocewicz@tu.koszalin.pl

Received: 13 November 2020; Accepted: 9 December 2020; Published: 11 December 2020

Abstract: This paper considers the dynamic vehicle routing problem where a fleet of vehicles deals with periodic deliveries of goods or services to spatially dispersed customers over a given time horizon. Individual customers may only be served by predefined (dedicated) suppliers. Each vehicle follows a pre-planned separate route linking points defined by the customer location and service periods when ordered deliveries are carried out. Customer order specifications and their services time windows as well as vehicle travel times are dynamically recognized over time. The objective is to maximize a number of newly introduced or modified requests, being submitted dynamically throughout the assumed time horizon, but not compromising already considered orders. Therefore, the main question is whether a newly reported delivery request or currently modified/corrected one can be accepted or not. The considered problem arises, for example, in systems in which garbage collection or DHL parcel deliveries as well as preventive maintenance requests are scheduled and implemented according to a cyclically repeating sequence. It is formulated as a constraint satisfaction problem implementing the ordered fuzzy number formalism enabling to handle the fuzzy nature of variables through an algebraic approach. Computational results show that the proposed solution outperforms commonly used computer simulation methods.

Keywords: dynamic vehicle routing problem; ordered fuzzy numbers formalism; declarative modelling; service delivery management

1. Introduction

The Industry 4.0, also referred to as "smart" factory, and including solutions such as smart networking, mobility, flexibility of industrial operations and their interoperability, integration with customers and suppliers [1] using the possibilities of modern IT technologies, enables to monitor physical processes and make smart decisions through real-time communication and cooperation with humans, machines, sensors, etc. In this context, the Maintenance 4.0, also known as predictive maintenance, seems to be its main application area [2]. This is because by using advanced Artificial Intelligent (AI) methods to predict disruptions in the functioning of technical systems, predictive maintenance enables the minimization of downtime, prolonging machine life, increasing production efficiency, resource utilization, and reducing costs [3–5]. There is no common definition of Maintenance 4.0 or Industry 4.0, however, a number of studies undertaking these issues are growing rapidly and are also witnessed by many taxonomies of problems identified in both these areas, which presented, inter alia, in the works [2,6–8].

Technological changes such as the high need for transparency (e.g., supply chain visibility) and integrity control (right products, at the right time, place, quantity, condition, and at the right cost) in the supply chains make it possible to improve the level of requested services ordered by geographically dispersed customers. By analogy to the names of the aforementioned areas, the expectations mentioned here underlie the new concept of Logistics 4.0 [1].

In the context of the last two of the aforementioned concepts, i.e., Maintenance 4.0 and Logistics 4.0, it is worth paying attention to the next one called Perfective Maintenance. The idea behind this approach is to strive to improve the functioning of the system by supplementing it with additional functionalities and properties that improve it, e.g., improve accuracy, increase resistance, decrease to cost, etc. The essence of this concept, derived from Perfective Software Maintenance, the aim of which is to improve the performance (e.g., updating the software according to changes in the user interface), maintainability, or other attributes of a computer program [9]; it can also be used in systems providing ordered services with transport to the customer. The presented idea can be used in the course of maintenance of dispatcher's functionality responsible for planning of cyclically repeated delivery/service missions servicing spatially dispersed customers. In the considered case, the functioning of the vehicle fleet planning system could be improved by supplementing it with additional functionalities enabling to react to ad hoc changes in the ordered services. Consequently, such a perfective-maintenance-based approach concerned with the functional enhancement of the vehicle fleet planning system or enhancing its user interface would be especially useful in situations connected with the dynamic planning of milk-run driven systems providing ordered services while taking into account the constraints imposed by customer requests' uncertainty.

The milk-run routing and scheduling problems are usually recognized and formulated as a special case of the vehicle routing problem (VRP), [10–13]. Just as some authors distinguish between the inbound logistics referring to the transport, storage and delivery of goods coming into a business, and the outbound logistics referring to the same for goods going out of business [14], other authors distinguish in-plant milk-run (referring to raw materials, work in process and finished goods distribution) and out-plant milk-run supporting commodities and products transport between manufacturers and customers as well as service visits [15,16]. In both cases, the decisions regarding the vehicles routing policies are considered, i.e., the determination of routes along which customers are visited, and the schedule guaranteeing the congestion-free movement of the vehicles.

Milk-run problems usually concern planning routes that are cyclically repeated according to a fixed schedule in a fixed sequence and with fixed arrival times to plan whom to serve, how much to deliver and which regularly repeated routes to travel on using which fleet of vehicles. Relevant examples are provided by public transport systems including rail transport, urban transport, and intercity bus transport etc. Rhythmic delivery, repeated at regular intervals, is also a feature of systems of the cyclic delivery of food products to distribution centers, waste, recycling and composting pickup, packs to parcel locker-machine points, periodic service inspections as well as restocking beverages in street vending machines.

Since VRPs, which are Non-deterministic Polynomial-time (NP)-hard problems, only approximate solutions with the help of heuristic methods can be obtained [17,18]. In real-live cases, these kinds of problems become more complex due to the necessity of taking into account the influences caused by disruptions (following changes in execution of already planned deliveries and the appearance of new requests/orders, congestion or accidents) and the fuzzy nature of the parameters determining the timeliness of the performed services/deliveries. From a dynamic perspective, arising from the fact that orders are revealed incrementally over time, the considered outbound dynamic routing problem (DRP) consists of designing the vehicle routes (determined by customers' visit sequences) in an online fashion, i.e., communicating to the vehicle which customer to serve next as soon as the visit is accomplished. All related decisions are made without the knowledge of future orders. The need to take DRP commonly arises in the area of maintenance operations, where the ability to redirect a moving vehicle to a new request nearby allows for additional savings [19–21]. However, the fulfillment of these expectations is conditioned by the ability to track the vehicle's position on an ongoing basis and communication ensuring the quick assignation of a new destination, i.e., with a guarantee of dynamically delivered services.

The uncertainty of DRP data due to traffic disruptions as well as changing the dates of the services completion implying the uncertainty of the final result force the necessity to adopt a model implementing the formalism of fuzzy sets. In turn, considering the necessity to take into account the aforementioned constraints of the nature of inequalities, implications and logical conditions, the declarative model seems to be best suited to guarantee these expectations. Therefore, the DRP can be formulated as a fuzzy constraint satisfaction problem and solved using both computer simulation and an analytical ordered-fuzzy-numbers-driven approach. It should be noted, that in opposition to standard fuzzy numbers, the support of the fuzzy number being a result of algebraic operations performed on ordered fuzzy numbers domain does not expand. This is the reason why the proposed use of the oriented fuzzy numbers algebra increases the competitiveness of the analytical approach in relation to the time-consuming computer simulation-based calculations of the feasible scenario of outbound mobile teams' dynamic rerouting.

In this context, the purpose of our research was to develop an ordered-fuzzy-numbers-driven declarative model, enabling to define the DRP subject to fuzzy maintenance time and transportation time constraints, the solution to which provides the possible dynamic rerouting scenarios. Unlike most of the problems discussed in the literature which focus on the search for solutions that optimize the path traveled or the cumulative cost of the mission carried out, in our approach, an answer to the following question was sought: can the newly reported delivery requests or the performance date correction of the already requested ones be accepted or not?

The present study is a continuation of our previous work that explored methods of the fast prototyping of solutions to the problems related to the routing and scheduling of tasks typically performed in batch flow production systems [22–28]. Its main contribution are threefold:

- Outbound mobile teams-driven maintenance services require taking into account disruptions occurring in road traffic (e.g., congestion-restricted delivery time) and the uncertainty of the delivery (e.g., unpacking and storage) operations or maintenance (e.g., repair or condition monitoring) services as well as changing the ordered dates of the service/delivery performance.
- Formulation of the DRP implementing the algebra of ordered fuzzy numbers allows one to plan mobile teams' operation, taking into account the uncertainty of their travel time and the time of conducted repairs.
- In opposition to standard fuzzy numbers, the support of the fuzzy number as a result of algebraic operations performed on ordered fuzzy numbers domain does not expand, which determines its dominance on the currently used computer simulation methods, the proposed algebraic approach allows for online vehicle rerouting and/or rescheduling forced by disturbances caused by ad hoc changes in the orders performed.

The structure of the paper is organized as follows. Section 2 includes the review of the literature. Section 3 provides preliminaries briefly referring to some known concepts from ordered fuzzy numbers theory and constraint programming techniques. The problem statement and the methodology used for its solution are described in Sections 4 and 5, respectively. Computational results are then reported and analyzed in Section 6, while conclusions and future directions of work are considered in Section 7.

2. Related Work

Most of the problems appearing in the milk-run systems are aimed at searching for an optimal periodic distribution policy. Examples of such problems [15,29] include both simple ones, e.g., Mix Fleet VRP, Multi-depot VRP, Split-up Delivery VRP, Pick-up and Delivery VRP, VRP with Time Windows, VRP with Backhauls, and more complex ones, e.g., VRP with multi-trip multi-traffic pick-up and delivery problem with time windows and synchronization being a combination of variants of the vehicle routing problem with multiple trips, a vehicle routing problem with a time window, and a vehicle routing problem with pick-up delivery. Since milk-run routing and scheduling problems follow VRPs which are NP-hard, hence their solutions derived from the milk-run distribution policy while, for instance, aimed at determining in what time windows parts, can be collected from suppliers, and how many logistic trains and along which routes they should run, can be obtained with the help of heuristic methods [17,18,20,30]. Regardless of the class of the problems whether typical for in-plan or out-plant milk-run systems [14] or accentuating either the dynamic or static character of vehicle routing [15,17,21,29,30], their goal is to search for optimal solutions. These studies implicitly assume that there exist admissible solutions, e.g., ones that ensure the congestion-free flow of concurrently executed transport processes [31,32] and/or that planned routings and schedules are robust to assumed disruptions [20,21]. The most studies, which address outbound milk-run systems, focus on the routings and schedules of the vehicle fleet used. Most of the implemented mathematical model-based frameworks employ heuristic approaches using different metaheuristics, such as hybrid ant colony optimization and Tabu search.

It is worth noting that among the aforementioned issues, relatively few studies are devoted to the problems of outbound milk-run dynamic routing and the systems in which services are provided by appointment. In systems of this type, the dynamic multi-period vehicle problem is solved, which boils down to services scheduling being implemented in a rolling horizon fashion, in which new requests are received while unfulfilled during the first period together with the set of customer requests preplanned for the next period constitute the new portfolio of orders to be considered for subsequent scheduling [13,16,33]. Mentioned approaches do not take into account many the practical requirements and limitations imposed by, for example, the need to take into account the specificity of the same services and the capabilities of the teams performing these services. In general, in addition to the need to balance the needs of the serviced customers with the capability of the team implementing the ordered services, the issues of the synchronization of works carried out for a given user by various service teams (e.g., in mutual exclusion or rendez-vous mode) should also be noticed. A broad review of VRP taxonomy-inspired problems formulated in the milk-run systems class are presented in the works [10–12,19].

In many real situations, DRP data uncertainty due to traffic disruptions (uncertain travel times caused by weather conditions, daily changes in traffic intensity etc.) as well as the degree of difficulty of the service provided (caused by intertwined overhauls, condition monitoring, product repairs operations, etc.) cannot be valued in a precise way. However, the minority of models of the so-called Fuzzy VRP only assume vagueness for fuzzy demands to be collected and fuzzy service or travel times. Literature on these issues is very scarce [34], despite the rapidly growing demand for predictive maintenance-oriented service providers [10]. The rapidly developing enterprise servitization indicates the growing demand for this type of services [35–37].

It is worth adding that the development of the servitization-based approach is determined by the ability to reconfigure a delivery/service system, e.g., by taking into account the change of used vehicles' number and their capacity, the number and location of refilling stations (concerning fuel, tools, materials) and so on. In that context, the reconfigurability of the outbound milk-run driven delivery/services system can be seen as the answer to expectations related to achieving the desired level of system flexibility as well as the requirements of the outbound logistics resilience (referring to maintaining the assumed system's stability and robustness levels). It is worth noting that such challenges fit into the concept of intertwined supply network viability, guaranteeing survival in a changing environment [38].

To summarize, the presented review shows that there is an urgent need to develop analytical methods that would replace the labor-intensive and time-consuming methods of the computer simulation-based assessment of possible maintenance service scenarios. The methods sought should take into account the fact that the mobile service missions carried out require taking into account the uncertainty factor resulting from the fuzzy nature of the vehicle movement and services period.

It seems that the requirements mentioned above meet our approach, which combines the declarative modeling paradigm (implemented through the constraints of programming techniques) with an algebra of ordered fuzzy numbers.

3. Preliminaries

3.1. An Ordered Fuzzy Numbers Framework

The routing and scheduling problems developed to date have limited use due to the data uncertainty observed in practice. The values describing parameters such as transport time or loading/unloading times depend on the human factor, which means they cannot be determined precisely. It is difficult to account for data uncertainty by using fuzzy variables due to the imperfections of the classical fuzzy numbers algebra [26]. Equations which describe the relationships between fuzzy variables (variables with fuzzy values) using algebraic operations (in particular, addition and multiplication) do not meet the conditions of the Ring (among others if the condition $\forall_{A\in\mathcal{F}}\ A+0=A$ is met, then condition $\forall_{A\in\mathcal{F}}\ \exists!_{B\in\mathcal{F}}\ A+B=0$ is not met). In addition, algebraic operations based on standard fuzzy numbers follow Zadeh's extension principle. In practice, this means that no matter what algebraic operations are used, the support of the fuzzy number, which is the result of these operations, expands. Consequently, it is impossible to solve algebraic equations with fuzzy variables. In particular, this means that for any fuzzy numbers a, b, c, the following implication $(a+b=c) \Rightarrow [(c-b=a) \wedge (c-a=b)]$ does not hold. This makes it impossible to solve a simple equation $A+X=C$. This fact significantly hinders the use of approaches based on declarative models, in which most of the relationships between decision variables are described as linear/nonlinear equations and/or algebraic inequalities.

We address these issues by proposing the formalism of ordered fuzzy numbers (OFNs) algebra [39]:

Definition 1. *An OFN is a pair of continuous real functions:*

$$\hat{A} = (f_A,\ g_A),\ where:\ f_A,\ g_A\colon\ [0,1] \to \mathbb{R}. \tag{1}$$

The functions f_A and g_A are called the up part and the down part of the OFN $\hat{A}$, respectively. The values of these continuous functions are limited ranges, which can be defined as the following bounded intervals: $UP_A = (l_{A0}, l_{A1})$ and $DOWN_A = (p_{A1}, p_{A0})$. Assuming that f_A is increasing and g_A is decreasing as well as that $f_A \le g_A$, the membership function μ_A of the OFN $\hat{A}$ is as shown in Figure 1a,b:

$$\mu_A(x) = \begin{cases} f_A^{-1}(x) & \text{when } x \in UP_A \\ g_A^{-1}(x) & \text{when } x \in DOWN_A \\ 1 & \text{when } x \in [l_{A1}, p_{A1}] \\ 0 & \text{otherwise} \end{cases} \tag{2}$$

A property called the orientation (direction) is defined for an OFN. There are two types of orientation: positive, when $\hat{A} = (f_A,\ g_A)$ the direction is consistent with the direction of the OX axis, and negative, when $\hat{A} = (g_A,\ f_A)$, the direction is opposite to the direction of the OX axis. Assuming that the values of all fuzzy variables may have a different orientation, the definitions of the algebraic operations used are as follows:

Definition 2. *Let $\hat{A} = (f_A,\ g_A)$ and $\hat{B} = (f_B,\ g_B)$ be OFNs. $\hat{A}$ is a number equal to $\hat{B}$ ($\hat{A} = \hat{B}$), $\hat{A}$ is a number greater than $\hat{B}$ or equal to or greater than $\hat{B}$ ($\hat{A} > \hat{B}$; $\hat{A} \ge \hat{B}$), $\hat{A}$ is less than $\hat{B}$ or equal to or less than $\hat{B}$ ($\hat{A} < \hat{B}$, $\hat{A} \le \hat{B}$) if: $\forall_{x\in[0,1]}\ f_A(x) * f_B(x) \wedge g_A(x) * g_B(x)$, where the symbol $*$ stands for: $=, >, \ge, <$, or $\le$.*

Definition 3. *Let $\hat{A} = (f_A,\ g_A)$, $\hat{B} = (f_B,\ g_B)$, and $\hat{C} = (f_C,\ g_C)$ be OFNs. The operations of addition $\hat{C} = \hat{A} + \hat{B}$, subtraction $\hat{C} = \hat{A} - \hat{B}$, multiplication $\hat{C} = \hat{A} \times \hat{B}$ and division $\hat{C} = \hat{A}/\hat{B}$ are defined as follows: $\forall_{x\in[0,1]}\ f_C(x) = f_A(x) * f_B(x) \wedge g_C(x) = g_A(x) * g_B(x)$, where the symbol $*$ stands for $+, -, \times$, or $\div$. The operation of division is defined for $\hat{B}$ such that $|f_B|>0$ and $|g_B|>0$ for $x \in [0,1]$.*

In recent years, the concept of OFNs has been continuously developed and used in various practical applications. Many publications have been devoted to the analysis of the OFN model in relation to convex fuzzy sets [40–43].

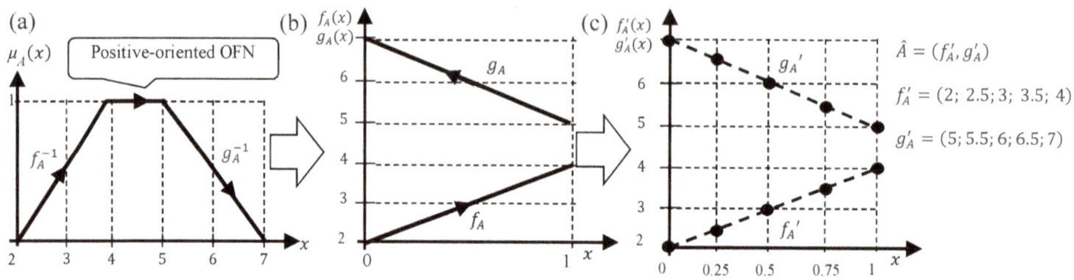

Figure 1. (**a**) Ordered fuzzy number (OFN) $\hat{A}$ represented as a convex fuzzy number; (**b**) functions f_A, g_A determining $\hat{A}$ (positive orientation); and (**c**) the discrete representation of $\hat{A}$ ($dx = 0.5$).

3.2. Illustrative Example

Let us consider the graph $G = (N, E)$ modelling a transportation network composed of $|N| = \omega = 11$ delivery points (hereinafter referred to as nodes), i.e., customers and the service base, as shown in Figure 2. The points include 1 node representing the service point N_1 and 10 nodes representing customers N_2–N_{11}. The customers N_2–N_{11} are cyclically serviced (with period $T = 2000$ u.t.) by the mobile service teams (MSTs) traveling form node N_1. The beginning moment of the node N_λ occupation (service) by team U_k is described by variable y_λ^k. The service is executed in intervals determined by the service deadline $\Delta_\lambda = [ld_\lambda; ud_\lambda] \in \Delta$ (see Table 1), i.e., $y_\lambda^k \geq ld_\lambda$ and $y_\lambda^k + t_\lambda \leq ud_\lambda$ (where t_λ is time of node N_λ occupation). Moreover, each node N_λ can be serviced by MSTs offering required qualifications and confirmed with the appropriate certificates. The considered sets of qualifications $\psi_\lambda \in \Psi$ which are required by customers N_λ are shown in Table 2.

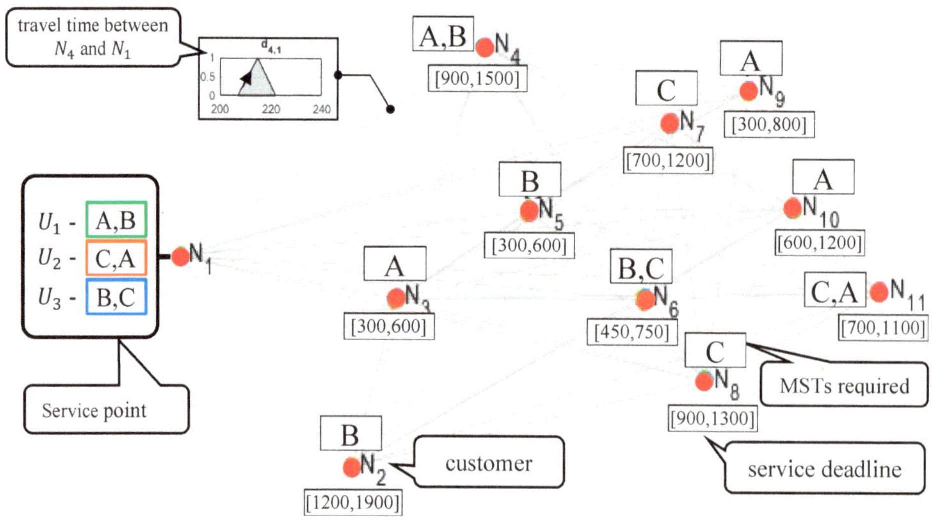

Figure 2. Graph G modeling transportation network.

Table 1. Service deadlines for customers N_2–N_{11}.

	N_2	N_3	N_4	N_5	N_6	N_7	N_8	N_9	N_{10}	N_{11}
ld_λ	1200	300	900	300	450	700	900	300	600	700
ud_λ	1900	600	1500	600	750	1200	1300	800	1200	1100

Table 2. Sets of required qualifications N_2–N_{11}.

	N_2	N_3	N_4	N_5	N_6	N_7	N_8	N_9	N_{10}	N_{11}
ψ_λ	$\{B\}$	$\{A\}$	$\{A,B\}$	$\{B\}$	$\{B,C\}$	$\{C\}$	$\{C\}$	$\{A\}$	$\{A\}$	$\{C,A\}$

For example, customer N_4 should be serviced within the interval time [900; 1500] by MSTs offering qualifications A and B (one MST offering set $\{A,B\}$ or two MSTs: the first offering A and the second offering B).

Each edge $(N_\beta, N_\lambda) \in E$ linking nodes N_β and N_λ is labelled with a fuzzy variable (in the OFN representation) representing the uncertainty of the traveling time $d_{\beta,\lambda}$ between the nodes N_β and N_λ (see Figure 3). Given is a set of MSTs $\mathcal{U} = \{U_1, \ldots, U_k, \ldots, U_K\}$ servicing customers spatially dispersed in network G.

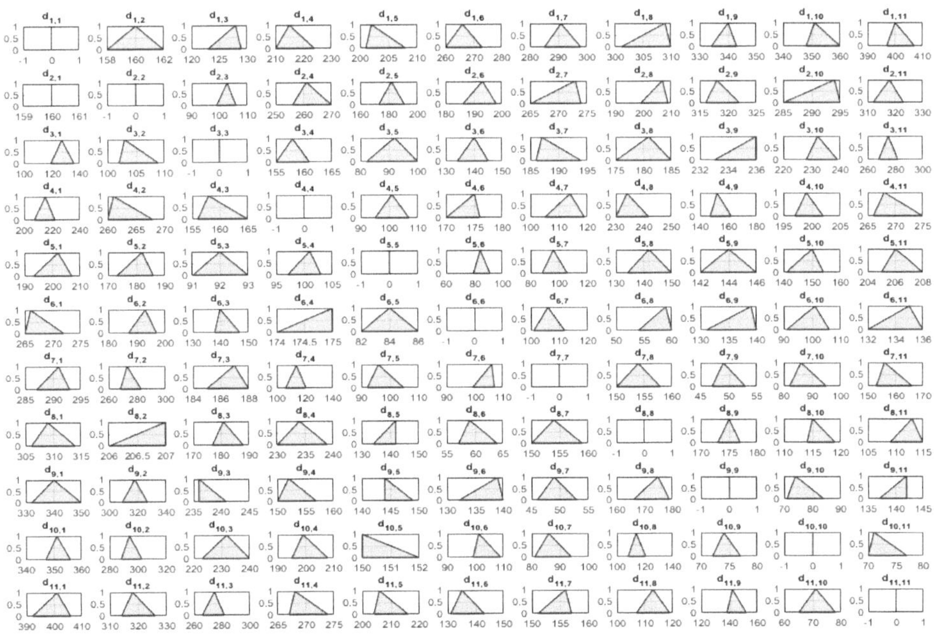

Figure 3. Traveling times between nodes for the networks from Figure 2 (an OFN representation).

For each U_k the set Φ_k of the offered qualifications is assigned. For example, the available set $\mathcal{U} = \{U_1, U_2, U_3\}$ in a network G (Figure 2) contains three MSTs offering the following qualifications: $\Phi_1 = \{A,B\}$; $\Phi_2 = \{C,A\}$; $\Phi_3 = \{B,C\}$. This means that:

- The team U_1 can completely satisfy the expectations of the nodes: N_2, N_3, N_4, N_5, N_9, N_{10}, and partially those of the nodes: N_6, N_{11};
- The team U_2 can completely satisfy the expectation of nodes: $N_3, N_7, N_8, N_9, N_{10}, N_{11}$, and partially of nodes: N_4, N_6;

- The team U_3 can completely satisfy the expectations of the nodes: N_2, N_5, N_6, N_7, N_8, and partially those of the nodes: N_4, N_5 and N_{11}.

The routes traveled by team U_k are denoted by sequences of nodes: $\pi_k = (N_{k_1}, \ldots, N_{k_i}, N_{k_{i+1}}, \ldots, N_{k_\mu})$, where $k_i \in \{1, \ldots, K\}$, $\forall_{k_i \neq k_j}\ N_{k_i} \neq N_{k_j}$, $\left(N_{k_i}, N_{k_{i+1}}\right) \in E$. Nodes representing the service point (e.g., N_1) appear along every route. Moreover, each route π_k consists of nodes in which customers N_λ assigned to them expect services that require qualifications ψ_λ, i.e., for each team U_k offering qualifications Φ_k the following condition holds $\Phi_k \cap \psi_\lambda \neq \varnothing$.

In this context, the problem of the proactive planning of service team trips boils down to the question: do the schedule and routings of MSTs guarantee the timely execution of the ordered services?

Given a set $\mathcal{U}$ of MSTs providing services (according to given qualifications Φ_k) to the customers allocated in a network G (ordering an assumed kind of services Ψ). Does there exist a set of routes Π guaranteeing the timely execution of the ordered services (according to given service deadlines Δ_λ)?

The examples of such routes Π and the associated fuzzy schedule for the network G for Figure 2 are illustrated in Figures 4 and 5. The routes are specified by the sequences of nodes: $\pi_1 = (N_1, N_9, N_{10}, N_4, N_1)$, $\pi_2 = (N_1, N_3, N_{11}, N_1)$, $\pi_3 = (N_1, N_5, N_6, N_7, N_8, N_2, N_1)$. It should be noted that in the presented solution, customer service is provided only by the necessary MSTs. Moreover, despite the uncertain (fuzzy) traveling times $d_{\beta,\lambda}$, it is also assumed that all customers are serviced cyclically (with period $T = 2000$) due to given service deadlines Δ_λ—see Figure 5.

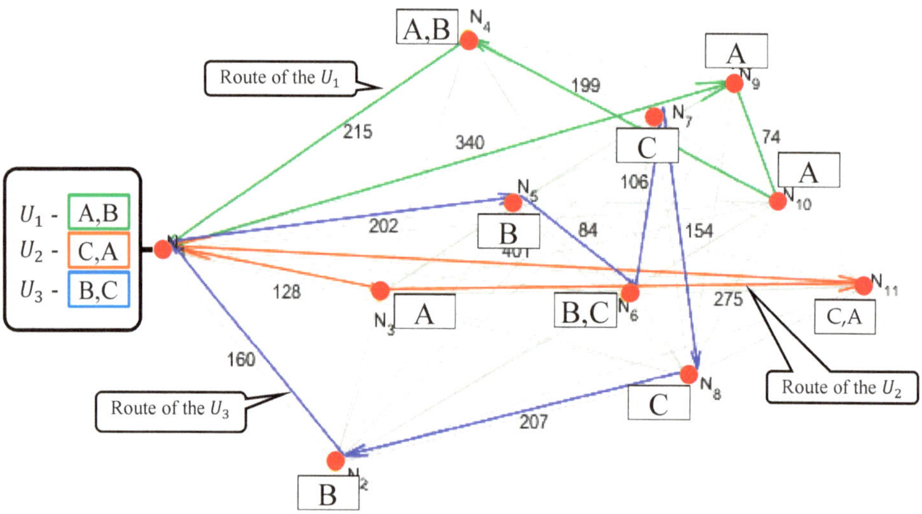

Figure 4. The routes Π of $\mathcal{U} = \{U_1, U_2, U_3\}$ guaranteeing the timely service of the customers $N_2 - N_{11}$.

Due to the occurrence of unforeseen disturbances, the implementation of proactively designated customers service plans becomes practically impossible. An example of such a disturbance are the unforeseen changes of service deadlines. Such a kind of disturbance is presented in Figure 5 where the dispatcher receives information about changing the date of the customer service being located at the node N_6 (from $\Delta_6 = [450; 750]$ to $\Delta_6^* = [650; 950]$), see the second window (moment $t^* = 2500$ when U_1 occupies N_9, U_2 occupies N_3 and U_5 occupies N_5). Due to this change, the adopted routes do not guarantee the implementation of maintenance services on the set dates—the handling of N_6 according to the new service deadline $\Delta_6^* = [650; 950]$ prevents the timely handling of the client N_8 and vice versa. In such a situation, it becomes necessary to answer the following question:

Given a set $\mathcal{U}$ of MSTs providing services (according to given qualifications Φ_k) to the customers allocated in a network (ordering assumed kind of services Ψ), MSTs move along a given set of routes Π according to a cyclic fuzzy schedule $\hat{\mathbb{Y}}$. Given is a disturbance changing Δ_λ toΔ^*_λ at the moment t^*. Does there exist a rerouting $^*\Pi$ and rescheduling $\widehat{^*\mathbb{Y}}$: of MSTs, which guarantee the timely execution of the ordered services?

The possibility of the reactive (dynamic) planning of MST missions in the event of the disruption of service deadlines is the subject of the following chapters.

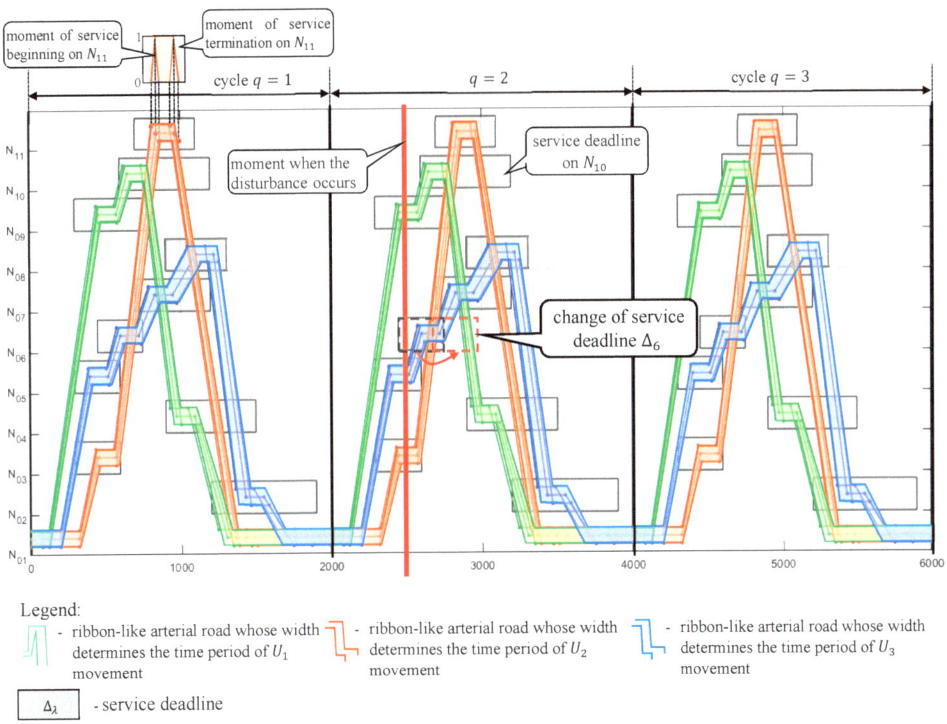

Figure 5. Fuzzy schedule for the implementation of maintenance services.

4. Problem Description

4.1. Assumptions

The following assumptions are met:

- Given is a network $G = (N, E)$,
- Each node $N_\lambda \in N$ is labelled with a fuzzy value $\widehat{t_\lambda}$ (represented in terms of OFN) denoting the duration of node occupation (service);
- Each edge $(N_\beta, N_\lambda) \in E$ is labeled with a fuzzy value $\widehat{d_{\beta,\lambda}}$ denoting the travel time between nodes N_β and N_λ;
- Given is a set $\mathcal{U}$ of MSTs, in which each MST U_k travels route π_k ($\pi_k \in \Pi$);
- Each U_k offers the set of qualifications Φ_k expected by the served customers;
- Node N_1 representing a service point occurs uniquely in all routes;
- Customer assigned to the node N_λ ($\lambda > 1$) expects services that require proper set of qualifications ψ_λ;

- Each route π_k consists of nodes in which customers N_λ assigned to them expect services that require qualifications ψ_λ following $\Phi_k \cap \psi_\lambda \neq \varnothing$;
- Customers are serviced cyclically in time windows repeated with period T;
- Customer assigned to the node N_λ is serviced by deadline $\Delta_\lambda = [ld_\lambda; ud_\lambda]$;
- Fuzzy beginning moments $\widehat{y_\lambda^k}$ (represented as OFN) of the node N_λ occupation make up final fuzzy cyclic schedule $\widehat{\mathbb{Y}}$;
- Disruption is understood as a change of the service deadlines from Δ to Δ^*;
- Moment t^* determines the disruption occurring.

It is also assumed that—values of the decision variables are represented as OFN, see Definition 1. Consequently, the OFN $\hat{A}$ is described by sequences $f_A{}'$ and $g_A{}'$ containing the values of functions f_A and g_A obtained by discretization of the interval $[0, 1]$:

$$f_A{}' = (f_A(0),\ f_A(dx),\ \ldots,\ f_A((M - 1)\,dx),\ f_A(1)), \tag{3}$$

$$g_A{}' = (g_A(1),\ g_A((M-1)dx),\ \ldots,\ g_A(1dx),\ g_A(0)),\ dx = \frac{1}{M}, \tag{4}$$

where $(M+1)$ is the number of samples (Figure 1c).

4.2. Declarative Model

Following the assumptions stated above, the proposed reference model consists of:

Parameters:

Crisp parameters:

G: graph of a transportation network $G = (N, E)$, where $N = \{N_1, \ldots, N_\lambda, \ldots, N_n\}$ is a set of nodes and $E = \{(N_i, N_j) \mid i,\ j \in N,\ i \neq j\}$ is a set of edges, n—the number of nodes;

$\mathcal{U}$: set of MSTs: $\mathcal{U} = \{U_1, \ldots, U_k, \ldots, U_K\}$, where U_k is the k-th MST;

K: size of the fleet;

Ψ: family of required sets of service qualifications: $\Psi = \{\psi_1, \ldots, \psi_\lambda, \ldots, \psi_n\}$, where ψ_λ is a set of qualifications required by customer N_λ (see example in Figure 3);

Φ: family of sets of offered qualifications: $\Phi = \{\Phi_1, \ldots, \Phi_k, \ldots, \Phi_K\}$, where Φ_k is a set of qualifications offered by U_k (see example in Figure 3);

Δ: set of service deadlines: $\Delta = \{\Delta_1, \ldots, \Delta_\lambda, \ldots, \Delta_n\}$, where $\Delta_\lambda = [ld_\lambda; ud_\lambda]$ is a deadline for service at the customer N_λ (see example in Figure 5);

IS: disturbance $IS = (M, \Delta^*)$ where: M is a state of fleet mission at the moment t^*: $M = ((\mu_1, \ldots, \mu_k, \ldots, \mu_K), t^*)$, where $\mu_k \in N$ is the node occupied by U_k (or the node the U_k is headed to) at time t^*, the information about the disturbance is received. For example, in the situation shown in Figure 5, the information about the disturbance IS was received at moment $t^* = 2500$ where the mission state is equal to: $M = ((N_9, N_3, N_5), 2500)$;

Δ^* is a set of changed service deadlines (caused by the appearance of disturbances): $\Delta^* = \{\Delta_1^*, \ldots, \Delta_\lambda^*, \ldots, \Delta_n^*\}$, where $\Delta_\lambda^* = [ld_\lambda^*;\ ud_\lambda^*]$ is a new deadline (after the occurrence of disturbance) for providing a service to customer N_λ;

T: window width, understood as a period, repeated at regular intervals, in which all nodes should be serviced (see Figure 5–$T = 2000$);

Π: set of routes π_k before the occurrence of the disturbance IS, where π_k is a route of U_k:

$$\pi_k = (N_{k_1}, \ldots, N_{k_i}, N_{k_{i+1}}, \ldots, N_{k_\mu}),\ \text{where } x_{k_i,k_{i+1}}^k = 1 \text{ for } i = 1, \ldots, \mu - 1 \text{ and } x_{k_\mu,k_1}^k = 1$$

$$x^k_{\beta,\lambda} = \begin{cases} 1 & \text{if } U_k \text{ travels from node } N_\beta \text{ to node } N_\lambda \\ 0 & \text{otherwise} \end{cases}$$

Imprecise parameters: (defined as positive-oriented OFNs and marked by "^"):

$\widehat{d_{\beta,\lambda}}$: traveling time along edge $\left(N_\beta, N_\lambda\right)$;
$\widehat{t_\lambda}$: time of node N_λ occupation;
$\widehat{\mathbb{Y}}$: fuzzy schedule of fleet $\mathcal{U}$, $\widehat{\mathbb{Y}} = \left(\widehat{Y}, \widehat{W}\right)$ before the disturbance IS:

$\widehat{Y}$: family of $\widehat{Y^k}$, where $\widehat{Y^k}$ is a sequence of moments $\widehat{y^k_\lambda}$: $\widehat{Y^k} = \left(\widehat{y^k_1}, \ldots, \widehat{y^k_\lambda}, \ldots, \widehat{y^k_n}\right)$, $\widehat{y^k_\lambda}$ is fuzzy time at which U_k arrives at node N_λ;
$\widehat{W}$: family of $\widehat{W^k}$, where $\widehat{W^k}$ is a sequence of laytimes $\widehat{w^k_\lambda}$: $\widehat{W^k} = \left(\widehat{w^k_1}, \ldots, \widehat{w^k_\lambda}, \ldots, \widehat{w^k_n}\right)$, $\widehat{w^k_\lambda}$ is laytime at node N_λ for U_k.

Variables:

Crisp variables:

${}^*x^k_{\beta,\lambda}$: binary variable indicating the travel of U_k between nodes N_β, N_λ after disturbance IS:

$${}^*x^k_{\beta,\lambda} = \begin{cases} 1 & \text{if } U_k \text{ travels from node } N_\beta \text{ to node } N_\lambda \\ 0 & \text{otherwise} \end{cases}$$

Imprecise variables (positive-/negative-oriented OFNs):

$\widehat{{}^*y^k_\lambda}$: fuzzy time at which U_k arrives at node N_λ, after occurrence of the disturbance IS;
$\widehat{{}^*w^k_\lambda}$: laytime at node N_λ for U_k, after occurrence of the disturbance IS;
$\widehat{{}^*s^k}$: take-off time of U_k.

Sets and sequences:

${}^*\pi_k$: route of U_k, after occurrence of the disturbance IS: ${}^*\pi_k = \left(N_{k_1}, \ldots, N_{k_i}, N_{k_{i+1}}, \ldots, N_{k_\mu}\right)$, where:

$${}^*x^k_{k_i,k_{i+1}} = 1 \text{ for } i = 1, \ldots, \mu - 1 \text{ and } {}^*x^k_{k_\mu,k_1} = 1;$$

${}^*\Pi$: set of routes ${}^*\pi_k$;
$\widehat{{}^*W^k}$: sequence of laytimes $\widehat{{}^*w^k_\lambda}$: $\widehat{W^k} = \left(\widehat{{}^*w^k_1}, \ldots, \widehat{{}^*w^k_\lambda}, \ldots, \widehat{{}^*w^k_n}\right)$;
$\widehat{{}^*W}$: family of $\widehat{{}^*W^k}$;
$\widehat{{}^*Y^k}$: sequence of moments $\widehat{{}^*y^k_\lambda}$: $\widehat{{}^*Y^k} = \left(\widehat{{}^*y^k_1}, \ldots, \widehat{{}^*y^k_\lambda}, \ldots, \widehat{{}^*y^k_n}\right)$;
$\widehat{{}^*Y}$: family of $\widehat{{}^*Y^k}$;
$\widehat{{}^*\mathbb{Y}}$: fuzzy schedule of fleet $\mathcal{U}$, after occurrence of the disturbance IS: $\widehat{{}^*\mathbb{Y}} = \left(\widehat{{}^*Y}, \widehat{{}^*W}\right)$.

Constraints:

Routes. Relationships between the variables describing MST take-off times/mission start times and the task order:

$$\widehat{{}^*s^k} \geq 0\,;\ k = 1 \ldots K, \tag{5}$$

$$\left(\widehat{s^k} \leq t^*\right) \Rightarrow \left(\widehat{{}^*s^k} = \widehat{s^k}\right);\ k = 1 \ldots K \tag{6}$$

$$\left(\widehat{y_j^k} \le t^*\right) \Rightarrow \left({}^*x_{i,j}^k = x_{i,j}^k\right);\ j = 1\ldots n;\ i = 2\ldots n;\ k = 1\ldots K, \tag{7}$$

$$\left(\widehat{y_j^k} \le t^*\right) \Rightarrow \left(\widehat{{}^*y_j^k} = \widehat{y_j^k}\right);\ j = 2\ldots n; k = 1\ldots K, \tag{8}$$

$$\left(\widehat{y_j^k} \le t^*\right) \Rightarrow \left(\widehat{{}^*w_j^k} = \widehat{w_j^k}\right);\ j = 2\ldots n; k = 1\ldots K, \tag{9}$$

$$\sum_{j=1}^{n} {}^*x_{1,j}^k = 1\ ;\ k = 1\ldots K, \tag{10}$$

$$\left({}^*x_{1,j}^k = 1\right) \Rightarrow \left(\widehat{{}^*y_j^k} = \widehat{{}^*s^k} + \widehat{d_{1,j}}\right);\ j = 1\ldots n;\ k = 1\ldots K, \tag{11}$$

$$\left(\widehat{{}^*y_j^k} > 0 \wedge \widehat{{}^*y_j^q} > 0\right) \Rightarrow \left(\left|\widehat{{}^*y_j^k} - \widehat{{}^*y_j^q}\right| \ge 0\right); i = 1\ldots n;\ k,q = 1\ldots K;\ k \ne q, \tag{12}$$

$$\left({}^*x_{i,j}^k = 1\right) \Rightarrow \left(\widehat{{}^*y_j^k} = \widehat{{}^*y_i^k} + \widehat{d_{i,j}} + \widehat{t_i} + \widehat{{}^*w_i^k}\right);\ j = 1\ldots n;\ i = 2\ldots n;\ k = 1\ldots K, \tag{13}$$

$$\left(\Phi_k \cap \psi_j = \varnothing\right) \Rightarrow \left(\sum_{i=1}^{n} {}^*x_{i,j}^k = 0\right),\ j = 2\ldots n; k = 1\ldots K, \tag{14}$$

$$\cup_{k\in\mathcal{X}_j} \Phi_k = \psi_j,\ j = 2\ldots n,\ \mathcal{X}_j = \{k : \sum_{i=1}^{n} {}^*x_{i,j}^k > 0\} \tag{15}$$

$$\widehat{{}^*s^k} + T = \widehat{{}^*y_1^k} + \widehat{t_1} + \widehat{{}^*w_1^k};\ k = 1\ldots K, \tag{16}$$

$$\widehat{{}^*y_j^k} \ge 0;\ i = 1\ldots n;\ k = 1\ldots K, \tag{17}$$

$$\sum_{j=1}^{n} {}^*x_{i,j}^k = \sum_{j=1}^{n} {}^*x_{j,i}^k;\ i = 1\ldots n;\ k = 1\ldots K, \tag{18}$$

$$\widehat{{}^*y_i^k} \le T,\ i = 1\ldots n;\ k = 1\ldots K, \tag{19}$$

$${}^*x_{i,i}^k = 0;\ i = 1\ldots n;\ k = 1\ldots K. \tag{20}$$

Service deadlines. All customers N_λ should be serviced by the given deadlines $\Delta_\lambda^* = \left[ld_\lambda^*;\ ud_\lambda^*\right]$:

$$\widehat{{}^*y_i^k} + \widehat{t_i} + c \times T\ \le\ ud_\lambda^*,\ i = 1\ldots n;\ k = 1\ldots K, \tag{21}$$

$$\widehat{{}^*y_i^k} + c \times T\ \ge\ ld_\lambda^*,\ i = 1\ldots n;\ k = 1\ldots K. \tag{22}$$

4.3. Fuzzy Constraint Satisfaction Problem

The model proposed above allows to define the problem under consideration in the following way:

Given a set $\mathcal{U}$ of MSTs servicing customers allocated in a network G (customers are serviced by prescheduled deadlines Δ), MSTs move along a given set of routes Π according to a cyclic fuzzy schedule $\widehat{\mathbb{Y}}$. Assuming that there occurs a disturbance IS which changes Δ to Δ^*, a feasible way of rerouting (${}^*\Pi$) and rescheduling ($\widehat{{}^*\mathbb{Y}}$) of MSTs, guaranteeing timely execution of the ordered services, is sought.

The response to the signaled disturbance *IS* is the rescheduling and rerouting of the MSTs resulting then in a new plan of service delivery. In that context, when disturbance *IS* occurs, the new set of routes ${}^*\Pi$ and a new schedule $\widehat{{}^*\mathbb{Y}}$, which guarantees the timely servicing of customers, are determined by solving the following fuzzy constraint satisfaction (FCS) problem (23):

$$\widehat{FCS}\left(\widehat{\mathbb{Y}}, \Pi,\ IS\right) = \left(\left(\widehat{\mathcal{V}}, \widehat{\mathcal{D}}\right), \widehat{C}\left(\widehat{\mathbb{Y}}, \Pi,\ IS\right)\right), \tag{23}$$

where:

$\widehat{\mathcal{V}} = \{\widehat{^*\mathbb{Y}}, {^*\Pi}\}$ is a set of decision variables: $\widehat{^*\mathbb{Y}}$—a fuzzy cyclic schedule guaranteeing the timely provision of service to customers in the case of disturbance IS, and $^*\Pi$—a set of routes determining the fuzzy schedule $\widehat{^*\mathbb{Y}}$;

$\widehat{\mathcal{D}}$— a finite set of decision variable domains: $\widehat{^*y^k_\lambda}, \widehat{^*w^k_\lambda} \in \mathcal{F}$ ($\mathcal{F}$ is a set of OFNs (1)), $^*x^k_{\beta,\lambda} \in \{0,1\}$;

$\widehat{C}$— a set of constraints which take into account the set of routes Π, fuzzy schedule $\widehat{\mathbb{Y}}$ and disturbance IS, while determining the relationships that link the operations occurring in MSTs cycles (5)–(22).

To solve $\widehat{FCS}$ (23), it is necessary to determine the values of the decision variables from the adopted set of domains for which the given constraints are satisfied. The implementation of $\widehat{FCS}$ in a constraint programming environment, such as IBM CPLEX ILOG, enables to find the solution.

5. Solution Methodology

The approach proposed assumes that the reaction to randomly occurring disruptions IS (resulting in, e.g., resignation from services and/or change of the dates of their implementation) takes place on an ongoing basis in the online mode. This is done through dynamic adaptation (i.e., the rerouting and rescheduling) of previously adopted routes Π, and schedules $\widehat{\mathbb{Y}}$, i.e., adjusting them (if possible) to the changes in services timetable.

It is understood that the considered output schedule $\widehat{\mathbb{Y}}$ sets the dates of periodically performed inspections/service repairs ordered by customers. Let $\widehat{\mathbb{Y}}(q)$ denote the fuzzy schedule of the q-th cycle defined as

$$\widehat{\mathbb{Y}}(q) = \left(\widehat{^*Y}(q), \widehat{^*W}(q)\right) \tag{24}$$

where $\widehat{^*Y}(q)$ and $\widehat{^*W}(q)$ are families of the following sets:

$$\widehat{^*Y^k}(q) = \left(\widehat{^*y^k_1}(q), \ldots, \widehat{^*y^k_\lambda}(q), \ldots, \widehat{^*y^k_n}(q)\right) \text{ and } \widehat{^*y^k_\lambda}(q) = \widehat{^*y^k_\lambda} + (q-1) \times T,\ q = 1,2\ldots,Q$$

$$\widehat{^*W^k}(q) = \left(\widehat{^*w^k_1}(q), \ldots, \widehat{^*w^k_\lambda}(q), \ldots, \widehat{^*w^k_n}(q)\right) \text{ and } \widehat{^*w^k_\lambda}(q) = \widehat{^*w^k_\lambda} + (q-1) \times T,\ q = 1,2\ldots,Q$$

The considered implementations of recurring service missions describe the routes Π and schedules: $\widehat{\mathbb{Y}}(1), \widehat{\mathbb{Y}}(2), \ldots, \widehat{\mathbb{Y}}(Q)$ sequences, where Q is the number of cycles performed. It is assumed that disturbance IS can occur in any cycle q.

An algorithm that supports dynamic planning, i.e., vehicle fleet rerouting and rescheduling, based on the proposed concept of $\widehat{FCS}$ (23), is shown in Figure 6. The algorithm processes the successive customer service cycles $q = 1,2\ldots,Q$. If there is a disturbance ($IS \neq \varnothing$) in a given cycle q (at moment t^*), then the problem $\widehat{FCS}$ is solved (*solve* function). The function *solve* represents algorithms implemented in declarative programming environments (responsible for the search for admissible solutions to the decision problems considered).

The existence of an admissible solution (i.e., $\left(\widehat{^*\mathbb{Y}} \neq \varnothing\right) \wedge ({^*\Pi} \neq \varnothing)$) means that there are routes which ensure that customers are serviced on time when the disturbance IS occurs in the cycle Q. If an admissible solution does not exist, then the currently used routes and the associated vehicle schedule should be modified (*reduce* function) in such a way as to remove the servicing operation at node N_λ at which disturbance IS occurs. The *reduce* function is responsible for modifying (rerouting) the routes. The proposed algorithm formulated in the constraints programming framework was implemented in the IBM CPLEX ILOG environment.

The presented algorithm generates in reactive mode (in situations of occurrence of service date change IS) alternative corrected versions of the assumed customer service plan. It needs to be highlighted that the proposed changes must not disrupt the timing of the customers' services to whom the disturbance does not apply to. Thus, there are situations in which such changes resulting in corrected versions of services delivery mission are not possible. In such cases, it is assumed that the affected customers will not be served in a given cycle (unhandled requests are not carried over to subsequent cycles).

The computational complexity of the algorithm from Figure 6 depends on the methods used to solve the problem $\widehat{FCS}$ (function *solve*). Due to the fact that the problem $\widehat{FCS}$ is an NP-hard possibility of the reactive change of assumed proactively scheduled services is limited to a small scale of problems. The assessment of the effectiveness of the proposed approach is the subject of the experiments described in the next section.

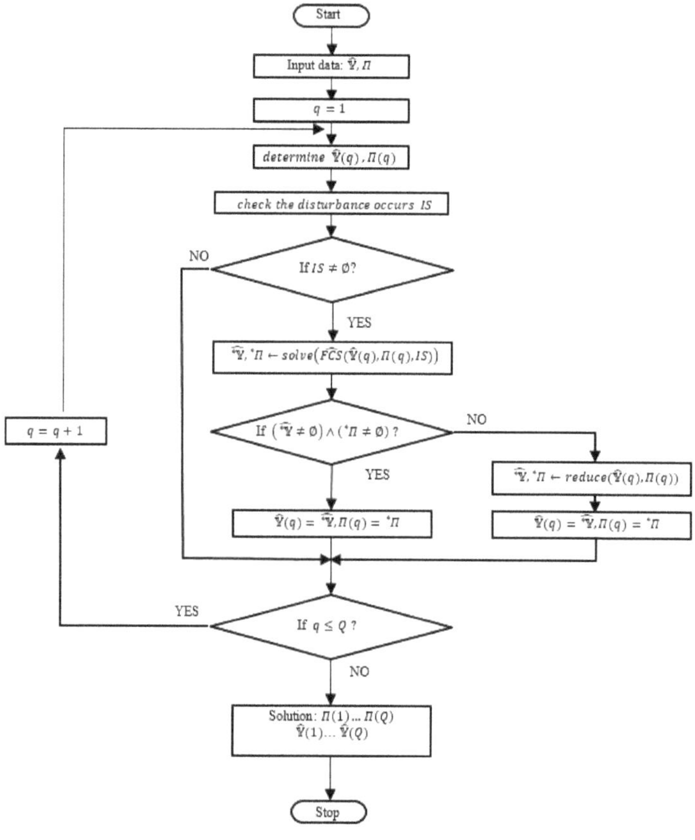

Figure 6. A dynamic rerouting and rescheduling algorithm.

6. Computational Experiments

Considering the graph model of the transportation network from Figure 2, in which three MSTs $\mathcal{U} = \{U_1, U_2, U_3\}$ periodically (with the period $T = 2000$ [u.t].) review the serviced stands owned, by using the customers located at nodes N_2–N_{11}, MSTs offer the following sets of qualifications: $\Phi_1 = \{A, B\}$; $\Phi_2 = \{C, A\}$; $\Phi_3 = \{B, C\}$. The assumed service deadlines Δ, required qualifications Ψ and fuzzy traveling times between the nodes $\widehat{d_{\lambda,\beta}}$ are collected in Tables 1 and 2, Figure 3, respectively. Routes $\pi_1 = (N_1, N_9, N_{10}, N_4, N_1)$, $\pi_2 = (N_1, N_3, N_{11}, N_1)$, $\pi_3 = (N_1, N_5, N_6, N_7, N_8, N_2, N_1)$ determine the fuzzy schedule $\widehat{\mathbb{Y}}$ of the service mission being carried out as shown in Figure 4. It is easy to see (Figure 4) that in the second cycle of the fuzzy schedule (in the state $M = ((N_9, N_3, N_5), 2500)$), an information about suddenly reported changes in the service deadline $\Delta_6^* = [450; 750]$ (instead $\Delta_6 = [650; 950]$) on node N_6 is announced. Given this, an answer to the following question is sought:

Does there exist a set of routes $^*\Pi$ operated by MSTs U_1, U_2 and U_3 for which the fuzzy cyclic schedule $^*\widehat{\mathbb{Y}}$ will guarantee that all customers are serviced on time when disturbance $IS = (S, \Delta^*)$ occurs?

In order to find the answer to this question, the algorithm shown in Figure 6 has been employed. The problem $\widehat{FCS}$ (23) was then implemented in IBM ILOG CPLEX (Windows 10, Intel Core Duo2 3.00 GHz, 4 GB RAM).

The solution time for the problems of this size does not exceed 60 s—see Figure 7c. The following routes were obtained: $^*\pi_1 = (N_1, N_9, N_6, N_{10}, N_4, N_1)$, $^*\pi_2 = (N_1, N_3, N_6, N_7, N_1)$, $^*\pi_3 = (N_1, N_5, N_{11}, N_6, N_8, N_2, N_1)$. It should be noted that the new routes provide simultaneous customer service N_6 by two MSTs: U_1, U_2 (whose qualifications meet the required service needs: $\psi_6 \subset \Phi_1 \cup \Phi_2$).

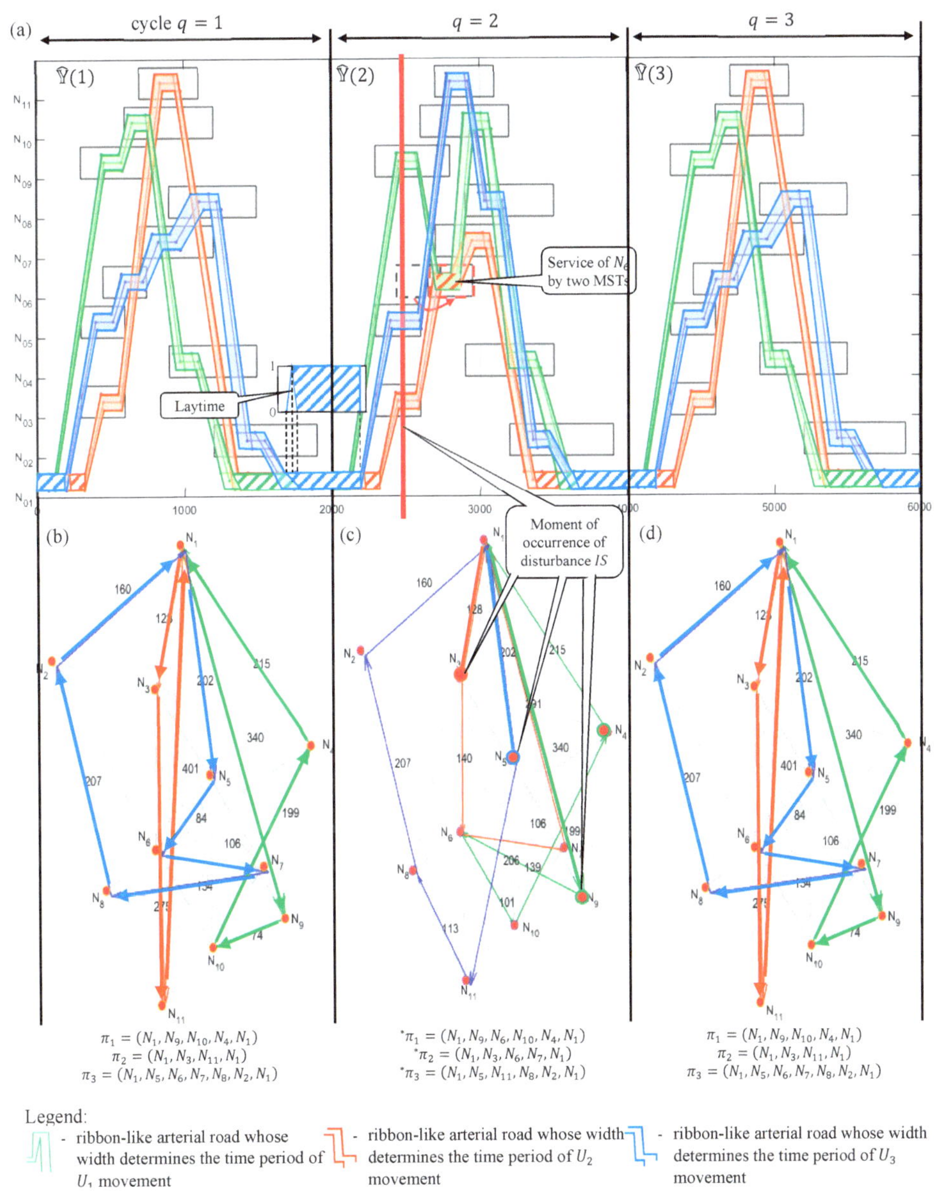

Figure 7. Cyclic fuzzy schedule (**a**); no disturbances (**b**); occurrence of the disturbance (**c**); and no disturbances (**d**).

In fuzzy schedule $^*\widehat{\mathbb{Y}}$ (Figure 7a), the operations are represented as ribbon-like "arterial roads", whose increasing width shows the time of vehicle movement resulting from the growing uncertainty. It is worth noting that the uncertainty is reduced at the end of each time window as a result of the operation of vehicles waiting at node N_1.The increasing uncertainty is not transferred to the subsequent cycles of the system. Uncertainty is reduced as a result of the implementation of OFN formalism. The MST waiting time at node N_1 has a negative orientation (laytimes $^*\widehat{w_1^1}$, $^*\widehat{w_1^2}$ and $^*\widehat{w_1^3}$). An example illustrating the use of standard fuzzy numbers for modeling the behavior of cyclic systems belonging to milk-run systems can be found in [44]. Taking the above into account, the proposed method of the dynamic planning of MSTs in cyclic maintenance delivery systems is unique, due to the possibility of taking into account the reduction in uncertainty in subsequent work cycles of the considered system.

Moreover, the routes $^*\pi_1$, $^*\pi_2$, $^*\pi_3$ remain unchanged (see routes $^*\pi_1$, $^*\pi_2$, $^*\pi_3$ in Figure 7a) until a disturbance occurs, and then they are rerouted, rescheduled and finally synchronized again so that all customers are serviced on time. This means that the model developed in this study allows to adjust the adopted delivery plans to disturbances changing the pre-established services timetable.

In addition to the above experiments, the effectiveness of the proposed approach was evaluated for the distribution networks of different sizes (different numbers of nodes and MSTs). The results are collected in Table 3.

Table 3. Results of the computational experiments carried out for the selected instances of distribution networks.

Number of Nodes n	Number of MSTs K	Calculation Time (s)
5	1	<1
5	2	<1
5	3	<1
5	4	<1
7	1	<1
7	2	<1
7	3	1
7	4	5
9	1	3
9	2	8
9	3	11
9	4	15
11	1	10
11	2	25
11 *	3	31
11	4	67
13	1	22
13	2	61
13	3	108
13	4	124
15	1	46
15	2	115
15	3	215
15	4	380
17	1	250
17	2	554
17	3	>900
17	4	>900
20	1	>900
20	2	>900
20	3	>900
20	4	>900

*—the solution from Figure 7.

To summarize, the experiments were carried out for networks containing 5–20 nodes in which services were made by sets consisting of 1–4 MSTs (the sizes of the instances considered correspond to the sizes of the networks encountered in practice [45]). The aim of the experiments was to estimate the time necessary to designate the routes to guarantee timely services in the case of disturbances *IS* occurrence. In all instances considered, the synthesis of routes required considerable time expenditure. This means that the problems considered can be solved online mode when the size of the service distribution network does not exceed 15 nodes. In the case of larger networks, the effect of combinatorial explosion becomes of significant importance and limits the practical use of this method to the offline prototyping of possible variants of service mission scenarios.

7. Conclusions

The novelty of this study is that it proposed ordered fuzzy numbers algebra framework aimed at the solution of the DMRP, which was stated in terms of the fuzzy constraint satisfaction problem. The specificity of the process involved in the course of the maintenance delivery schedule planning results in the need to determine the sequentially cumulative uncertainty in the performance of the operations involved in it. In other words, the accumulation of uncertainties of previously performed operations result in the increasing uncertainty of the timely execution of subsequent operations. The question that arises in this context concerns the method for preventing additional uncertainty introduced by the combinations of summing up uncertainties of successively summed uncertain deadlines for the implementation of operations. In this context, in contrast to standard fuzzy numbers, the support of a fuzzy number obtained by algebraic operations performed on the ordered fuzzy numbers domain does not expand. In turn, however, the possibility of carrying out algebraic operations is limited to select domains of the computability of these supports. For this reason, sufficient conditions implying the calculability of arithmetic operations guarantee interpretability of the results obtained are proposed. Their use confirms the competitiveness of the analytical approach in relation to the time-consuming computer-simulation-based calculations of MST schedules.

The proposed framework enhanced by modern IT technology, e.g., Internet-of-Things, enables the digital integration of a vehicle fleet providing maintenance services to geographically dispersed customers, and provides feasible solutions forced by ad hoc emerging disturbances, i.e., delivering near-optimal schedules prioritizing the just-in-time performance of maintenance services and the execution of a maximum of the many orders among those reported during the mission. The results of the conducted tests demonstrate that the proposed analytical approach enables to cope with the problems of dynamic routing and scheduling of mobile teams servicing customer requests while taking into account the uncertainty of the travel time and provided maintenance times. In this sense, the paper presents the method enabling to generate alternative MST routing scenarios to customer request change. Its implementation in DSS will support decision-making activities undertaken by service MSTs dispatcher.

The results of the conducted experiments indicate the implementation of the relevant methods in systems supporting the reactive scheduling of MSTs following the milk-run driven manner. In this context, the use of available environments, such as IBM ILOG CPLEX, ECLiPSe, Gurobi, etc., which make it possible to tackle the practical-scale problems, can be viewed as an attractive solution for problem-oriented DSS. It is also worth noting that the research conducted, being in line with the concept of Maintenance 4.0 which stresses the need to seek solutions that allow information systems to create a virtual copy of the physical world, and provides a programming framework for context-aware information model design.

In future work, some additional factors including the impatient customer concept [46], refilling stops, and synchronization of works carried out for a given user by various service teams, will be recorded and streamlined into the proposed approach. Furthermore, the currently studied problem will be extended to the dynamic planning of multi-period outbound MST-driven services, delivery aimed at scheduling being implemented in a rolling horizon approach [47].

Author Contributions: Z.B. and M.J.-K. prepared the state of art, G.B. formulated the problem and proposed the model, G.B. and P.N. conceived and designed the experiments; Z.B. and G.B. prepared the dynamic algorithm; M.J.-K. analyzed the data; G.B., M.J.-K., P.N. and Z.B. wrote the paper. All authors have read and agreed to the published version of the manuscript.

Funding: This research received no external funding.

Acknowledgments: This research was carried out under the internship: Declarative models of the vehicles fleet mission planning (Aalborg University, 23 November 2020–31 March 2021).

Conflicts of Interest: The authors declare no conflict of interest.

References

1. Barreto, L.; Amarala, A.; Pereira, T. Industry 4.0 implications in logistics: An overview, Manufacturing Engineering Society International Conference 2017, MESIC 2017. *Procedia Manuf.* **2017**, *13*, 1245–1252. [CrossRef]
2. Zschech, P. A Taxonomy of Recurring Data Analysis Problems in Maintenance Analytics. *Res. Pap.* **2018**, *197*. Available online: https://aisel.aisnet.org/ecis2018_rp/197 (accessed on 28 November 2018).
3. Cachada, A.; Moreira, C.; Romero, L.; Barbosa, J.; Leitno, P.; Gcraldcs, C.A.S.; Deusdado, L.; Costa, J.; Teixeira, C.; Teixeira, J.; et al. Maintenance 4.0: Intelligent and Predictive Maintenance System Architecture. In Proceedings of the 2018 IEEE 23rd International Conference on Emerging Technologies and Factory Automation (ETFA), Turin, Italy, 4–7 September 2018; pp. 139–146. [CrossRef]
4. Jasiulewicz-Kaczmarek, M.; Gola, A. Maintenance 4.0 Technologies for Sustainable Manufacturing—An Overview. *IFAC Pap.* **2019**, *52*, 91–96. [CrossRef]
5. Patalas-Maliszewska, J.; Pajak, I.; Skrzeszewska, M. AI-based Decision-making Model for the Development of a Manufacturing Company in the context of Industry 4.0. In Proceedings of the 2020 IEEE International Conference on Fuzzy Systems (FUZZ-IEEE), Glasgow, UK, 19–24 July 2020. [CrossRef]
6. Oztemel, E.; Gursev, S. *A Taxonomy of Industry 4.0 and Related Technologies*; IntechOpen: London, UK, 2020. [CrossRef]
7. Nazarov, D.; Klarin, A. Taxonomy of Industry 4.0 research: Mapping scholarship and industry insights. *Syst. Res. Behav. Sci.* **2020**, *37*, 535–556. [CrossRef]
8. Franciosi, C.; Pasquale, V.; Iannone, R.; Miranda, S. A taxonomy of performance shaping factors for human reliability analysis in industrial maintenance. *J. Ind. Eng. Manag.* **2019**, *12*, 115–132. [CrossRef]
9. Rine, D. Software perfective maintenance: Including retrainable software in software reuse. *Inf. Sci.* **1993**, *75*, 109–132. [CrossRef]
10. Pillac, V.; Gendreau, M.; Guéret, C.; Medaglia, A.L. A review of dynamic vehicle routing problems. *Eur. J. Oper. Res.* **2013**, *225*, 1–11. [CrossRef]
11. Braekers, K.; Ramaekers, K.; Nieuwenhuyse, I.V. The vehicle routing problem: State of the art classification and review. *Comput. Ind. Eng.* **2016**, *99*, 300–313. [CrossRef]
12. Psaraftis, H.N. Dynamic vehicle routing: Status and prospects. *Ann. Oper. Res.* **1995**, *61*, 143–164. [CrossRef]
13. Subramanyam, A.; Gounaris, C.E.; Laínez-Aguirre, J.M.; Pinto, J.M. Routing: Construction of Uncertainty Sets and Evaluation Via lling-Horizon Simulations. In Proceedings of the 2018 AlChE Annual Meeting Group: Computing and Systems Technology Division, Pittsburgh, PA, USA, 28 October 2018.
14. Purba, H.H.; Fitra, A.; Nindiani, A. Control and integration of milk-run operation in Japanese automotive company in Indonesia. *Manag. Prod. Eng. Rev.* **2019**, *10*, 79–88. [CrossRef]
15. Holborn, P.L. Heuristics for Dynamic Vehicle Routing Problems with Pickups and Deliveries and Time Windows. Ph.D. Thesis, School of Mathematics, Cardiff University, School of Mathematics, Cardiff, UK, 2013. Available online: http://orca.cf.ac.uk/47742/ (accessed on 30 March 2016).
16. Xi, C. Multi-Period Dynamic Technician Routing and Scheduling Problems with Experience-Based Service Times and Stochastic Customers. Ph.D. Thesis, University of Iowa, Iowa City, IA, USA, 2016. [CrossRef]
17. Okulewicz, M.; Mańdziuk, J. A metaheuristic approach to solve Dynamic Vehicle Routing Problem in continuous search space. *Swarm Evol. Comput.* **2019**, *48*, 44–61. [CrossRef]
18. Montemanni, R.; Gambardella, L.M.; Rizzoli, A.E. Ant Colony System for a Dynamic Vehicle Routing Problem. *J. Comb. Optim.* **2005**, *10*, 327–343. [CrossRef]
19. Ritzinger, U.; Puchinger, J.; Hartl, R.F. A survey on dynamic and stochastic vehicle routing problems. *Int. J. Prod. Res.* **2016**, *54*, 215–231. [CrossRef]

20. Sáeza, D.; Cortésb, C.E.; Núñez, A. Hybrid adaptive predictive control for the multi-vehicle dynamic pick-up and delivery problem based on genetic algorithms and fuzzy clustering. *Comput. Oper. Res.* **2008**, *35*, 3412–3438. [CrossRef]
21. Pavone, N.; Bisnik, E.; Frazzoli, V.; Isler, A. Stochastic and Dynamic Vehicle Routing Problem with Time Windows and Customer Impatience. *Comput. Sci. Eng.* **2009**, *14*, 350–364. [CrossRef]
22. Bożejko, W.; Gnatowski, A.; Pempera, J.; Wodecki, M. Parallel tabu search for the cyclic job shop scheduling problem. *Comput. Ind. Eng.* **2017**, *113*, 512–524. [CrossRef]
23. Pempera, J.; Smutnicki, C. Open shop cyclic scheduling. *Eur. J. Oper. Res.* **2018**, *269*, 773–781. [CrossRef]
24. Sitek, P.; Wikarek, J. Capacitated Vehicle Routing Problem with Pick-up and Alternative Delivery (CVRPPAD)—Model and implementation using hybrid approach. *Ann. Oper. Res.* **2017**, *273*, 257–277. [CrossRef]
25. Smutnicki, C. Minimizing cycle time in manufacturing systems with additional technological constraints. In Proceedings of the 22nd International Conference on Methods and Models in Automation & Robotics, Miedzyzdroje, Poland, 28–31 August 2017; pp. 463–470. [CrossRef]
26. Bocewicz, G.; Nielsen, I.; Banaszak, Z. Production flows scheduling subject to fuzzy processing time constraints. *Int. J. Comput. Integr. Manuf.* **2016**, *29*, 1105–1127. [CrossRef]
27. Bocewicz, G.; Bożejko, W.; Wójcik, R.; Banaszak, Z. Milk-run routing and scheduling subject to a trade-off between vehicle fleet size and storage capacity. *Manag. Prod. Eng. Rev.* **2019**, *10*, 41–53. [CrossRef]
28. Thibbotuwawa, A.; Nielsen, P.; Banaszak, Z.; Bocewicz, G. Energy consumption in un-manned aerial vehicles: A review of energy consumption models and their relation to the UAV routing. *Adv. Intell. Syst. Comput.* **2019**, *853*, 173–184. [CrossRef]
29. Yang, Z.; Osta, J.; Veen, B. Dynamic vehicle routing with time windows in theory and practice. *Nat. Comput.* **2017**, *16*, 119–134. [CrossRef] [PubMed]
30. Hanshar, F.; Ombuki-Berman, B. Dynamic vehicle routing using genetic algorithms. *Appl. Intell.* **2007**, *27*, 89–99. [CrossRef]
31. Kim, G.; Ong, Y.S.; Cheong, T.; Tan, P.S. Solving the Dynamic Vehicle Routing Problem Under Traffic Congestion. *IEEE Trans. Intell. Transp. Syst.* **2016**, *17*, 2367–2380. [CrossRef]
32. Bocewicz, G.; Nielsen, I.; Gola, A.; Banaszak, Z. Reference model of milk-run traffic systems prototyping. *Int. J. Prod. Res.* **2020**, 1–18. [CrossRef]
33. Wen, M.; Cordeau, J.-F.; Laporte, G.; Larsen, J. The dynamic multi-period vehicle routing problem. *Comput. Oper. Res.* **2010**, *37*, 1615–1623. [CrossRef]
34. Brito, J.; Moreno-Pérez, J.A.; Verdegay, J.L. Fuzzy optimization in vehicle routing problems. In Proceedings of the Joint 2009 International Fuzzy Systems Association, World Congress and European Society of Fuzzy Logic and Technology, Lisbon, Portugal, 20–24 July 2009; pp. 1547–1552.
35. Baines, T.; Ziaee Bigdeli, A.; Bustinza, O.F.; Shi, V.G.; Baldwin, J.; Ridgway, K. Servitization: Revisiting the state-of-the-art and research priorities. *Int. J. Oper. Prod. Manag.* **2017**, *37*, 256–278. [CrossRef]
36. Mastrogiacomo, L.; Barravecchia, F.; Franceschini, F. Definition of a conceptual scale of servitization: Proposal and preliminary results. *Cirp J. Manuf. Sci. Technol.* **2020**, *29*, 141–156. [CrossRef]
37. Kryvinska, N.; Kaczor, S.; Strauss, C. Enterprises' Servitization in the First Decade-Retrospective Analysis of Back-End and Front-End Challenges. *Appl. Sci.* **2020**, *10*, 2957. [CrossRef]
38. Ivanov, D.; Dolgui, A. Viability of Intertwined Supply Networks: Extending the Supply Chain Resilience Angles Towards Survivability. A Position Paper Motivated by COVID-19 Outbreak. *Int. J. Prod. Res.* **2020**, *58*, 2904–2915. [CrossRef]
39. Kosinski, W.; Prokopowicz, P.; Slezak, D. On algebraic operations on fuzzy numbers. In *Intelligent Information Processing and Web Mining*; Physica Verlag: Zakopane, Poland; Heidelberg, Germany, 2003; pp. 353–362.
40. Chwastyk, A.; Kosiński, W. Fuzzy calculus with applications. *Math. Appl.* **2013**, *41*, 47–96.
41. Czerniak, J.M.; Dobrosielski, W.T.; Apiecionek, Ł.; Ewald, D.; Paprzycki, M. Practical application of OFN arithmetics in a crisis control center monitoring. *Stud. Comput. Intell.* **2016**, *655*, 51–64.
42. Rudnik, K. Transport trolley control in a manufacturing system using simulation with the FSAW, FWASPAS and FTOPSIS methods. *Adv. Intell. Syst. Comput.* **2018**, *637*, 440–449. [CrossRef]
43. Rudnik, K.; Serafin, R. Probabilistic fuzzy approach to assessment of supplier based on de-livery process. *Adv. Intell. Syst. Comput.* **2019**, *835*, 254–266.

44. Bocewicz, G.; Banaszak, Z.; Rudnik, K.; Witczak, M.; Smutnicki, C.; Wikarek, J. Milk-run routing and scheduling subject to fuzzy pickup and delivery time constraints: An ordered fuzzy numbers approach. *IEEE Int. Conf. Fuzzy Syst.* **2020**, 1–10. [CrossRef]
45. Fedorko, G.; Molnár, V.; Honus, S.; Neradilova, H.; Kampf, R. The application of simulation model of a milk run to identify the occurrence of failures. *Int. J. Simul. Model.* **2018**, *17*, 444–457. [CrossRef]
46. Daugherty, P.J.; Bolumole, Y.; Grawe, S.J. The new age of customer impatience: An agenda for reawakening logistics customer service research. *Int. J. Phys. Distrib. Logist. Manag.* **2019**, *49*, 4–32. [CrossRef]
47. Angelelli, E.; Savelsbergh, T.; Speranza, G.M. Competitive Analysis for Dynamie Multi-Period Uncapacitated Routing Problems. *Networks* **2007**, *49*, 308–317. [CrossRef]

MDPI AG
Grosspeteranlage 5
4052 Basel
Switzerland
Tel.: +41 61 683 77 34

Applied Sciences Editorial Office
E-mail: applsci@mdpi.com
www.mdpi.com/journal/applsci